Advances in Polymer Science

Volume 294

Aims and Scope

The series Advances in Polymer Science presents critical reviews of the present and future trends in polymer and biopolymer science. It covers all areas of research in polymer and biopolymer science including chemistry, physical chemistry, physics, and material science.

The thematic volumes are addressed to scientists, whether at universities or in industry, who wish to keep abreast of the important advances in the covered topics.

Advances in Polymer Science enjoys a longstanding tradition and good reputation in its community. Each volume is dedicated to a current topic, and each review critically surveys one aspect of that topic, to place it within the context of the volume. The volumes typically summarize the significant developments of the last 5 to 10 years and discuss them critically, presenting selected examples, explaining and illustrating the important principles, and bringing together many important references of primary literature. On that basis, future research directions in the area can be discussed. Advances in Polymer Science volumes thus are important references for every polymer scientist, as well as for other scientists interested in polymer science - as an introduction to a neighboring field, or as a compilation of detailed information for the specialist.

Review articles for the individual volumes are invited by the volume editors. Single contributions can be specially commissioned.

Readership: Polymer scientists, or scientists in related fields interested in polymer and biopolymer science, at universities or in industry, graduate students.

Gert Heinrich • Reinhold Kipscholl •
Jean-Benoît Le Cam • Radek Stoček
Editors

Advances in Understanding Thermal Effects in Rubber

Experiments, Modelling, and Practical Relevance

With contributions by

K. Akutagawa · S. An · C. Breitkopf · J. J. C. Busfield · S. Dedova · T. G. Ebbott · E. Euchler · G. Heinrich · R. Hentschke · C. R. Herd · M. Kaliske · R. Kipscholl · O. Kratina · W. A. Kyei-Manu · J.-B. Le Cam · W. V. Mars · O. Peter · A. K. Sambale · K. Schneider · M. Schwartzkopf · R. Stoček · B. L. Su · F. Teng · L. B. Tunnicliffe · J. Vaněk · Y. S. Wang · S. Wießner · J. Wu · B. Yin · P. Zádrapa · L. Zhang

Springer

Editors
Gert Heinrich
TUD Dresden University of Technology
Institute of Textile Machinery and High Performance Material Technology
Dresden, Germany

Reinhold Kipscholl
Coesfeld GmbH & Co. KG
Dortmund, Germany

Jean-Benoît Le Cam
Institut de Physique de Rennes
Université de Rennes
Rennes Cedex, France

Radek Stoček
PRL Polymer Research Lab; Centre of polymer systems
Tomas Bata University in Zlín
Zlín, Czech Republic

ISSN 0065-3195 ISSN 1436-5030 (electronic)
ISBN 978-3-031-71055-1 ISBN 978-3-031-71056-8 (eBook)
https://doi.org/10.1007/978-3-031-71056-8

This Springer imprint is published by the registered company Springer Nature Switzerland AG
The registered company address is: Gewerbestrasse 11, 6330 Cham, Switzerland

Preface

In the case of an ideal rubber, one often thinks only of the linear dependence of the shear modulus on temperature as an expression of the typical entropy elasticity of polymer networks (see, e.g., [1]). Thermoelasticity studies on rubbers have a long history. The earliest qualitative investigation was published by J. Gough in 1805 on unvulcanized natural rubber [2]. He observed that a strip of rubber warms on stretching, but cools when allowing to contract. Under conditions of constant loading the length of the rubber strip shrinks on heating. This is different to a strip of solid metal which would expand on heating under a constant load. This effect of a negative linear coefficient of expansion in axial direction was investigated further with vulcanized rubber by James Joule in 1859 [3] and became known as the *Gough–Joule effect*. Recommended references on the fundamental thermodynamics of elastomers, which also contain descriptions of many experimental results, are, for example, the book by Treloar [4], review articles by Price [5], Shen and Chroucher [6] and Mark [7, 8] as well as review articles on thermoelasticity in [9, 10] with many more references.

However, it is known that the effects of temperature and especially temperature changes on typical technical rubber materials and rubber components are much more complicated. This has consequences for the practical behaviour of rubber-elastic components. For example, the temperatures that rubber seals are exposed to play a significant role in how well they function both immediately and over time. If the temperature is too low, then it can cause the elastomeric materials in the seal to contract. This directly results in the decrease of interference force due to reduction of compression, exposing the seal to a possible leak. We recall the dramatic space shuttle Challenger disaster in year 1986. The rubber used to seal the solid rocket booster joints with O-rings did not expand at temperatures of 0 °C or below, resulting in an opening in the solid rocket booster joint through which gas attempted to escape. Much like low temperatures, high temperatures also have an effect on rubber seals. The rubber on a seal is prone to deterioration at higher temperatures. This can cause the rubber to initially soften. However, the opposite effect can also occur. If the rubber is exposed to the higher temperatures for too long, it can cause the rubber seal to become hard. Furthermore, it is known that self-heat build-up (HBU) in a rubber occurs as this material is exposed to stresses in service. Data from HBU testing are commonly used to estimate the service quality and expected service life of different rubber components. We also remind of the link between HBU and rolling resistance (RR) of a rolling tire. There are a variety of factors that can cause tire HBU, including elevated air and road temperatures, improperly inflated tires, overloading, and unsafe driving habits such as excessive speed and hard braking. HBU can lead to uneven wear, shorter product lifespan, and possible tire failure. Fatigue and failure phenomena in rubbers are specifically described in our earlier book [11]. Of course, in connection with vehicle tires, one should not forget to mention that ambient temperatures influence the braking behaviour of tires. This opens up the topic of friction between rubber and road and its temperature dependence. For the engineers and chemists, the requirement is to relate the mentioned phenomena to the individual rubber properties.

The main physical reason for the heat generation processes is the hysteresis of rubber materials due to deformation and due to the viscoelastic nature of this kind of soft materials. Most elastomers therefore change significantly over time when exposed to heat (and also light or oxygen (ozone), as described in many examples in our earlier book [12]). These changes can have a dramatic effect on the service life and properties of the elastomers. Particularly pronounced heat development in a rubber occurs when it is subjected to a variety of compressive stresses in service. Heat evolution tests are commonly performed to estimate the quality of use and expected service life of various compounds or material options for end-product applications. New developments in recent years on test methods in this direction constitute an important part of this book. At the same time, corresponding simulation and modelling methods have been developed that contribute to a better understanding and enable the predictive simulation of self-heating and the kinetics of temperature fields in complex cyclically loaded rubber components. Specifically, finite-strain thermal viscoelastic damage models for predicting the cyclic thermomechanical response of rubber specimens under fatigue are therefore presented and analytical models for heat diffusion in stressed rubbers. The reader will find more recent information on all these phenomena in the form of original reports or overview articles within this book. The overview chapter by *C. Breitkopf* aims to summarize different approaches to the calculation of thermal conductivities of polymers and elastomers from the macroscopic to the microscopic level. Both approaches – macroscopic and microscopic – are presented in combination with their historical development, and complementary experimental methods are described. The chapter of *S. Dedova et al.* provides a new approach to study double crack behaviour in elastomer materials under complex loading. Particular attention is paid to the correlation of thermal, optical, and mechanical measurements to determine different energy components in the rubber material. The chapter by *E. Euchler et al.* shows combined *in-situ* experiments by synchronizing wide-angle X-ray scattering (WAXS) and infrared thermography (IR) on uniaxially deformed NR vulcanizates reinforced with different carbon black contents. Thus, experimental results are available to characterize the structural evolution under deformation by a crystallization index and a temperature change as a function of mechanical stress and strain. *R. Hentschke* provides in his chapter a comprehensive overview of all temperature-dependent results obtained so far for rubbers. It provides the reader with an understanding of how to use simulations creatively in the context of reinforced rubber materials and in combination with a physical theory when required. The chapter of *O. Kratina et al.* describes the kinetics of HBU in natural rubber reinforced with 10 different types of carbon blacks and under varying cyclic loading conditions. Simple mathematical functions were found that describe the heat development as a function of the type of carbon black and the load frequency.

In the chapter of *W. A. Kyei-Manu et al.* the HBU and energy dissipation of carbon black reinforced natural rubber compounds are studied using a rebound resilience tester, Goodrich flexometer, and a Coesfeld heat build-up analyser. It gives an understanding of how carbon black properties can be leveraged to optimize desired HBU and energy dissipation properties depending on the type of deformation during compound development. *J. B. Le Cam* discusses in detail the use of the surface calorimetry approach to study rubber deformation. He introduces into the corresponding theoretical framework and shows and analyses several calorimetric signatures of thermoelasticity, viscosity, strain-induced crystallization, stress softening, and energy storage. The review of *W. V. Mars et al.* considers the characterization and modelling of the impact of temperature on durability of rubber materials and parts. The chapter includes examples computed with the *Endurica* fatigue solvers. The explanations in this chapter are of great interest for industrial specialists. *J. Vaněk et al.* deal with the derivation of an analytical equation for predicting the heat generation in a rubber cylinder under dynamic cyclic loading. The analytical expression contributes to the effective prediction of the evolution of heat production in rubber. The chapter of *J. Wu et al.* combines experimental and simulation analyses to investigate the heat generation behaviour of aircraft tire rubber materials under different slip conditions. A high-speed sliding friction test was conducted to investigate the temperature distribution law generated mainly by frictional heat, and tests were conducted on the heat generation of the rubber wheel when hysteresis and frictional losses interact at different sliding speeds. In the contribution of *B. Yin et al.* a thermo-viscoelastic rheological model based on multiplicative decomposition of the deformation gradient is coupled to phase-field modelling to investigate temperature-dependent and rate-dependent fracture within polymeric materials. Several representative numerical studies demonstrate the capabilities of the present numerical model.

The book thus provides a comprehensive overview of the latest developments in the field of thermal effects in rubber. It will be of interest to both academic researchers with an application focus and industrial specialists in this field.

Dresden, Germany — Gert Heinrich
Dortmund, Germany — Reinhold Kipscholl
Rennes, France — Jean-Benoît Le Cam
Zlin, Czech Republic — Radek Stoček

References

1. Mark JE, Erman B (2007) Rubberlike elasticity: a molecular primer. Cambridge University Press
2. Gough J (1805) Mem Lit Philos Soc Manchester 1:288
3. Joule J (1859) Philos Trans R Soc Lond A 149:91–131
4. Treloar LRG (1975) The physics of rubber elasticity. Clarendon Press, Oxford
5. Price C (1976) Proc R Soc Lond A 351:331–350
6. Shen M, Chroucher M (1975) J Macromol Sci – Revs Macromol Chem C12 (2):287–329
7. Mark JE (1973) Rubber Chem Technol 46:593–618
8. Mark JE (1993) The rubber elastic state. In: Mark JE (ed) Physical properties of polymers, 2nd edn. American Chemical Society, Washington, DC
9. Erman B, Mark JE (1997) Structure and properties of rubberlike networks. Oxford University Press
10. Heinrich G (2014) Thermoelasticity of rubbers. Encyclopedia of polymeric nanomaterials. Springer, Heidelberg. https://doi.org/10.1007/978-3-642-36199-9_309-1
11. Heinrich G, Kipscholl R, Stoček R (eds) (2021) Fatigue crack growth in rubber materials, vol 286. Advances in polymer science. Springer Nature Switzerland AG
12. Heinrich G, Kipscholl R, Stoček R (eds) (2023) Degradation of elastomers in practice, experiments and modeling, vol 289. Advances in polymer science. Springer Nature Switzerland AG

Acknowledgement

G. H. acknowledges the DFG (German Research Foundation) project 380321452/GRK2430 for financial support.

R.S. acknowledges the Ministry of Education, Youth and Sports of the Czech Republic—DKRVO (RP/CPS/2022/006) for financial support.

Contents

Adv Polym Sci (2024) 294: 1–30
https://doi.org/10.1007/12_2024_174

Published online: 26 May 2024

Thermomechanical and Energetic Behavior of Elastomers Investigated with IR Thermography-Based Surface Calorimetry: Revisiting Rubber Elasticity, Viscosity, Stress Softening, Strain-Induced Crystallization, and Energy Stored

J.-B. Le Cam

Contents

J.-B. Le Cam (✉)
Univ Rennes, CNRS, IPR (Institut de Physique de Rennes) – UMR 6251, Rennes, France
e-mail: jean-benoit.lecam@univ-rennes.fr

Abstract This chapter discusses the use of the surface calorimetry approach to study rubber deformation. It introduces the theoretical framework for determining heat source (or heat power density) from full temperature field measurements and the heat diffusion equation. The determination of the mean intrinsic dissipation and the strain-induced crystallinity from the heat sources is presented and discussed. Typical results obtained under homogeneous tensile loadings are reviewed and analyzed according to four types of calorimetric responses. The types of responses depend on whether the rubber is filled or not, and whether it is crystallizing or not. The calorimetric signatures of thermo-elasticity, viscosity, strain-induced crystallization, stress softening, and energy storage are analyzed and discussed.

Keywords Surface calorimetry · Thermomechanics · Elastomers · Large deformations · Infrared thermography

1 Introduction

During deformation, materials produce or absorb heat due to reversible processes such as energetic elasticity, entropic elasticity, and latent heat of solid–solid phase transition, as well as irreversible processes like viscosity, plasticity, and damage such as cavitation, crack growth, and aging. The specific type of material and loading conditions will determine the extent of these processes. Investigating these processes through their thermal and calorimetric signatures provides therefore complementary information to purely mechanical approaches. This helps physicists and mechanicians to better link microstructure and thermomechanical properties.

A significant amount of research has been conducted on metallic materials in this field since the beginning of the last century, particularly in the field of plasticity. A typical illustration is the studies due to Farren and Taylor [1] and Taylor and Quinney [2] who measured the latent energy remaining after plastic deformation through "cold working" in a metal under quasi-static monotonous loadings. This work led to the consideration of the fraction of the anelastic deformation energy rate irreversibly converted into heat through a ratio, which is now known as the Taylor–Quinney ratio [3–7]. The studies examined the mechanical energy used to deform the material and how it is stored in the microstructure and raises its internal energy, rather than being converted into heat. In their review dedicated to the stored energy of cold work, Bever et al. [8] explained that under certain favorable conditions, cold-worked metals undergo recovery processes during which part of the stored energy can be released. This information is of primary importance to better understand the energetic behavior of rubbers. Such a thermodynamic approach of the deformation processes requires the measurement of the temperature, which is influenced by heat conduction as well as heat exchange with the ambient air and the grips of the testing machine. Even under adiabatic conditions at the scale of the tested samples, strong temperature gradients within the sample tested can induce significant heat diffusion by conduction. A typical example would be the crack tip zone in a stretched sample. This is the reason why the heat sources are preferred to temperature for studying the thermomechanical behavior of materials [9–12]. To a lesser extent than for metals, polymers also benefited from this calorimetric approach; see for example the work by Rittel et Rabin [13], who studied the heat generated during cyclic loading of two glassy polymers, and Chrysochoos and coworkers who studied the kinetics of stored and dissipated energy during cyclic loading of dry polyamide 6.6 samples [14] and the influence of the relative humidity and loading frequency on their thermomechanical cyclic behavior [15]. Despite the interest in this type of thermomechanical characterization, papers on calorimetry applied to stretched rubbers are still scarce, while the natural rubber (NR) was the first material whose thermosensitivity was already highlighted in the nineteenth century by Gough (1805) and Joule (1857).[1] In fact, most of the rubber experiments performed in the twentieth century were purely mechanical: see studies on the stress softening initiated as early as 1903 by Bouasse and Carrière [17], and continued by Holt [18] and Mullins [19]; studies on the fatigue properties of carbon black filled NR with the pioneering work of Cadwell et al. [20] and Fielding [21] for crack initiation, and Rivlin and Thomas [22] and Thomas [23] for crack growth resistance. In addition, since the 1970s onward, the mechanics of elastomers has focused more on behavior modeling than on the physical phenomena of deformation, which made numerical simulation and structural design possible with the advent of computational mechanics in the 1980s.

Nevertheless, physical models should benefit from a better understanding of deformation and damage processes, for example in the area of the fatigue life reinforcement in crystallizing rubbers [20, 24, 25], the importance and the physics of the mechanical hysteresis [26], and the temperature dependence of the mechanical response [27], to name a few. It was not until the early 1970s that the first calorimetric studies of elastomer deformation were carried out by Göritz and coworkers (see, e.g., Ref. [28]). They showed that the thermal energy of strain-induced crystallization (SIC) could be accurately quantified using a "stretch calorimeter." In fact, crystallization is strongly exothermic and the corresponding crystallinity can therefore be evaluated from the part of the total heat source due to SIC only. This technique offered a simpler way than the X-ray diffraction (XRD) technique to evaluate the crystallinity.[2] Nevertheless, calorimetry under stretch was no longer used to measure crystallinity nor to investigate physical phenomena involved in the rubber deformation. Among the possible reasons is the fact that the measurement is made at the scale of the sample and not at the local scale. The point is that the large and possibly heterogeneous strains experienced by rubbers make the local temperature measurement with a thermocouple impossible. This is probably why the advent of infrared cameras has led to new calorimetric studies in the field of

[1] Even before Clausius introduced the concept of entropy (page 391 in Ref. [16]).

[2] It should be noted that the crystallinity measurement with different techniques than the XRD was investigated as soon as 1968 from temperature measurement by Mitchell and Meier [29].

experimental thermomechanics of rubbers. This type of camera has also provided a wealth of information on thermal gradients in the presence of strong geometric singularities such as cracks [30]. Before going any further, it is worth focusing on the concept of the rubber elasticity, which is often considered as an entropy-driven elasticity. Nevertheless, deviations to this theory have been experimentally evidenced by several authors, see for example Allen et al. [31–33], Shen [34], and Shen and Croucher [35]. As detailed and discussed by Treloar in Refs. [36] and [37], non-entropic (energetic) effects contribute to rubber elasticity. They are due to change in the network chain conformations, which affects the intramolecular internal energy. This is confirmed by the fact that the energetic contribution to rubber elasticity does not change when diluents are incorporated [38]. Change in the chain packing may also affect the intermolecular internal energy. Therefore, both energetic and entropic elasticity contribute to the elastic reaction force. In addition to the book of Treloar [37], the reader can refer to Refs. [39–42] to find out more about experimental and theoretical thermodynamics of elastomers. To sum up, rubber elasticity, which induces a reaction force and a variation in temperature under stretching, is the result of two contributions: a change in internal energy, which increases the reaction force and decreases the temperature, and a change in entropy, which also increases the reaction force but decreases the temperature. This clearly justifies the use of thermal and calorimetric measurements to deconvolute these effects, for example by comparing the evolution of the strain and the heat power densities as done in Ref. [43]. In 2006, Honorat provided the first surface calorimetric characterization of cyclic deformation of an unfilled NR with an infrared camera under relaxing and non-relaxing uniaxial tensile loading [44]. He showed the competition between entropic and non-entropic (energetic) elasticity in the calorimetric response and proposed a simple model able to predict the well-known thermoelastic inversion in the thermal response [45, 46]. Caborgan completed the work by Honorat using the same technique and applying a larger number of deformation cycles [47]. In the same period, Samaca Martinez et al. carried out an exhaustive study on the calorimetric signature of the phenomena involved in the rubber deformation: the nature of the elasticity [48], the viscosity [49], the stress softening induced by fillers [50], and the crystallization [48]. A summary is proposed in Refs. [43] and [51]. Finally, Lachhab et al. [52] and Loukil et al. [53] defined an energetic ratio, called γ_{se}, which allowed them to identify the part of the mechanical energy stored and released during a thermodynamical cycle, due to the crystallization/melting process in the polyurethane studied in Lachhab et al. [52] or by the filler network in the nitrile rubber studied in Loukil et al. [53] This ratio highlights and quantifies the part of the hysteresis loop in rubbers that is not only due to viscosity or damage, and which can be significant. In addition, pushing forward the processing of the heat source, Le Cam [54] proposed to couple the work by Göritz and coworkers on stretch calorimetry and the IR thermography-based surface calorimetry approach to evaluate the strain-induced crystallinity. The method was applied to an unfilled NR in Le Cam [54] and successfully validated by comparison with the XRD technique in Le Cam et al. [55] Recently, the measurement has been extended to NR filled with different carbon black filler types in Le Cam et al. [56] and to the characterization of heterogeneities in the strain-induced crystallinity fields [57]. In the above-mentioned studies, the mechanical tests were performed under a homogeneous strain field. It should be noted that a few studies have investigated the calorimetric response under heterogeneous mechanical states, with very different objectives: inverse identification of the constitutive parameters without knowing boundary conditions [58], characterization of the effect of the multiaxial strain state on the calorimetric response with a 3-branch sample [59] and a cruciform sample [60], and investigation of the calorimetric behavior of the crack tip zone where the mechanical field is highly heterogeneous [61, 62]. This chapter reviews the calorimetric signatures of the main phenomena involved in the deformation processes from work done in our group at the Institute of Physics Rennes (University of Rennes, France). In Sect. 2, the thermodynamic framework of the surface calorimetry in the case of large deformations is presented. In particular, the simplification of the heat diffusion equation to be applied to homogeneous (or heterogeneous without heat conduction) heat source fields is described. The results presented in Sects. 3 and 4 summarize the results obtained in our group at the Institute of Physics of Rennes (University of Rennes, France). Section 3 recalls the experimental setups used (sample geometries, loading conditions, imaging techniques) and, more importantly, the experimental precautions required. The results obtained are presented in Sect. 4. They show that several families can be distinguished in terms of the calorimetric response, depending on whether the material is filled or not or whether crystallizing or not. Concluding remarks close the paper.

2 Theoretical Framework for Quantitative Surface Calorimetry

Surface calorimetry lies in the reconstruction of the heat power density, also called the heat source, from temperature measurements at the surface of a material and the heat diffusion equation. When the thermal measurement provides a two-dimensional field, typically obtained with an infrared (IR) camera at a flat surface, the three-dimensional formulation of the heat diffusion equation must be simplified to process the thermal measurement. This section is based on the theoretical framework presented in Ref. [60].

2.1 Tri-dimensional Heat Diffusion Equation in the Case of Finite Strains

In the case where the constitutive state equations are derived from the Helmholtz free energy function and the heat conduction follows Fourier's law, the formulation of the local heat diffusion equation writes as follows with respect to the reference configuration:

$$\rho_0 C\dot{T} + DivQ - R = \underbrace{\mathcal{D}_{int} + T\frac{\partial P}{\partial T} : \dot{F}}_{\widehat{S}} \tag{1}$$

with ρ_0 the mass density, C the heat capacity, T the absolute temperature, Q the heat influx per unit reference area,[3] R the external heat source (due to radiation, for instance), F the deformation gradient tensor, P the first Piola–Kirchhoff stress tensor (or nominal stress tensor) and $\mathcal{D}_{int}$ the intrinsic dissipation (also named mechanical dissipation). $\widehat{S} = \mathcal{D}_{int} + T\frac{\partial P}{\partial T} : \dot{F}$ denotes the heat source. The superposed dot stands for the material time derivative.

2.2 *Simplification of the Heat Diffusion Equation*

2.2.1 2D Formulation

IR cameras provide two-dimensional thermal fields. Reconstructing the two-dimensional heat source field from these fields requires therefore the formulation of a two-dimensional version of the heat diffusion equation. In this approach, several assumptions have to be made: (i) the material is incompressible, (ii) the temperature is constant over the sample thickness,[4] and (iii) the heat conduction is isotropic. The sample is therefore thin and the plane stress state can be generally assumed. Under such conditions, Eq. (1) can be integrated over the thickness, which leads to the two-dimensional formulation of the heat diffusion equation:

---

[3] $\mathbf{Q} = -\mathbf{K_0}\mathbf{C}^{-1}\, det\mathbf{F}\mathrm{Grad}T$, where $\mathbf{K_0}$ is the thermal conductivity tensor and $\mathbf{C} = \mathbf{F}^{\mathrm{T}}\mathbf{F}$ is the right Cauchy–Green tensor.

[4] This assumption can be verified for the test configuration considered by calculating the Biot number B_i. This number is the ratio between the thermal resistance for conduction inside a body and the resistance for convection at the surface of the body. It characterizes how significant the temperature inside a body varies in space when the body is submitted to a heat flux at its surface. In the case of a rectangular thin sample:

$$B_i = \frac{h}{k_0} \cdot \frac{e}{2} \tag{2}$$

where h is the convection coefficient and e is the thickness. If $B_i < 0.1$, the temperature variation in the thickness is less than 5% [63], which is typically the case for a 2 mm thick rectangular rubber sample.

$$\rho_0 C\left(\frac{T-T_{amb}}{\tau_{2D}}\right)-\mathrm{Div}_{2D}\left(k_0\mathbf{C}^{-1}\mathrm{Grad}_{2D}T\right)-R=\widehat{S}, \tag{3}$$

where T_{amb} is the ambient temperature, k_0 is the thermal conductivity coefficient, Div_{2D} and Grad_{2D} are, respectively, the divergence and gradient operators in the sample plane calculated according to the reference (Lagrangian) configuration, and τ_{2D} is a time representing the heat exchanges by convection with the air at the sample surface along the thickness direction.

By considering the temperature variation $\theta(X,Y,t) = T(X,Y,t) - T_{ref}(X,Y)$ instead of the temperature itself. It should be noted that T_{ref} is the ambient temperature if it is constant over the test. In this case, Eq. (3) writes:

$$\rho_0 C\left(\dot{\theta}+\frac{\theta}{\tau_{2D}}\right)-\mathrm{Div}_{2D}\left(k_0\mathbf{C}^{-1}\mathrm{Grad}_{2D}\theta\right)=\widehat{S}. \tag{4}$$

This two-dimensional formulation of the heat diffusion equation has recently been applied to the thermomechanical characterization of moderate heterogeneities in the mechanical strain state [58, 59] or strong ones in the influence zone of crack tips in filled rubber samples [61, 62, 64].

To apply Eq. (4), τ_{2D} has to be characterized according to the stretch λ, defined as the ratio of the current length to the initial length in the stretch direction, and the biaxiality coefficient B, as defined in Refs. [65, 66]. In other words,

as τ varies spatially according to the change in the thickness, i.e., to the stretch λ and to the biaxiality ratio B, $\tau_{2D}= \tau(\lambda,B)$. $\tau(\lambda,B)$ can be assessed in two different ways:

- By heating (or cooling) the sample and measuring the temperature field during the return to thermal equilibrium and then by fitting the curves with an exponential function.
- By considering that the thickness changes are derived by assuming that the material is incompressible:

$$\tau(\lambda,B)=\tau_0\lambda^{-B-1}. \tag{5}$$

τ_0 is the value of τ determined from the thermalization in the undeformed state.

The accurate evaluation of $\tau(\lambda,B)$ is all the more important that the test is performed at a low loading rate, which leads to significant heat diffusion by convection and thus to small temperature variations. In Charlès and Le Cam [58], the two methods have been compared and a good correlation was found.

Counterintuitively, measuring τ for different global stretches is not systematically the more accurate method. Typically, for stretches activating SIC in crystallizable rubbers, the return to the thermal equilibrium is disturbed by the competition between heat production due to recrystallization and heat diffusion due to convection.

2.2.2 "0D" Formulation

In practice, for materials with low thermal diffusivity such as rubber, the only thermal gradient obtained under homogeneous loading is limited to the area near the grips; i.e., heat conduction in the sample plane is negligible and heat exchanges with the grips of the testing machine do not induce temperature gradient in the measurement zone, generally in the sample middle. This is also true for heterogeneous heat source fields if the diffusion length is small compared to the spatial resolution of the thermal measurement. In this case, the two-dimensional heat diffusion equation can be simplified [67, 68] and would be a "0D"-like formulation of the heat diffusion equation[5]:

$$\rho_0 C\left(\dot{\theta}+\frac{\theta}{\tau_{2D}}\right)=\widehat{S}. \tag{6}$$

Such simplification has previously been done for metallic materials even under heterogeneous strain states [9, 12, 69–71] and more recently applied for characterizing the calorimetric signature of the different mechanisms in play in the homogeneous [43, 44, 47–51] and heterogeneous deformations [58, 60] of rubber.

2.3 *Energy Balance*

The calorimetric response accompanying the mechanical response gives access to numerous additional information about the deformation process. For instance, the difference between the heat absorbed and produced over one thermodynamic cycle provides the mean intrinsic dissipation, i.e., the mechanical energy converted into heat, which can be compared to the energy involved in the hysteresis loop. This provides therefore a direct evaluation of the mean intrinsic dissipation responsible for the self-heating. It should be noted that two types of dissipation can be obtained, the thermal one due to heat diffusion and the intrinsic one due to irreversible deformation processes. The intrinsic dissipation is a power density, not an energy. The mean intrinsic dissipation is an energy rate. The reader can refer to Ref. [72] for further information on the thermodynamics of irreversible processes.

The difference between the energy involved in the hysteresis loop and the mean intrinsic dissipation corresponds to the mechanical energy stored in the material and that changes its internal energy. The continuum quantities required for such an energy balance are listed below:

[5]The relevancy of this assumption for a given situation can be evaluated by calculating the Biot number B_i [63].

- The total strain energy density (per unit reference volume) W_{strain} is the energy supplied mechanically to the material during the loading W_{strain}^{load} and the unloading W_{strain}^{unload},

$$W_{strain}^{load} = \int_{loading} \pi d\lambda \text{ and } W_{strain}^{unload} = \int_{unloading} \pi d\lambda, \text{respectively,} \tag{7}$$

where π denotes the first component of Piola–Kirchhoff stress tensor $\mathbf{P}$ in the loading direction and λ is the corresponding stretch.

- The mechanical energy involved in the hysteresis loop W_{hyst}^{cycle} and the corresponding energy rate P_{hyst}^{cycle} are determined as follows:

$$W_{hyst}^{cycle} = W_{strain}^{load} - W_{strain}^{unload}, P_{hyst}^{cycle} = \frac{W_{hyst}^{cycle}}{t_{cycle}} \tag{8}$$

where t_{cycle} denotes the cycle duration.

- Integrating the heat source over time of each cycle gives the mean intrinsic dissipation $\widetilde{\mathcal{D}}_{int}$:

$$\widetilde{\mathcal{D}}_{int} = \frac{1}{t_{cycle}} \int_{cycle} S dt. \tag{9}$$

- The difference between P_{hyst}^{cycle} and $\widetilde{\mathcal{D}}_{int}$ represents the energy stored rate P_{stored}^{cycle} during each cycle:

$$P_{stored}^{cycle} = P_{hyst}^{cycle} - \widetilde{\mathcal{D}}_{int}, \tag{10}$$

and the corresponding mean energy is therefore:

$$W_{stored}^{cycle} = \frac{P_{stored}^{cycle}}{t_{cycle}}. \tag{11}$$

- The stored energy ratio γ_{se} is defined as

$$\gamma_{se} = \frac{W^{cycle}_{stored}}{W^{cycle}_{hyst}} \tag{12}$$

and further characterizes the ability of the material to store energy:

- If γ_{se} tends to 0, no energy is stored during the deformation. The whole hysteresis loop is due to the intrinsic dissipation.
- If γ_{se} tends to 1, the whole hysteresis loop is due to stored energy and no intrinsic dissipation is detected. This is typically the case in unfilled natural rubber [48]. It should be noted in this case that this energy is released with a different kinetics during the unloading.

This ratio is influenced by both the strain range and the strain rate applied to the sample [52].

2.4 Strain-Induced Crystallinity Evaluation

The strain-induced crystallinity of rubber is classically evaluated by the XRD technique, which has the advantage of providing additional information of importance on the crystalline phase structure [73–76], chain orientation [77], and crystallization kinetics [78, 79]. Since SIC is highly exothermic, alternative methods based on temperature or calorimetry measurements can be used to determine the crystallinity. To this end, Le Cam proposed a technique to evaluate the thermal energy and the corresponding temperature variation due to SIC (denoted ΔT_{cryst} in the following) from surface calorimetry [54], without using the mechanical response, in particular the strain power density. The crystallinity is obtained by considering that the crystallization energy of natural rubber can be approximated by the fusion enthalpy ΔH_{cryst}, which is assumed to be independent of temperature and stretch. Here, the value determined in Ref. [80] and set $\Delta H_{cryst} = 59.9\ J.\ cm^{-3}$ is used. The crystallinity can be expressed in terms of ΔT_{cryst} as follows:

$$\chi(\mathrm{t}) = \frac{\rho C \Delta T_{cryst}(t)}{\Delta H_{cryst}} \tag{13}$$

For filled materials, this ratio has to be weighted by a factor $\frac{1}{1-\Phi_{filler}}$, where Φ is the filler volume fraction.

The methodology is briefly recalled by the diagram in Fig. 1. It is composed of four steps:

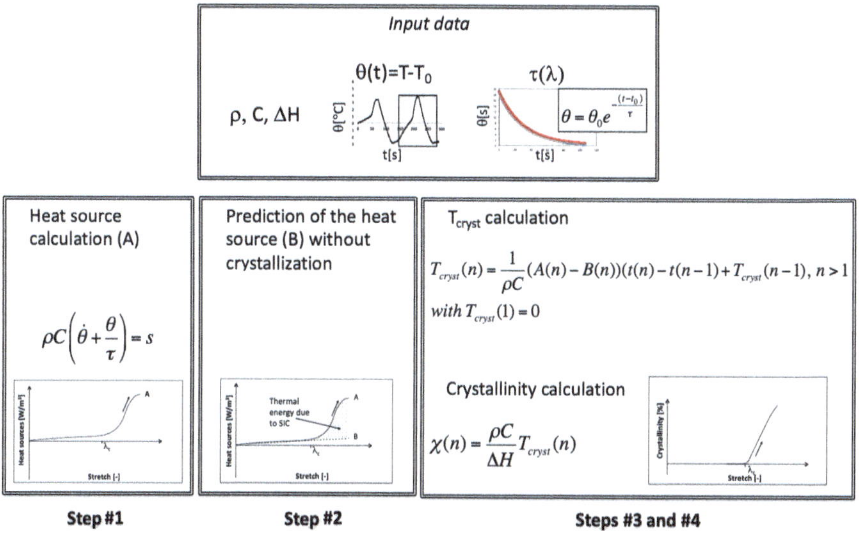

Fig. 1 Methodology for evaluating the crystallinity from the calorimetric response

- Step #1: the temperature variation $\theta(t)$ and parameter $\tau(\lambda)$ are processed with the 0D formulation of the heat diffusion equation to calculate the heat source.
- Step #2: the thermal energy due to SIC is determined from the area between the heat source measured and the heat source prediction due to thermoelastic (entropic and energetic) couplings only.
- Steps #3 and #4: the corresponding temperature variation due to SIC is calculated for a numerical scheme of integration and the crystallinity is finally obtained with Eq. (13).

Results obtained with this method show very good agreement with those obtained with the XRD technique when applied to the same unfilled NR [55].

Remark on the heat source prediction due to thermoelastic couplings only: for unfilled NR, the thermal energy of SIC is well approximated whatever the polynomial chosen to predict the heat source due to thermoelastic couplings only. Indeed, the exothermal peak due to SIC is of high intensity and of narrow width, so the degree of the polynomial has very few effects on the prediction. Recently, Le Cam et al. studied the crystallinity measurement in an NR unfilled and filled with different types of fillers [56]. By comparison with XRD measurements, the authors showed that even the tangent to the heat source before SIC starts can be used in the case of unfilled NR. However, in the case of filled NR, the calorimetric response exhibits a lower in intensity and larger peak. Therefore, the degree of the polynomial chosen has a strong influence on the results. The authors demonstrated that the tangent or a Neo-Hookean-like expression fitted before SIC starts provides accurate results and applies for both filled and unfilled NR.

3 Experimental Setup

Numerous experimental setups have been developed to characterize the calorimetric response of elastomers. They differ in terms of materials tested, loading conditions, and sample dimensions.

The aim of this section is to describe to the reader how the calorimetric response can be characterized in the simplest way and what are the different families to be distinguished in terms of the calorimetric response. This section is based on tests performed in our group at the Institute of Physics of Rennes (University of Rennes, France), which are summarized in Table 1. Only homogeneous strain fields at the macroscopic scale are considered here.

3.1 *Materials*

Several elastomers have been characterized with the stretched surface calorimetry technique. So far, crystallizing NR and thermoplastic polyurethane (TPU) and non-crystallizing styrene butadiene rubber (SBR) and nitrile butadiene rubber (NBR) have been studied. In addition to SIC, other factors affect the calorimetric response, especially the addition of fillers. Fillers induce both viscosity and stress softening (the Mullins effect) that have specific calorimetric signatures. Therefore, more than the chemical composition itself, what distinguishes the different materials according to their calorimetric response is their belonging to one of the four main families of compounds: filled or unfilled, and crystallizing or not crystallizing under strain.

From a thermodynamic point of view, the required material parameters are the density, the thermal conductivity when heat conduction occurs, and the heat capacity. Generally, volume variations are neglected, which simplifies the prediction of $\tau(\lambda)$ (see Sect. 2.2.1). The heat capacity can be measured as a function of temperature (see, e.g., Ref. [57]) or a mean value can be chosen (see, e.g., Ref. [48]).

3.2 *Sample Geometries and Loading Conditions*

Sample geometries used for uniaxial loadings are thin; 2 mm maximum (to meet conditions exposed in Sect. 2.2.1), widths between 4 and 20 mm, and lengths of 10 to ca 40 mm. Obviously, the smaller the length, the smaller the width must be to meet uniaxial tension loading condition. The purpose here is not to provide an exhaustive list of the various dimensions reported in the literature, but rather to discuss some experimental precautions. Typically, the initial sample length has an effect on the measurement: if the loading is applied with only one jaw, the longer the sample, the greater the displacement in the air of the surfaces to achieve the same

Table 1 Summary of the setups used for surface calorimetry with an IR camera, with t: thickness, w: width, l: length, S: symmetrical tension, NS: non-symmetrical tension, MC: motion compensation of the measurement zone in the IR images, CJ: Cedip Jade III-MWIR 320 × 240 px, FX6: FLIR X6540sc InSb 640 × 512 px, phr: part per hundred of rubber in weight

Reference	Material	Filler type and amount (phr)	t × w × l [mm^3]	Loading rate [mm/min]	Strain rate ± [s^{-1}]	Loading conditions	MC	IR camera
[48, 81]	NR	0	1.4 × 5 × 10	100, 300	0.167, 0.500	NS, 3 cycles at $\lambda = 2, 5, 6, 7.5$	X	CJ
[49]	SBR	5 CB (N347)	1.4 × 5 × 10	300	0.500	NS, 3 cycles at $\lambda = 2, 3, 3.5, 4$	X	CJ
	SBR	50 CB (N347)	1.4 × 5 × 10	300	0.500	NS, 3 cycles at $\lambda = 2, 3, 4, 4.5$	X	CJ
[50]	NR	50 CB (N347)	1.4 × 5 × 10	300	0.500	NS, 3 cycles at $\lambda = 1.4, 2, 4, 6$	X	CJ
	SBR	50 CB (N347)	1.4 × 5 × 10	300	0.500	NS, 3 cycles at $\lambda = 2, 3, 4, 4.5$	X	CJ
[51]	SBR	5 CB (N347)	1.4 × 5 × 10	300	0.500	NS, 3 cycles at $\lambda = 2, 3, 3.5, 4$	X	CJ
	SBR	50 CB (N347)	1.4 × 5 × 10	300	0.500	NS, 3 cycles at $\lambda = 2, 3, 4, 4.5$	X	CJ
	NR	0	1.4 × 5 × 10	300	0.500	NS, 3 cycles at $\lambda = 2, 5, 6, 7.5$	X	CJ
	NR	50 CB (N347)	1.4 × 5 × 10	300	0.500	NS, 3 cycles at $\lambda = 1.4, 2, 4, 6$	X	CJ
[43][a]	NR	0	1.4 × 5 × 10	100	0.167	NS, 3 cycles at $\lambda = 2, 5, 6, 7.5$	X	CJ
[52]	TPU[b]	0	5 × 9 × 20	100, 300	0.083, 0.250	NS, 5 cycles at $\lambda = 1.5, 2, 2.5, 3$	X	FX6
[53]	NBR	0	1.9 × 7.25 × 20	50, 100, 200	0.042, 0.083, 0.167	NS, 5 cycles at $\lambda = 1.5, 2, 2.5, 3, 3.5, 4$	X	FX6
	NBR	35 CB (N347)	1.9 × 7.25 × 20	50, 100, 200	0.042, 0.083, 0.167	NS, 5 cycles at $\lambda = 1.5, 2, 2.5, 3, 3.5, 4$		FX6
[54][a]	NR	0	1.4 × 5 × 10	100	0.167	NS, 3 cycles at $\lambda = 2, 5, 6, 7.5$	X	CJ
[55]	NR	0	1.4 × 7.2 × 19.3	100, 200	0.086, 0.173	S, 3 cycles at $\lambda = 7$		FX6
[60]	NR	20–30 CB	2 × 10 × 24	300	0.208	S, 3 cycles at $\lambda = 2.5, 4, 6$		FX6
[57]	NR	0	2 × [8.36–18.8] × 12	300	0.417	S, 3 cycles at $\lambda = 7$		FX6
[56]	NR	0	2 × 5 × ca 40[c]	300	ca 0.125	S, 5 cycles at $\lambda = 6$		FX6
	NR	50, 3 different CB types	2 × 5 × ca 40	300	ca 0.125	S, 5 cycles at $\lambda = 2, 3, 4, 5, 6$		FX6

[a] Experimental data from Refs. [48, 81]
[b] Irogran A87H4615 Huntsman Corporation (The Woodlands, Texas)
[c] The sample ends were wound around a metal axle. The initial length was measured once the sample was fixed and not buckled

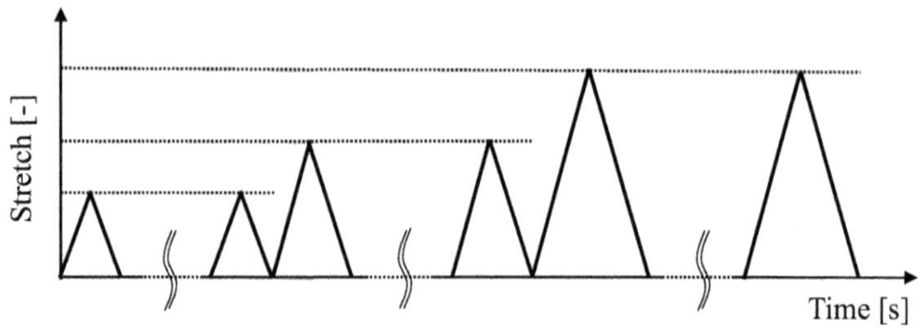

Fig. 2 Typical loading profile under uniaxial tension. Several cycles of increasing maximum strain are applied at the same strain rate

stretch, and the more the convection to be forced. This is a key point because τ is evaluated from the thermalization of a static surface. From this point of view, it is preferable to carry out symmetrical tension test whenever possible with the equipment used [54, 55, 60]. It should be noted that we will not deal with relaxation tests or cyclic tests with relaxation phases, but the reader can refer to Refs. [48, 60] on these subjects. Nor will we discuss tests under heterogeneous loading, which are carried out with 3- or 4-branch samples and are aimed more at developing methods for motion compensation [59] or inverse identification of constitutive parameters without boundary conditions [58], or to study the effect of multiaxiality on the calorimetric response [60] or to characterize the functioning of the crack-laying zone [61, 62, 64].

The loading conditions generally consist of several cycles of increasing strain amplitude. These cycles are also repeated even if the material does not exhibit significant accommodation (stress softening), in order to stabilize the thermal response, which is a key point for the proper assessment of the intrinsic dissipation [12]. Figure 2 shows a typical loading case, where several cycles (usually 3 to 5 maximum) of increasing maximum strain are applied. The different geometries and loading and thermal measurement conditions are summarized in Table 1.

3.3 *Temperature Measurement*

Thermal measurements are performed with IR cameras. The temperature fields obtained allow us to study heterogeneous heat source fields, a typical example being a crack tip zone. The IR cameras are usually turned on several hours before the test to stabilize the temperature of the IR detector matrix. Acquisition frequencies are generally a few tens of Hertz to reduce measurement noise by using filters before deriving temperatures with respect to time. With cooled cameras, the thermal resolution or noise equivalent temperature difference (NETD) is about 20 mK for a temperature range between 5 and 40°C. The spatial resolution of the thermal field ranges generally between 200 and 500 μm/px. The surface emissivity of the tested materials is set at 0.92 ± 0.02. This value was verified with NR samples, filled and unfilled by comparing the temperature of the sample surfaces placed in an oven and heated to 50°C with that of a black body regulated at the same temperature. It should be noted that an optical infrared pyrometer could be used for the temperature measurement, especially if the loading is applied symmetrically.

4 Typical Calorimetric Responses under Uniaxial Tensile Loading

This section introduces the different types of calorimetric response. They are reviewed from the results reported in Ref. [51]. The crystallizable and non-crystallizable rubbers are those among the most produced each year, namely NR and SBR, respectively. The filler type considered was the carbon black, in a quantity of 50 phr. It should be noted that the unfilled SBR was filled with 5 phr to improve its manufacturing. Figure 3 presents their mechanical responses in terms of the nominal stress versus the stretch. The loading was applied under prescribed displacement, at a rate of 300 mm/min (a strain rate of 0.5 s^{-1}).

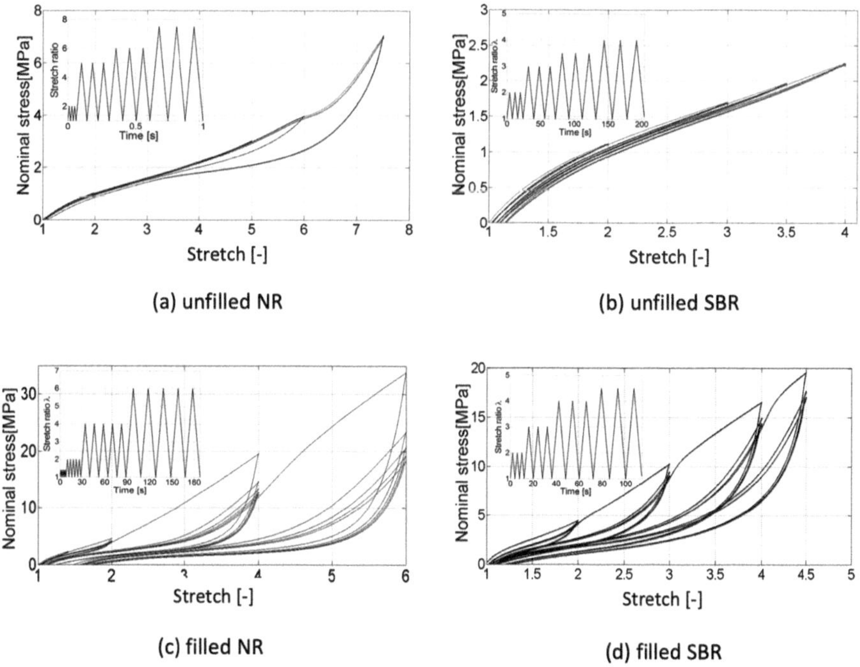

Fig. 3 Mechanical response of NR and SBR, unfilled and filled with carbon black (from Ref. [51])

For the unfilled rubbers, the mechanical cycles, i.e., the number of cycles at a given stretch, have no effect on the mechanical response. No significant permanent strain was observed in unfilled NR. In SBR, the 5 phr of carbon black may be responsible for the slight residual stretch observed. No hysteresis loop is observed for the SBR. This is also the case with NR if the maximum stretch does not significantly exceed that of the onset of crystallization (approx. 4). For higher stretches, a hysteresis loop is formed. This phenomenon is due to the difference in the kinetics of crystallization and crystallite melting, which was first instituted by Clark as soon as 1940 [26] and demonstrated in Refs. [47, 52, 82]. As expected, the addition of fillers induces a hysteresis loop (whatever the stretch reached) that decreases with the number of cycles and stabilizes at a given maximum stretch, which is accompanied by a significant permanent stretch and stress softening. These phenomena were first observed by Bouasse and Carrière [17] and extensively studied by Mullins [19, 83, 84].

Figure 4 presents the calorimetric responses determined from the temperature measurements and the simplified heat diffusion equation (Eq. (6)).

In the literature, the calorimetric response is given either in W/m^3, the unit of $\widehat{S}$, or in K/s, the unit of $\frac{\widehat{S}}{\rho_0 C}$. This latter unit has a more intuitive meaning, i.e., the amount of Kelvin gained or lost per unit of time.

For all the compounds, an increase in the stretch induces an increase in the heat source. This is a consequence of the entropic elasticity, which is preponderant in rubbers compared to non-entropic (energetic) elasticity. It should be noted that for the lowest stretches, a thermoelastic inversion is generally observed, which corresponds to an inversion in the heat source sign. This is not discussed in this chapter, but the reader can refer to Refs. [46, 48, 81, 85] for further information on this phenomenon that is due to non-entropic effects.

In filled NR, the stabilized heat source is close to the maximum heat source obtained in unfilled NR at a given maximum stretch, which was not the case for SBR. In fact, at $\lambda = 4$, the heat source in filled SBR is approximately 4 times that in unfilled SBR. A strong difference (a factor 3 at $\lambda = 4$) was also observed with a non-crystallizing NBR in Loukil et al. [53], unfilled and filled with 35 phr of carbon black. This difference was shown not being affected by the loading rates that were applied (see the loading rate range in Table 1). It should be noted that for the unfilled and the stabilized response of filled SBR, the heat source is symmetrical between the loading and unloading. For the unfilled rubber, it is observed when the maximum stretch applied is inferior to λ_C.

In the next paragraphs, special attention is paid to the calorimetric signature of the different phenomena involved in the rubber deformation.

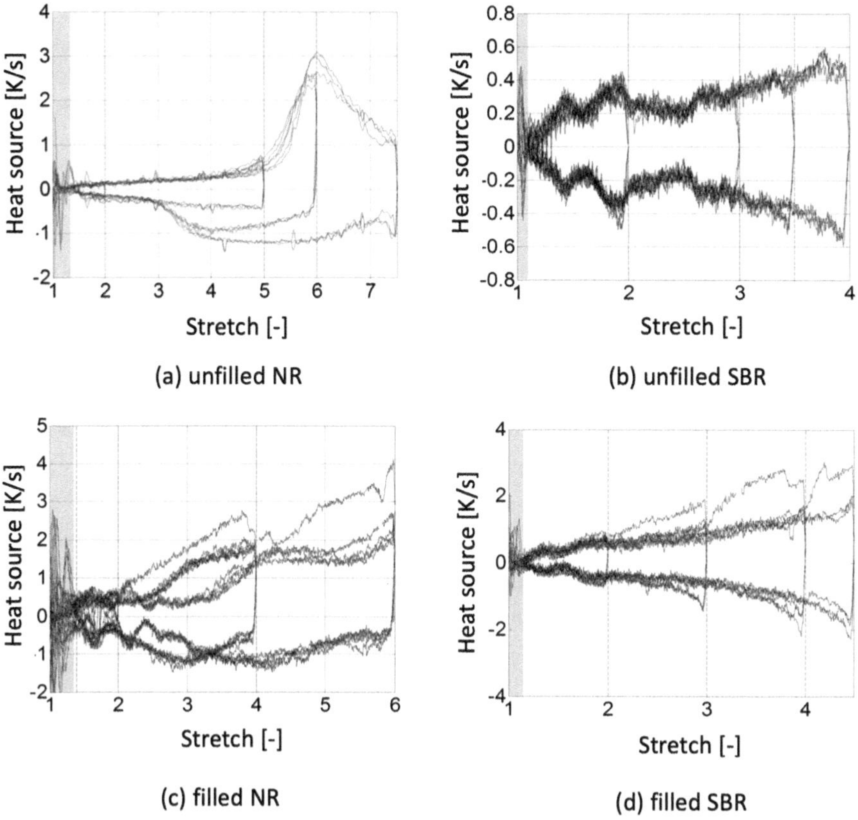

Fig. 4 Calorimetric response of NR and SBR, unfilled and filled with carbon black (extracted from data appearing in Ref. [51])

4.1 Heat Sources Due to SIC

4.1.1 Qualitative Analysis on How SIC and Crystallites Melting Affect the Calorimetric Response

The effects of SIC and crystallite melting can be discussed from the calorimetric response of unfilled NR, whose hysteresis loop is due neither to viscosity nor stress softening. Figure 4a highlights strong singularities in the calorimetric response of unfilled NR compared to the unfilled non-crystallizing SBR (see Fig. 4b). As the stretch rate is constant over the test, these singularities can only be due to strain-induced crystallization and crystallite melting. When the mechanical cycles are performed at maximum stretches inferior to λ_C, the heat source evolves symmetrically between loading and unloading with respect to the zero heat source axis. This indicates that no intrinsic dissipation has occurred. The symmetry is lost when maximum stretch equal to 5 is applied. During loading, the heat source evolves quasi-linearly until reaching a stretch close to λ_C, from which an exothermal peak forms. Moreover, Samaca Martinez et al. [48] found that the areas under the loading and unloading curves were equal, which confirms that no heat was produced due to intrinsic dissipation for the loading conditions tested. When the maximum applied stretch increases, loading–unloading dissymmetry increases as the maximum crystallinity reached increases. From a mechanical point of view, the area of the hysteresis loop also increased with the maximum stretch reached. As no mechanical dissipation was detected, this is the demonstration that a hysteresis loop can form without any detectable intrinsic dissipation. From stretches equal to 6 during the loading phase, the heat source decreased instead of increasing continuously. As the heat source remains positive, this means that heat continues to be produced, but at a lower rate. Several causes were addressed in Le Cam et al. [51]: (i) the crystallinity saturation if any is approached; (ii) the contribution of internal energy increases due to crystallite deformation; and (iii) a less exothermal crystallization process for the highest stretches.

4.1.2 Quantitative Analysis: Determination of the Crystallinity from the Calorimetric Response

As the crystallization is strongly exothermal and that a link is established between the enthalpy of fusion, the density, the heat capacity, the temperature variation due to SIC, and the crystallinity, the crystallinity can be measured from the calorimetric response by applying the method described in Sect. 2.4.

First of all, Fig. 5a shows the variations of the power densities during the test: the strain power density and the heat source (curve A), as well as a prediction of the heat power density without crystallization (curve B) for an unfilled NR different from the

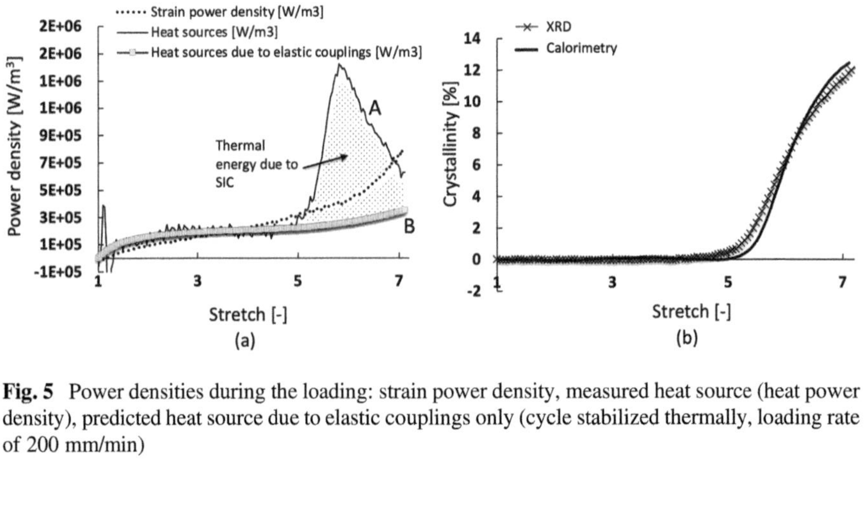

Fig. 5 Power densities during the loading: strain power density, measured heat source (heat power density), predicted heat source due to elastic couplings only (cycle stabilized thermally, loading rate of 200 mm/min)

previous one (information on the material is given in Le Cam et al. [55]). This prediction is made using the method described in Ref. [54]. The area between curves A and B corresponds to the thermal energy of crystallization. This energy is used to derive the crystallinity. The result obtained is compared with the XRD measurement in Fig. 5b. This figure shows the surface calorimetry technique provides an accurate crystallinity measurement. We recall that for unfilled NR, the tangent or Neo-Hookean-like prediction fitted before SIC starts provides an accurate measurement of the crystallinity (see Le Cam et al. [56]).

It should be noted that if the material were purely entropic, the thermal energy of crystallization would be the area between curve A and the curve in dotted line. Moreover, the curves would superimpose before crystallization elongation was reached. This is not the case. Furthermore, if we were to determine the thermal energy of crystallization as the area between these two curves, we would obtain a negative crystallinity from lambda equal to 3.5 and consequently underestimate the crystallinity, especially for the highest elongations. This shows once again that (i) the elasticity of unfilled NR is not purely entropic and (ii) crystallinity cannot be systematically determined by coupling thermal (or calorimetric) measurement and stress (or strain power density) measurement, since increasing the stretch always increases the stress, but can induce a temperature decrease (negative heat source due to energetic elasticity such as in metals), a temperature increase (positive heat source due to entropic elasticity), or both (in rubber-like materials).

Remark on SIC heterogeneity under homogeneous loadings: a recent study by Le Cam et al. [57] investigated heterogeneities in the crystallinity field in unfilled NR under uniaxial tension. The authors regularly observed temperature fields on cruciform samples in areas where the mechanical field should be quasi-homogeneous (uniaxial tension). The areas that were hottest during loading were those that absorbed the most heat when discharged; the maximum local stretch reached being greater than λ_C. This phenomenon is highlighted in Fig. 6a. The authors undertook a study to investigate this heterogeneity using homogeneous uniaxial tensile tests. Their measurements, shown in Fig. 6b show highly heterogeneous thermal fields and crystallinities along the width at half-height of the sample that can vary by almost 3 %. This was the first evidence of SIC heterogeneity under a homogeneous loading, which could be explained by strain/stress concentrations that initiate SIC (i) the distribution in the active chain density and (ii) the presence of particles.

4.2 *Heat Sources Due to Viscosity and Stress Softening in Filled Rubbers: Determination of the Mean Intrinsic Dissipation*

When fillers were added to the compounds, the first mechanical cycle at a given maximum stretch ratio resulted in a higher maximum heat source and a decreasing value until stabilization for the following cycles. All the cycles, even the first one,

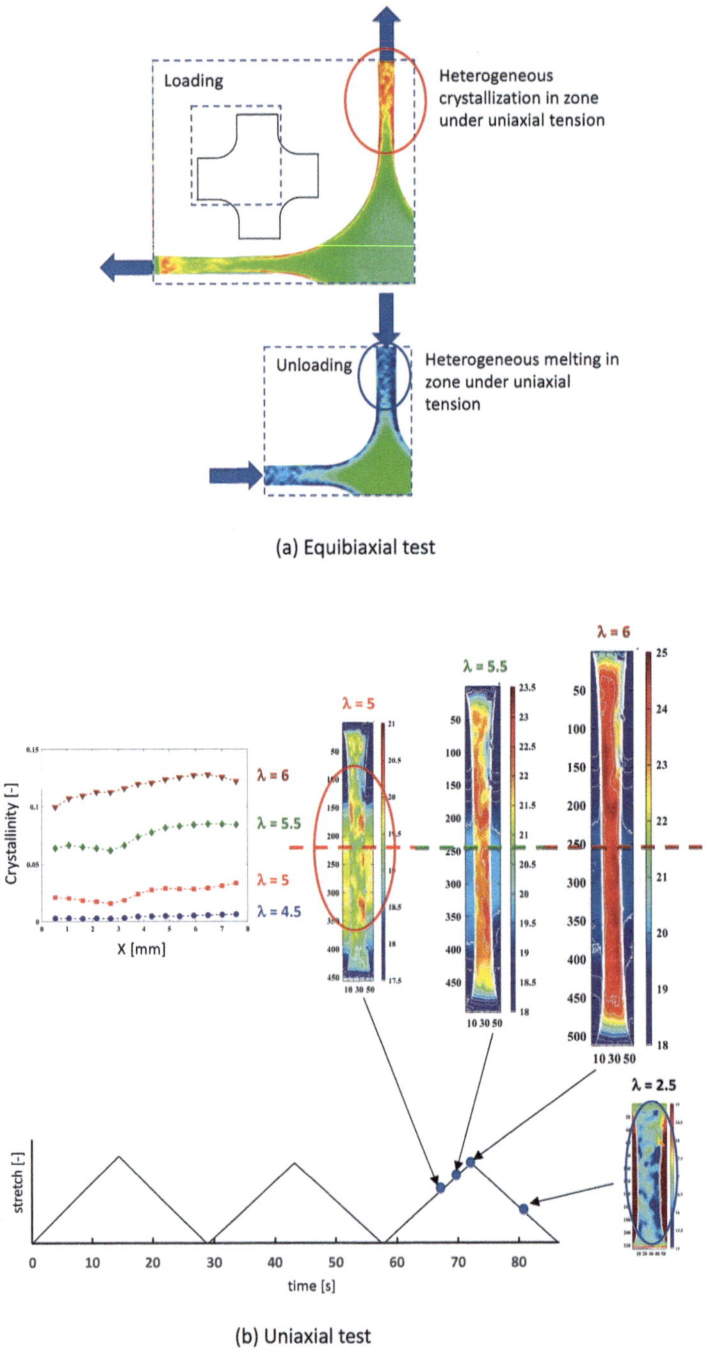

Fig. 6 Heterogeneous crystallization and melting (**a**) in a cruciform sample submitted to equibiaxial tension loading (**b**) under uniaxial tension loading

show the same heat source evolution during the unloading. This is clearly shown in Fig. 4c, d. Therefore, the calculation of the mean intrinsic dissipation $\widetilde{\mathcal{D}}_{int}$ applied to the first cycle provides an evaluation of the mean intrinsic dissipation due to the stress softening and to the viscosity, and when applied to the stabilized (mechanical and thermal) cycles, the calculation provides the mean intrinsic dissipation due to viscosity.[6]

Thus, the mean intrinsic dissipation due to viscosity can be removed from the mean intrinsic dissipation of the first cycles in order to evaluate the mean intrinsic dissipation due to the stress softening. Figure 7 gives an example of the mean intrinsic dissipation due to Mullins effect and to viscosity in the filled NR tested in Khiem et al. [60] We can observe that $\widetilde{\mathcal{D}}_{int}$ is almost linear with stretch for the applied loading conditions. This is an important information for modeling the viscosity and the Mullins effect in filled rubbers.

4.3 *Investigating How Rubbers Use the Mechanical Energy to Deform: An Energetic Approach*

The previous results clearly demonstrate that the mechanical energy involved in the hysteresis loop is not systematically dissipated into heat by viscous processes, the typical example being the hysteresis loop observed in the mechanical response of unfilled NR only when SIC occurs. The fact that the hysteresis can be partly or even not due to viscosity explains why it cannot be time-dependent [53, 86–89] and why the self-heating under repeated cycles is generally much less than the one predicted by assuming that all the energy involved in the hysteresis loop is converted into heat. This is generally observed by measuring the self-heating during fatigue-like test loadings [90–92].

Thus, other factors affect the mechanical hysteresis:

(a) The thermal dissipation (under non-adiabatic test conditions). In most of the homogeneous (in terms of heat source field) tests performed, considering the thermal properties of elastomers and the loading rate relatively high, the thermal dissipation does not contribute significantly to the mechanical hysteresis.
(b) The change in microstructure. In this case, as observed by Mott et al. [93] for polyurea, all the work done to the system is not measured as an apparent temperature change. The authors concluded that a significant part of the mechanical energy is used by the material to reorganize, which changes its internal energy. The authors also related this mechanical energy "lost" in the microstructure to the residual strain. This result could be discussed in terms of energy stored

[6]The mean intrinsic dissipation has to be calculated over a thermodynamic cycle, i.e., with the same temperature and strain between the loading beginning and the unloading end [12]. This means that only the mechanical and thermally stabilized cycle provides quantitative data on the heat source produced by the viscosity.

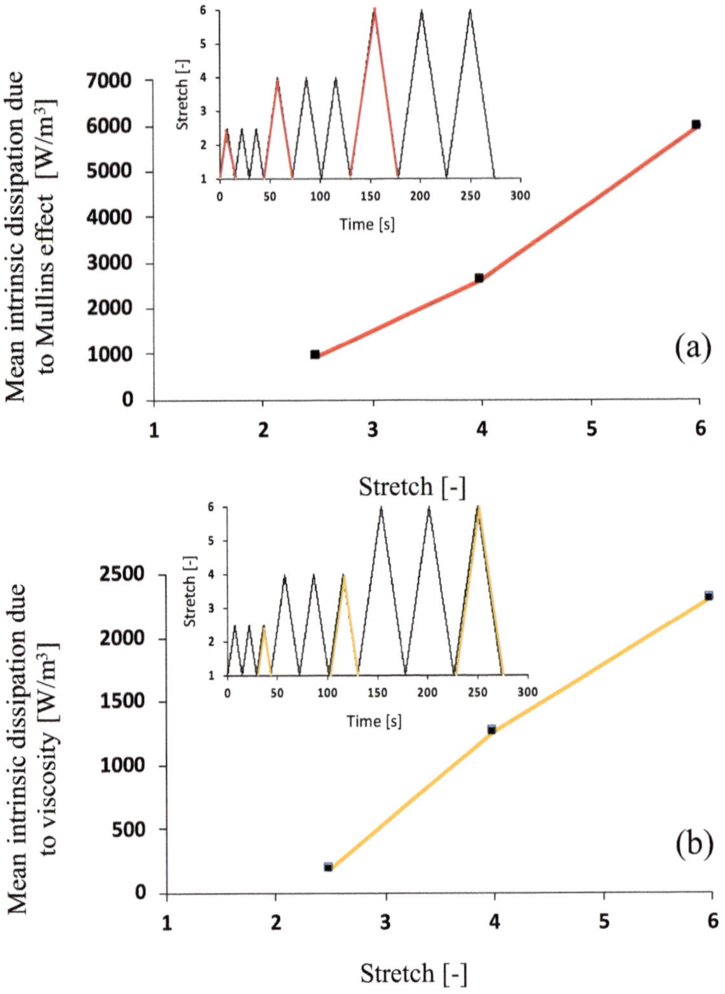

Fig. 7 Mean intrinsic dissipation ($\widetilde{\mathcal{D}}_{int}$) versus stretch due to (**a**) the Mullins effect and (**b**) the viscosity. The lines are guides to the eye

and not released. Le Cam [43] showed that the mechanical hysteresis of the unfilled NR he studied was entirely due to the difference in kinetics between crystallization and crystallite melting. No residual strain accompanied the occurrence of the hysteresis loop. Therefore, energy is stored in the material during the crystallization process and is released at a different rate during melting. This energy storage may be due not only to the SIC but also to the filler network. In fact, Loukil et al. [53] demonstrated that filled NBR he studied stored part of the mechanical energy in the filler network reorganization and released it at a different rate during unloading. This should be due to the fact that the viscous effect due to fillers change the kinetics of energy storage and release between loading and unloading.

Therefore, the amount of self-heating, or more relevant, intrinsic dissipation, must be put into perspective in relation to the energy rate involved in the hysteresis loop. Even more interesting is the part of the mechanical energy brought that is stored and released by reversible microstructure changes and is not dissipated as heat or used to damage the material, for example to propagate cracks. This is the aim of the stored energy ratio γ_{se} defined in Sect. 2.3, which enables us to characterize and to compare the energetic behavior of rubbers. An illustration is given in Fig. 8 for 3 different materials. The data are available in Lachhab et al. [52] for the TPU, in Loukil et al. [53] for the filled NBR, and in Samaca Martinez et al. [48] for the unfilled NR. Only the zones containing the responses are represented here. It is shown that the energetic response of unfilled NR does not depend on the stretch, nor on the stretch rate (in the range investigated). Similar to unfilled NR, TPU response is not stretch-dependent. Nevertheless, it depends on the stretch rate and on the density, but the responses are included in a range where γ_{se} is close to one. For the filled NBR, γ_{se} depends on both stretch and stretch rate. This type of energetic map brings new insights for understanding how rubbers use the mechanical energy to deform and motivate the systematic measurement of the temperature during mechanical test.

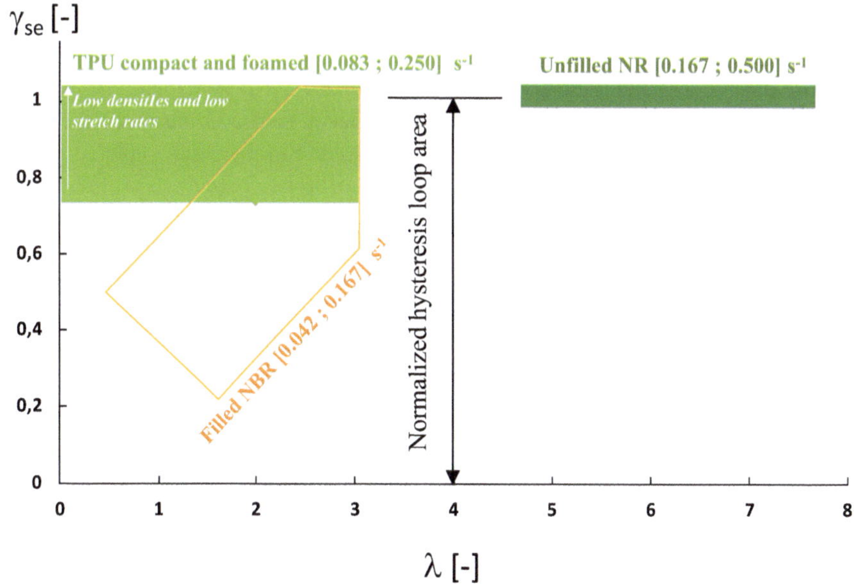

Fig. 8 γ_{se} evaluated from thermodynamic cycles

5 Conclusion

This chapter reviews the calorimetric signatures of the main phenomena involved in the rubber deformation, i.e., entropic and energetic elasticity, reinforcement by fillers, strain-induced crystallization, stress softening, and energy storage.

For this purpose, the thermodynamic framework to determine the heat source from full temperature field measurements processed with the heat diffusion equation is recalled, as well as the methodology to evaluate the crystallinity from the calorimetric response.

The main conclusions are summarized as follows:

- When the rubber is stretched (relaxed), entropic elasticity is the major contributor to the calorimetric response and leads to heat production (absorption). The load–unload heat source curves are symmetrical in an unfilled and non-crystallizing rubber. This does not mean that the material behaves as a pure entropic material. In fact, the strain power density and the heat power density do not superimpose, which means that changes in the internal energy also contribute to the calorimetric response.
- The symmetry in the heat source between loading and unloading in unfilled NR is no longer observed if crystallization starts. The areas under the calorimetric load–unload curves are equal, meaning that crystallization and melting occur without detectable intrinsic dissipation.

Addition of fillers induces viscosity and stress softening, which is accompanied by the production of intrinsic dissipation that can be quantified and related to the energy rate of the mechanical hysteresis with the γ_{se} ratio.

- The mechanical and the calorimetric responses do not systematically stabilize at the same number of cycles. This has to be taken into account since the mean intrinsic dissipation must be determined from a thermodynamic cycle.
- The crystallinity in unfilled NR was found to be heterogeneous at the macroscopic scale. Differences in the order of magnitude of one percent can be found from one point to another at the surface of a stretched sample. Since fillers concentrate the strain and the stress much more than active chain density variations or particles, experiments are currently being performed in our group in order to define whether or not the addition of fillers, i.e., the addition of much stronger stress/strain concentrators, smooths this heterogeneity.

In conclusion, surface calorimetry offers a wide range of complementary information to better understand the deformation and damage mechanisms in rubbers. The perspectives offered by the extension of this technique to heterogeneous heat source field are significant, typically to study the calorimetric behavior of the crack tip zone in crystallizing rubbers.

6 Acknowledgments

The author acknowledge the Association nationale de la recherche et de la technologie (ANRT) for the financial support (CIFRE grant) of the PhD theses of Dr. Samaca Martinez and Dr. Lachhab, and the "Programme doctoral normalien. nes" from Ecole Normale Supérieure de Rennes (CDSN grant) for the PhD thesis of Dr. Charlès. The author thanks his various industrial partners for their financial support: PCM (France), Cooper Standard (France), Continental (France), and La Manufacture des Pneumatiques Michelin (France). The author acknowledges his academic partners for the fruitful discussions and their significant contribution to the advancement of the surface calorimetry in elastomeric materials by means of full thermal field measurements: Dr. Robin, Dr. Miroir, Mr. Le Fur, and Mr. Burgaud from the University of Rennes (France), Prof. Balandraud from Sigma Clermont (France), Prof. Toussaint and Prof. Grédiac from the University Clermont-Auvergne (France), and Prof. Itskov and Dr. Khiem from the University of Aachen (Germany). Full thermal and kinematic measurements at the University of Rennes were performed with cameras purchased thanks to the financial support of the Région Bretagne, Rennes Métropole, the Département d'Ille et Vilaine, and the European Union (CPER-FEDER 2015–2023 "Institut du Verre").

References

1. Farren WS, Taylor GI (1925) The heat developed during plastic extension of metals. Proc R Soc Lond A Math Phys Eng Sci 107(743):422–451
2. Taylor GI, Quinney H (1934) The latent energy remaining in a metal after cold working. Proc R Soc Lond A Math Phys Eng Sci 143(849):307–326
3. Chrysochoos A (1985) Energy balance for elastic plastic deformation at finite strain (in French). J Méc Appl 5:589–614
4. Chrysochoos A, Maisonneuve O, Martin G, Caumon H, Chezeau JO (1989) Plastic and dissipated work and stored energy. Nucl Eng Des 114:323–333
5. Mason JJ, Rosakis AJ, Ravichandran G (1994) On the strain and strain rate dependence of the fraction of plastic work converted to heat: an experimental study using high speed infrared detectors and the Kolsky bar. Mech Mater 17(2):135–145
6. Rittel D (1999) On the conversion of plastic work to heat during high strain rate deformation of glassy polymers. Mech Mater 31(2):131–139
7. Oliferuk W, Maj M, Raniecki B (2004) Experimental analysis of energy storage rate components during tensile deformation of polycrystals. Mater Sci Eng A 374(1):77–81
8. Bever MB, Holt DL, Titchener AL (1973) The stored energy of cold work. Prog Mater Sci 17: 5–177
9. Chrysochoos A, Louche H (2001) Thermal and dissipative effects accompanying Lüders band propagation. Mat Sci Eng A 307:15–22
10. Berthel B, Wattrisse B, Chrysochoos A, Galtier A (2007) Thermoelastic analysis of fatigue dissipation properties of steel sheets. Strain 43:273–279
11. Wattrisse B, Chrysochoos A, Muracciole J-M, Nèmoz Gaillard M (2001) Analysis of strain localization during tensile tests by digital image correlation. Exp Mech 41:29–39

12. Chrysochoos A, Huon V, Jourdan F, Muracciole J-M, Peyroux R, Wattrisse B (2010) Use of full-field digital image correlation and infrared thermography measurements for the thermomechanical analysis of material behaviour. Strain 46:117–130
13. Rittel D, Rabin Y (2000) An investigation of the heat generated during cyclic loading of two glassy polymers. Part II: thermal analysis. Mech Mater 32(3):149–159
14. Benaarbia A, Chrysochoos A, Robert G (2014) Kinetics of stored and dissipated energies associated with cyclic loadings of dry polyamide 6.6 specimens. Polym Test 34:155–167
15. Benaarbia A, Chrysochoos A, Robert G (2015) Influence of relative humidity and loading frequency on the pa6.6 thermomechanical cyclic behavior: part ii. energy aspects. Polym Test 41:92–98
16. Clausius R (1865) Sur diverses formes facilement applicables qu'on peut donner aux equations fondamentales de la théorie mécanique de la chaleur (translated from German by H.-FBessard). Journal de mathématiques pures et appliquées 2ème série 10:361–400
17. Bouasse H, Carrière Z (1903) Courbes de traction du caoutchouc vulcanisé. Annales de la Faculté des Sciences de Toulouse 5:257–283
18. Holt WL (1931) Behavior of rubber under repeated stresses. J Ind Eng Chem 23:1471–1475
19. Mullins L (1948) Effect of stretching on the properties of rubber. Rubber Chem Technol 21: 281–300
20. Cadwell SM, Merril RA, Sloman CM, Yost FL (1940) Dynamic fatigue life of rubber. Ind Eng Chem 13:304–315. 12:19–23, reprinted in Rubber Chem and Tech
21. Fielding JH (1943) Flex life and crystallisation of synthetic rubber. Ind Eng Chem 35(12): 1259–1261
22. Rivlin RS, Thomas AG (1953) Rupture of rubber. I. Characteristic energy for tearing. J Polym Sci 10:291–318
23. Thomas AG (1960) Rupture of rubber – vi – further experiments on the tear criterion. J Polym Sci 3:168–174
24. Stoček R, Heinrich G, Kipscholl R, Kratina O (2021) Cut & chip wear of rubbers in a range from low up to high severity conditions. Appl Surf Sci Adv 6(1):100152. https://doi.org/10.1016/j.apsadv.2021.100152
25. Saintier N, Cailletaud G, Piques R (2011) Cyclic loadings and crystallization of natural rubber: an explanation of fatigue crack propagation reinforcement under a positive loading ratio. Mater Sci Eng A 528(3):1078–1086
26. Clark GL, Kabler M, Blaker E, Ball JM (1940) Hysteresis in crystallization of stretched vulcanized rubber from X-ray data. Ind Eng Chem 32:1474–1477
27. Rey T, Chagnon G, Le Cam J-B, Favier D (2013) Influence of the temperature on the mechanical behaviour of filled and unfilled silicone rubbers. Polym Test 32:492–501
28. Göritz D, Müller FH (1970) Die kalorimetrische erfassung der dehnungskristallisation polymerer. Kolloid-Zeitschrift und Zeitschrift für Polymere 241(20):1075–1079
29. Mitchell JC, Meier DJ (1968) Rapid stress-induced crystallization in natural rubber. J Polym Sci Part A-2 Polym Phys 6(10):1689–1703
30. Dedova S, Schneider K (2023) Characterisation of crack propagation in natural rubber under complex loading conditions using thermographic methods. In: Advances in polymer science. Springer, Berlin. https://doi.org/10.1007/12_2023_164
31. Wolf FP, Allen G (1975) Polymer 16:209
32. Allen G, Bianchi U, Price C (1963) Trans Faraday Soc 59:2493
33. Allen G, Kirkham MJ, Padget J, Price C (1971) Trans Faraday Soc 67:1278
34. Shen MC (1969) Macromolecules 2:358–364
35. Shen M, Croucher M (1975) Contribution of internal energy to the elasticity of rubberlike materials. Polym Rev 12:287–329
36. Treloar LRG (1973) The elasticity and related properties of rubbers. Rep Prog Phys 36(7):755
37. Treloar LRG (1975) The physics of rubber elasticity. Oxford University Press, Oxford
38. Heinrich G, Kaliske M, Klüppel M, Mark JE, Straube E, Vilgis TA (2003) The thermoelasticity of rubberlike materials and related constitutive laws. J Macromol Sci A 40:87–93

39. Price C (1976) Thermodynamics of rubber elasticity. Proc R Soc Lond A 351:331–350
40. Mark JE (1973) Thermoelastic properties of rubber like networks and their thermodynamic and molecular interpretation. Rubber Chem Technol 46(3):593–618
41. Gedde UW (1999) Chapter 3: the rubber elastic state. Springer, Dordrecht, pp 39–53
42. Heinrich G (2014) Thermoelasticity of rubbers. Springer, Berlin
43. Le Cam J-B (2017) Energy storage due to strain-induced crystallization in natural rubber: the physical origin of the mechanical hysteresis. Polymer 127:166–173
44. Honorat V (2006) Analyse thermomécanique par mesure de champs des élastomères. PhD thesis, Sciences et techniques du Languedoc. Montpellier, Université Montpellier 2
45. Joule JP (1857) On some thermodynamic properties of solids. Phil Mag 14:227
46. Anthony RL, Caston RH, Guth E (1942) Equations of state for naturals and synthetic rubber like materials: unaccelerated natural soft rubber. J Phys Chem 46:826
47. Caborgan R (2011) Contribution à l'analyse expérimentale du comportement thermomécanique du caoutchouc naturel. PhD thesis, Mécanique et Génie civil Montpellier 2
48. Samaca Martinez JR, Le Cam J-B, Balandraud X, Toussaint E, Caillard J (2013) Mechanisms of deformation in crystallizable natural rubber. part 2: quantitative calorimetric analysis. Polymer 54:2727–2736
49. Samaca Martinez JR, Le Cam J-B, Balandraud X, Toussaint E, Caillard J (2013) Filler effects on the thermomechanical response of stretched rubbers. Polym Test 32:835–841
50. Samaca Martinez JR, Le Cam J-B, Balandraud X, Toussaint E, Caillard J (2014) New elements on Mullins effect: a thermomechanical analysis. Eur Polym J 55:98–107
51. Le Cam JB, Martinez JRS, Balandraud X, Toussaint E, Caillard J (2015) Thermo-mechanical analysis of the singular behavior of rubber: entropic elasticity, reinforcement by fillers, strain-induced crystallization and the Mullins effect. Exp Mech 55(4):771–782
52. Lachhab A, Robin E, Le Cam J-B, Mortier F, Tirel Y, Canevet F (2018) Energy stored during deformation of crystallizing tpu foams. Strain:e12271
53. Loukil MT, Corvec G, Robin E, Miroir M, Le Cam J-B, Garnier P (2018) Stored energy accompanying cyclic deformation of filled rubber. Eur Polym J 98:448–455
54. Le Cam J-B (2018) Strain-induced crystallization in rubber: a new measurement technique. Strain 54(1):e12256
55. Le Cam JB, Albouy P-A, Charlès S (2020) Comparison between X-ray diffraction and quantitative surface calorimetry based on IR thermography to evaluate strain-induced crystallinity in natural rubber. Submitted to. Rev Sci Instrum 91:044902
56. Le Cam J-B, Kyei-Manu WA, Tayeb A, Albouy P-A, Busfield JJC, Tunnicliffe LB (2023) Strain-induced crystallisation of reinforced elastomers using surface calorimetry, vol 131, p 108341
57. Le Cam J-B, Tayeb A, Charlès S (2022) First evidence and characterization of strain-induced crystallization heterogeneity in natural rubber under homogeneous strain states. Polymer 255: 125120
58. Charlès S, Le Cam J-B (2020) Inverse identification of constitutive parameters from heat source fields: a local approach applied to hyperelasticity. Strain 56(2)
59. Toussaint E, Balandraud X, Le Cam J-B, Grédiac M (2012) Combining displacement, strain, temperature and heat source fields to investigate the thermomechanical response of an elastomeric specimen subjected to large deformations. Polym Test 31(7):916–925
60. Khiem VN, Le Cam J-B, Charles S, Itskov M (2022) Thermodynamics of strain-induced crystallization in filled natural rubber under uni- and biaxial loadings, part 1: complete energetic characterization and crystallinity evaluation. J Mech Phys Solids 159:104701
61. Pottier T, Moutrille M-P, Le Cam J-B, Balandraud X, Grédiac M (2009) Study on the use of motion compensation technique to determine heat sources. application to large deformations on cracked rubber specimens. Exp Mech 49:561–574
62. Samaca Martinez JR, Balandraud X, Toussaint E, Le Cam JB, Berghezan D (2014) Thermomechanical analysis of the crack tip zone in stretched crystallizable natural rubber by using infrared thermography and digital image correlation. Polymer 55(24):6345–6353

63. Louche H (2009) Etudes de certains phénomènes de localisation à partir de champs thermomécaniques. Habilitation thesis,. Savoie University
64. Samaca Martinez JR, Toussaint E, Balandraud X, Le Cam JB, Berghezan D (2015) Heat and strain measurements at the crack tip of filled rubber under cyclic loadings using full-field techniques. Mech Mater 81:62–71
65. Verron E, Andriyana A (2008) Definition of a new predictor for multiaxial fatigue crack nucleation in rubber. J Mech Phys Solids 56(2):417–443
66. Baaser H, Hopmann C, Schobel A (2013) Reformulation of strain in variants at incompressibility. Arch Appl Mech 83:273–280
67. Chrysochoos A (1995) Analyse du comportement des matériaux par thermographie infra rouge. In: Colloque Photomécanique, vol 95, pp 201–211.
68. Berthel B, Chrysochoos A, Wattrisse B, Galtier A (2008) Infrared image processing for the calorimetric analysis of fatigue phenomena. Exp Mech 48:79–90
69. Chrysochoos A, Louche H (2000) An infrared image processing to analyse the calorific effects accompanying strain localisation. Int J Eng Sci 38:1759–1788
70. Balandraud X, Chrysochoos A, Leclercq S, Peyroux R (2001) Influence of the thermomechanical coupling on the propagation of a phase change front. C R Acad Sci Ser IIB Mech 329:621–626
71. Maquin F, Pierron F (2009) Heat dissipation measurements in low stress cyclic loading of metallic materials: from internal friction to micro-plasticity. Mech Mater 41:928–942
72. Maugin GA, Muschik W (1994) Thermodynamics with internal variables. Part I. General concepts. J Non-Equilib Thermodyn 19:217–249
73. Bunn CW (1942) Proc R Soc London Ser 1(180):40
74. Takahashi Y, Kumano T (2004) Crystal structure of natural rubber. Macromolecules 37:4860
75. Immirzi A, Tedesco C, Monaco G, Tonelli AE (2005) Crystal structure and melting entropy of natural rubber. Macromolecules 38:1223
76. Rajkumar G, Squire JM, Arnott S (2006) Macromolecules 39:7004
77. Toki S, Sics I, Ran SF, Liu LZ, Hsiao BS, Murakami S, Tosaka M, Kohjiya S, Poompradub S, Ikeda Y, Tsou AH (2004) Rubb Chem Technol 42:956–964
78. Toki S, Fujimaki T, Okuyama M (2000) Strain-induced crystallization of natural rubber as detected real-time by wide-angle x-ray diffraction technique. Polymer 41:5423–5429
79. Trabelsi S, Albouy P-A, Rault J (2003) Effective local deformation in stretched filled rubber. Macromolecules 36:9093–9099
80. Roberts DE, Mandelkern L (1955) Thermodynamics of crystallization in high polymers: Natural rubber. Rubber Chem Technol 28(3):718–727
81. Samaca Martinez JR, Le Cam J-B, Balandraud X, Toussaint E, Caillard J (2013) Mechanisms of deformation in crystallizable natural rubber. part 1: thermal characterization. Polymer 54:2717–2726
82. Le Cam J-B (2022) Fast evaluation and comparison of the energy performances of elastomers from relative energy stored identification under mechanical loadings. Polymers 14(3)
83. Mullins L (1949) Permanent set in vulcanized rubber. Rubb Chem Technol 22(4):1036–1044
84. Mullins L (1969) Softening of rubber by deformation. Rubb Chem Technol 42:339–362
85. Holzapfel G-A (2000) Non linear solid mechanics – a continuum approach for engineering. Wiley
86. D'Ambrosio P, De Tommasi D, Ferri D, Puglisi G (2008) A phenomenological model for healing and hysteresis in rubber-like materials. Int J Eng Sci 46(4):293–305
87. Dorfmann A, Ogden RW (2003) A pseudo-elastic model for loading, partial unloading and reloading of particle-reinforced rubber. Int J Solids Struct 40(11):2699–2714
88. Rey T, Chagnon G, Favier D, Le Cam J-B (2014) Hyperelasticity with rate-independent microsphere hysteresis model for rubberlike materials. Comput Mater Sci 90:89–98
89. Vandenbroucke A, Laurent H, Ait Hocine N, Rio G (2010) A hyperelasto-visco-hysteresis model for an elastomeric behaviour: experimental and numerical investigations. Comput Mater Sci 48(3):495–503

90. Le Saux V, Marco Y, Calloch S, Doudard C, Charrier P (2010) Fast evaluation of the fatigue life-time of rubber-like materials based on a heat build-up protocol and micro-tomography measurements. Int J Fatigue 32(10):1582–1590
91. Ovalle Rodas C, Zaïri F, Naït-Abdelaziz M, Charrier P (2016) A thermo-visco-hyperelastic model for the heat build-up during low-cycle fatigue of filled rubbers: formulation, implementation and experimental verification. Int J Plasticity 79:217–236
92. Peter O, Stoček R, Kratina O (2023) Experimental and numerical description of the heat build-up in rubber under cyclic loading. Springer, Cham
93. Mott PH, Giller CB, Fragiadakis D, Rosenberg DA, Roland CM (2016) Deformation of polyurea: where does the energy go? Polymer 105:227–233

Adv Polym Sci (2024) 294: 31–52
https://doi.org/10.1007/12_2023_165

Published online: 7 October 2023

2D Transient Thermal Analytical Solution of the Heat Build-Up in Cyclically Loaded Rubber Cylinder

Jiří Vaněk, Ondřej Peter, Gert Heinrich, and Radek Stoček

Contents

Abstract Predicting the heat build-up in rubber under dynamic cyclic loading is still challenging due to its complexity. The simplest way to make such a prediction is

---

J. Vaněk
Faculty of Technology, University Institute, Tomas Bata University in Zlín, Zlín, Czech Republic

O. Peter
Centre of Polymer Systems, University Institute, Tomas Bata University in Zlín, Zlín, Czech Republic

Bonatrans Group a.s, Bohumín, Czech Republic

G. Heinrich
Institut für Textilmaschinen und Textile Hochleistungswerkstofftechnik, Technische Universität Dresden, Dresden, Germany

R. Stoček (✉)
Centre of Polymer Systems, University Institute, Tomas Bata University in Zlín, Zlín, Czech Republic

PRL Polymer Research Lab., s.r.o., Nad Ovčárnou 3685, Zlín, Czech Republic
e-mail: stocek@utb.cz

to use numerical methods like the Finite Element Method (FEM), which is preferable especially for geometrically complex rubber components. However, for components with simple geometry, analytical methods may be more appropriate as they better reflect the physical background and allow an assessment of the accuracy of the FE methods through comparison. This paper deals with the detailed derivation of the analytical equations for predicting the heat generation from the heat equation in a rubber cylinder under dynamic cyclic loading. The work focuses exclusively on a thermal solution assuming known rubber material parameters and corresponding loading conditions of the cylinder. The results of the equations derived in this way are compared and discussed with the results of a Finite Element Analysis (FEA) in the radial and axial directions and in the time domain. The analytical expression represents a significant contribution to the effective prediction of the evolution of heat generation in rubber without meeting the requirements of the FEA implementation.

Keywords Cylinder · Heat build-up · Heat equation · Rubber · Transient thermal analytical solution

1 Introduction

Rubber is an indispensable material widely used in engineering practice, especially for its viscoelastic properties. The amount and number of industrial and everyday applications of these classes of polymeric materials have been steadily increasing in recent decades. Typical examples are synthetic and natural covalently cross-linked rubbers integrated into automobile tires, or other applications such as drive belts, seismic isolators, fuel system hoses, seals, turbocharger hoses, cooling system hoses, engine bushing, spring pads, and more. Other technological applications are even emerging today in new fields such as soft robotics, stretchable electronics, tissue engineering, and more. Most of these products can perform very well without detailed analysis and prediction of their service life. However, a potential durability problem arises with dynamically cyclically loaded components that may experience temperature rise due to dissipation of mechanical energy [1]. Since these materials are specific in their low thermal conductivity [2], they cannot easily transfer the accumulated heat to the environment, and so the temperature in the component increases [3], causing thermal degradation [4] and changes in mechanical [5] and fatigue [6] behavior.

The temperature rise, known in the literature as Heat Build-Up (HBU) [3], together with fatigue should be reliably predicted in order to guarantee the required service life when designing such loaded parts. There are several studies [7–13] that focus on the prediction of either HBU alone or on HBU together with material fatigue. An analytical solution was used in [7], where the heat equation was simplified to a 1D steady-state solution and the results were compared with an

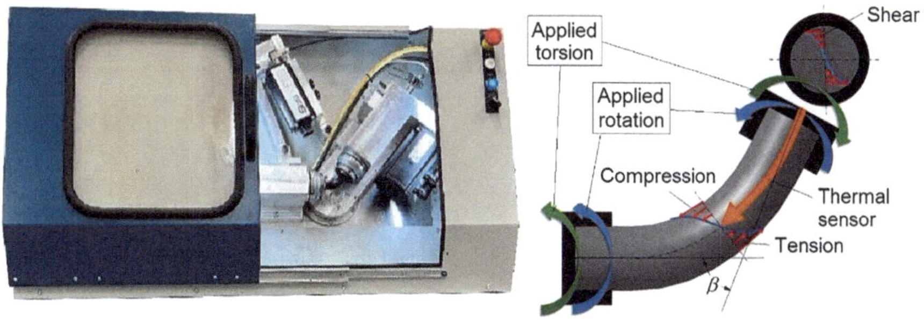

Fig. 1 Photograph of the HBuA

experiment performed on a planar strain with the so-called pure-shear sample. In [8] the storage and loss modulus together with Bessel functions were used to calculate the temperature rise and in [9] a numerical approach was taken. In both cases, the calculated results were compared with those of a rubber cylinder under cyclic compression. A finite strain thermo-viscoelastic constitutive model was presented in [10] and validated by tensile experiments. This model was later extended by the use of a stretch amplification factor to account for the effect of carbon-black filler on HBU [11]. Fatigue along with HBU was studied in [12, 13] using a numerical approach. Almost all these studies predicted and measured the temperature evolution individually in either tension or compression, but not in combined tension/ compression.

In order to measure the HBU on a specimen loaded in both tension and compression, Coesfeld GmbH & Co. Kg, Germany (see Fig. 1, left) developed a test device called Heat Build-up Analyzer (HbuA) [14, 15], which can in-situ measure the increase in surface and internal temperature with respect to number of loading cycles or time. The measurement is performed on a rubber cylinder, both ends of which are clamped in the spindles and then one of the spindles is turned to the desired angle to bend the rubber cylinder. The two spindles are then rotated around their axis, so that tension and compression alternate in the cylinder (see Fig. 1, right), creating the HBU.

The HBU measurement results in a temperature rise on the outer surface and inside the cylinder as a function of load cycles or time. An example of such a result is presented in Fig. 2, where the kinetics of HBU on the surface and inside the cylinder are shown. It can be seen that the heat starts to be induced first on the outer surface of the cylinder, which is characterized by a higher increase in the temperature curve compared to the inner surface. After a certain number of load cycles/time, $n(E_{qv})$, the temperature reaches an equilibrium state, $T_{HBU}(E_{qv})$. The term equilibrium state, E_{qv}, is defined as the point at which, for a given number of cycles/time, n, the temperature of the T_{HBU} reaches the same value on the surface and inside the cylinder. With further loading beyond the equilibrium state, the temperature inside the sample is higher and the kinetics of the evolution of the two monitored temperatures starts to stabilize at constant values without further increase.

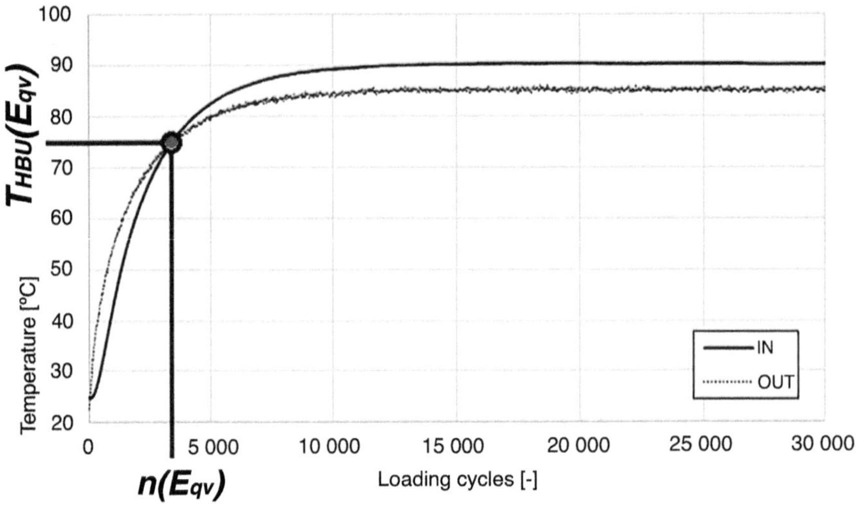

Fig. 2 Example of HBU results, the inner (line) and outer (dot) measured temperatures

The explanation for this phenomenon is based on the fact that the greatest deformation occurs at the surface of the cylinder and therefore the greatest heat increase, while the deformation decreases from the surface to zero deformation in the axis of the cylinder. However, during cyclic loading, heat is transferred from the outer surface of the cylinder to the surrounding area and at the same time, temperature starts to accumulate in the center of the cylinder due to the low thermal conductivity of the rubber.

The simulation of the HBU in such a loaded cylindrical specimen was firstly presented in [16]. The authors used a viscohyperelastic material model whose constants vary with temperature. Based on this model, they are able to obtain hysteretic loops in all elements of the finite element mesh on the cylinder using a mechanical solver in the FEM software Abaqus. According to these hysteretic loops, they determined the amount of dissipated energy, which was converted into heat flux. The result is a steady-state temperature value across all elements of the cylinder, which is determined by iterative calculations.

A similar approach was used in [17], but the viscoelastic properties of the material were replaced by the so-called dissipation constant, D, defined as a ratio of the dissipated energy density, W_{DED}, to the strain energy density, W_{SED}. Another modification concerns the thermal simulation, which is already transient, not just steady-state. However, the assumption that the amount of dissipated energy does not vary with temperature is used, which may lead to less accurate results.

This approach is currently being further developed, particularly with regard to the methodology for measuring the dissipation constant, D, which has not yet been fully established. In this paper, we follow the methodology of this approach and assume that the dissipation constant, D, is well known along with other material parameters.

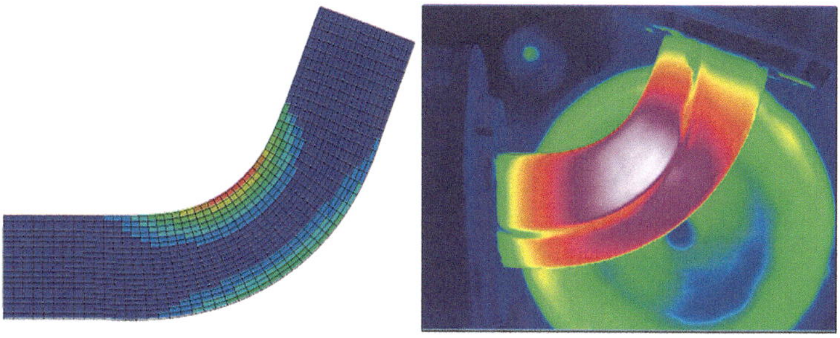

Fig. 3 Example of deformed shape of rubber cylinder in FEM (left) and during the test (right)

The methodology itself can be divided into three parts. First, it is necessary to simulate the structural part of the problem. This step is solved by FEM, because the analytical solution of this problem is complicated due to the material nonlinearity, large deformations, etc. For these reasons, the commercial software Ansys Workbench Mechanical 2021R2 is used as the FEM solver. One end of the cylinder is bent to the desired angle, β (Fig. 1, right), while the other end of the cylinder is fixed. The cylinder is then rotated about its neutral axis as in the experimental test, resulting in a strain energy density, W_{SED}, in all elements of the finite element mesh as a function of time, t. An example of the comparison of the deformed cylinder shape according to the FEA and the real one is shown in Fig. 3, where the figure on the left shows the strain energy density, W_{SED}, mapped to the mesh and the figure on the right shows the temperature distribution on the identically deformed real body.

Second, the internal heat generator, $\dot{q}$, must be determined. During cycling, two phases of deformation, stretching and un-stretching, occur. When a force is applied, the specimen starts to stretch, work is stored as strain energy and the temperature increases. After the specimen has been unloaded, the reversible part of the temperature can be recovered, but irreversible temperature occurs [18]. This means that part of the stored strain energy is converted to heat. In our case, the strain energy density, W_{SED}, is used instead of strain energy. The cylinder is loaded by bending and the strain energy density, W_{SED}, is distributed unevenly along both the length and across the cross-section (see Fig. 3, left). The amount of strain energy density, W_{SED}, converted to heat is calculated for each element, i, of the finite element mesh using the dissipation constant, D, and is called dissipative energy density, W_{DED}.

$$W_{\mathrm{DED},i} = D.W_{\mathrm{SED},i} \tag{1}$$

The dissipative power density, W_{DPD}, is then obtained as the derivative of the dissipative energy density, W_{DED}, with respect to time, t.

$$W_{\mathrm{DPD},i} = \frac{dW_{\mathrm{DED},i}}{dt} \tag{2}$$

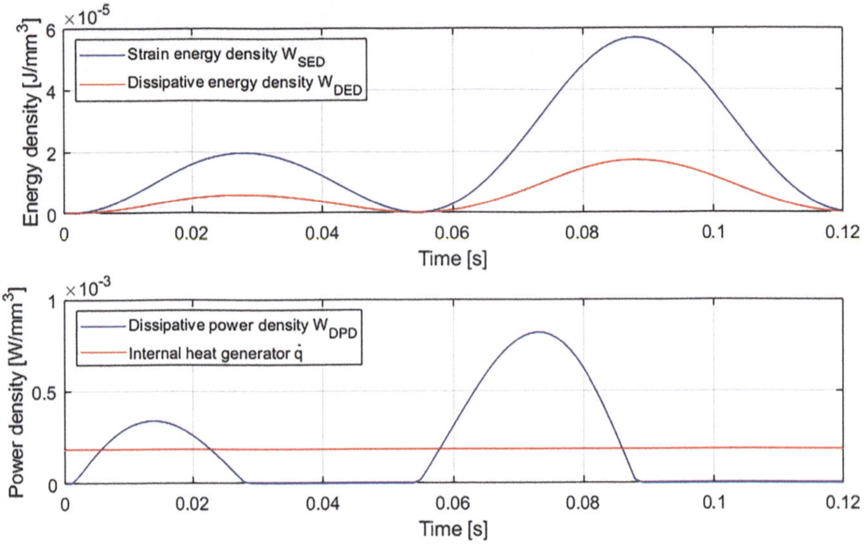

Fig. 4 Energies density, power density and internal heat generator of i_{th} element with respect to time – per one cycle

In general, the dissipative power density, W_{DPD}, has positive values, which means that heating occurs when stretching and negative values, which means that cooling occurs when un-stretching. Because we have separated this irreversible part of the heating from the reversible part using Eq. 1 and assuming that the heating only occurs when stretching, the negative values of the dissipative power density, W_{DPD}, can be set to zero. Then the average value over time is taken as the internal heat generator $\dot{q}$. Performing these operations for all elements, i, of the finite element mesh yields the thermal loading surface required for either the FEA or the analytical thermal solution. An example of the strain energy density, W_{SED}, dissipative energy density, W_{DED}, and dissipative power density, W_{DPD}, curve is shown in Fig. 4.

$$\dot{q}_i = \text{mean}(W_{\text{DPD},i}) \tag{3}$$

Finally, a transient thermal calculation can be performed by using FEM, but also analytically by deriving relations from the general heat equation. The analytical solution has already been presented in [19], where the heat equation was simplified into a 1D steady-state thermal solution and the resulting temperature across the cross-section was compared with FEA results. Furthermore, a 2D transient thermal solution is derived in this paper.

2 Computational Section

For the analytical and FEA calculation of the HBU in the cylinder, it is necessary to know the course of the internal heat generator, $\dot{q}$, which is part of the general heat equation in the cylindrical coordinate system [20], see Fig. 5.

$$\frac{\partial^2 T}{\partial r^2} + \frac{1}{r}\frac{\partial T}{\partial r} + \frac{1}{r^2}\frac{\partial^2 T}{\partial \phi^2} + \frac{\partial^2 T}{\partial z^2} + \frac{\dot{q}}{k} = \frac{\rho C_p}{k}\frac{\partial T}{\partial t} \tag{4}$$

Where T is the temperature, r is the radius variable, Φ is the rotation variable along the cross-section, z is the length position variable, k is the thermal conductivity, ρ is the density, C_p is the specific heat, and t is the time (Fig. 5).

2.1 Internal Heat Generator

In order to use the analytical approach, it is necessary to find a function for the internal heat generator, $\dot{q}$, determined by FEA. The following general polynomial function provides a good fit.

$$\dot{q} = \dot{q}(r, z) = ar^{\gamma} + bz^{\delta} + cr^{\gamma}z^{\delta} + d \tag{5}$$

where a, b, c, d are fitting constants and γ, δ are fitting polynomials.

For the determination of the fitting constants, the values of $\dot{q}(r,z)$ can be assumed to be located in the red points marked in Fig. 6. The cylinder is divided into two parts in terms of loading and symmetry is also used, as can be seen in the scheme. In the left part, where the cylinder is bent, the load course is assumed according to Eq. (5). In the right part, i.e. the part fixed under the spindle, the value of $\dot{q}(r,z)$ is assumed to be zero.

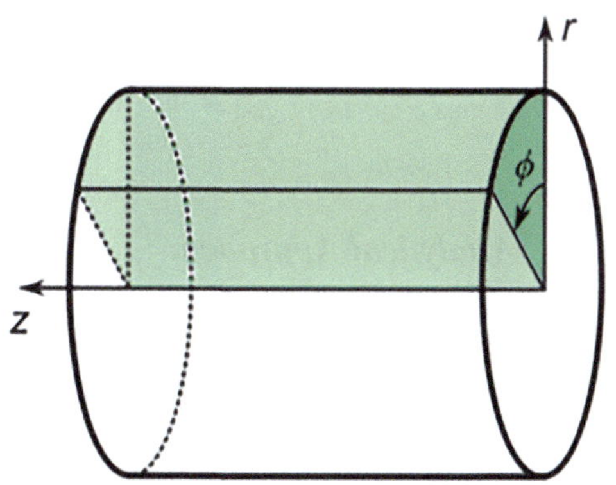

Fig. 5 Scheme of cylindrical coordinate system

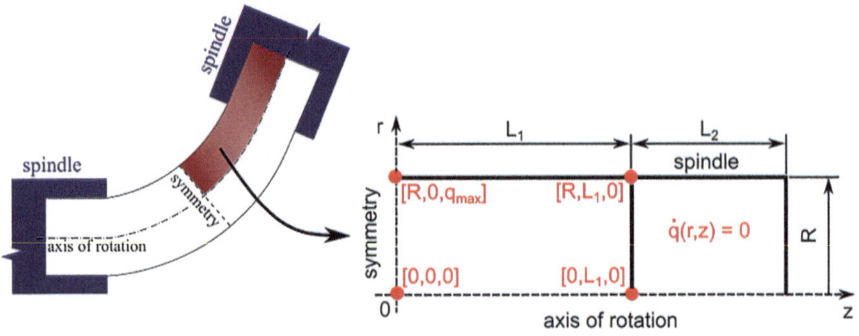

Fig. 6 Scheme of internal heat generation in cylinder

$$[0,0,0] \Rightarrow 0 = a0^{\gamma} + b0^{\delta} + c0^{\gamma}0^{\delta} + d \Rightarrow d = 0$$
$$[0,L_1,0] \Rightarrow 0 = a0^{\gamma} + bL_1^{\delta} + c0^{\gamma}L_1^{\delta} + d \Rightarrow b = 0$$
$$[R,0,\dot{q}_{\max}] \Rightarrow \dot{q}_{\max} = aR^{\gamma} + cR^{\gamma}0^{\delta} \Rightarrow a = \frac{\dot{q}_{\max}}{R^{\gamma}}$$
$$[R,L_1,0] \Rightarrow 0 = \frac{\dot{q}_{\max}}{R^{\gamma}}R^{\gamma} + cR^{\gamma}L_1^{\delta} \Rightarrow c = -\frac{\dot{q}_{\max}}{R^{\gamma}L_1^{\delta}} \tag{6}$$

where according to Fig. 6, right, R is the outer radius, L_1 is the length of the free surface, and L_2 is the length of the clamped surface in the spindle.

A suitable polynomial function is therefore of the form:

$$\dot{q}(r,z) = \frac{\dot{q}_{\max} r^{\gamma}}{R^{\gamma}} - \frac{\dot{q}_{\max} r^{\gamma} z^{\delta}}{R^{\gamma} L_1^{\delta}} \tag{7}$$

Now the values of the fitting polynomials γ, δ and the maximum value of the internal heat generator $\dot{q}_{\max}$ are required to be found, so that the fitted course of $\dot{q}(r,z)$ corresponds to the one obtained from the FEA. For the current solution, these values are $\gamma = 2$, $\delta = 2$ and $\dot{q}_{\max} = 2.1\text{e-}4\ \text{W/mm}^3$ provides a good fit, whereas a comparison of the internal heat generator $\dot{q}(r,z)$ from the FEA and the analytical solution is presented in Fig. 7, left, and the difference between the both thermal loadings $\dot{q}_{\text{diff}} = \dot{q}_{\text{FEA}} - \dot{q}_{\text{anl}}$ is shown in Fig. 7, right.

2.2 Analytical Approach

The analytical solution is based on the general heat equation in the cylindrical coordinate system (4). This equation can be simplified by introducing the assumption of temperature homogeneity at each rotation of the cross-section (ϕ-axis), so $\partial^2 T/\partial \phi^2 = 0$. At the same time, the assumption that the material and load parameters

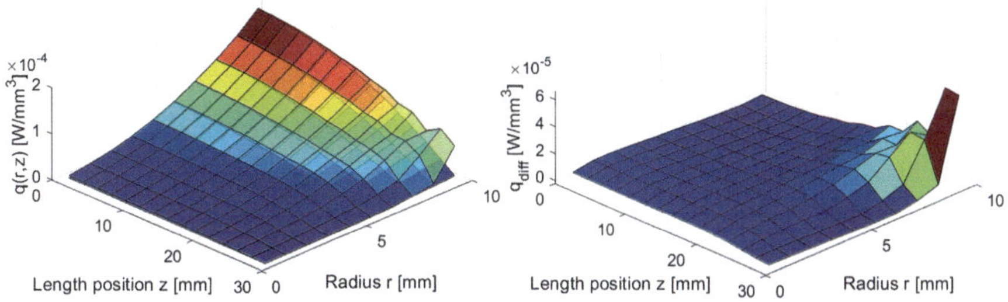

Fig. 7 Internal heat generator $\dot{q}(r,z)$ from the FEA (transparent surface with coarse mesh) and the analytical (fine mesh) solution (left), the difference between both solutions $\dot{q}_{\mathrm{diff}}$ (right)

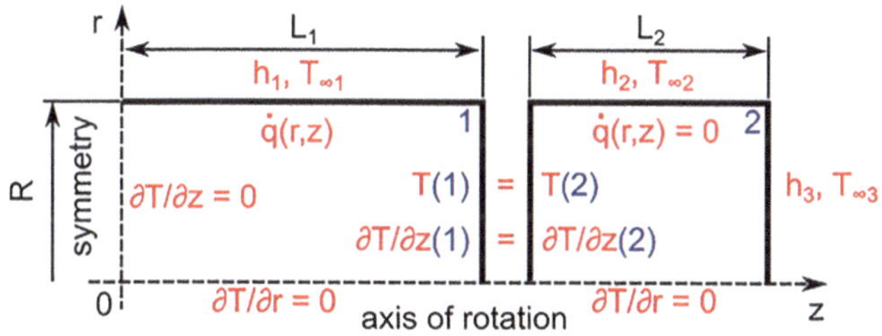

Fig. 8 Split of the cylinder to two bodies + boundary conditions

do not change as a function of temperature was introduced. Subsequently, a 2D transient heat equation was obtained which is analytically solvable.

$$\frac{\partial^2 T}{\partial r^2}+\frac{1}{r}\frac{\partial T}{\partial r}+\frac{\partial^2 T}{\partial z^2}+\frac{\dot{q}(r,z)}{k}=\frac{\rho C_p}{k}\frac{\partial T}{\partial t} \tag{8}$$

Due to the differences in convection between the free surface and the surface clamped in the spindle, the body can be divided into two individual and separate parts for future calculation. Symmetry has been taken into account, since the body is symmetrical in terms of shape and load. A graphical representation of the cylinder split along with the boundary conditions is shown in Fig. 8, where h_1, h_2, h_3 are surface convections and $T_{\infty 1}$, $T_{\infty 2}$, $T_{\infty 3}$ are the ambient temperatures at these surfaces.

The two bodies divided in this way will first be dealt with separately. Finally, we will combine two solutions using boundary conditions. The solution for body 1 will be derived first.

Note that:

$$\frac{\rho C_p}{k}=\alpha^{-1}\Rightarrow\frac{k}{\rho C_p}=\alpha$$

$$r\frac{\partial^2 T}{\partial r^2}+\frac{\partial T}{\partial r}=\frac{\partial}{\partial r}\left(r\frac{\partial T}{\partial r}\right) \tag{9}$$

where α is called thermal diffusivity. By substituting and reducing Eq. (8) the following term is obtained:

$$\frac{\partial}{\partial r}\left(r\frac{\partial T}{\partial r}\right)=-\frac{\dot{q}_{\max}r^{(\gamma+1)}}{kR^{\gamma}}+\frac{\dot{q}_{\max}r^{(\gamma+1)}z^{\delta}}{kR^{\gamma}L_1^{\delta}}-r\frac{\partial^2 T}{\partial z^2}+\frac{r}{\alpha}\frac{\partial T}{\partial t} \tag{10}$$

Then Eq. (10) is integrated over the radius r using indefinite integral.

$$\int_r \frac{d}{dr}\left(r\frac{\partial T}{\partial r}\right)dr=\int_r\left(-\frac{\dot{q}_{\max}r^{(\gamma+1)}}{kR^{\gamma}}+\frac{\dot{q}_{\max}r^{(\gamma+1)}z^{\delta}}{kR^{\gamma}L_1^{\delta}}-r\frac{\partial^2 T}{\partial z^2}+\frac{r}{\alpha}\frac{\partial T}{\partial t}\right)dr \tag{11}$$

$$\frac{\partial T}{\partial r}=-\frac{\dot{q}_{\max}r^{(\gamma+1)}}{(\gamma+2)kR^{\gamma}}+\frac{\dot{q}_{\max}r^{(\gamma+1)}z^{\delta}}{(\gamma+2)kR^{\gamma}L_1^{\delta}}-\frac{r}{2}\frac{\partial^2 T}{\partial z^2}+\frac{r}{2\alpha}\frac{\partial T}{\partial t}+\frac{C_1}{r} \tag{12}$$

Then Eq. (12) is again integrated over the radius r using indefinite integral to obtain the temperature relation just with respect to the dependence on the radius.

$$T=-\frac{\dot{q}_{\max}r^{(\gamma+2)}}{(\gamma+2)^2kR^{\gamma}}+\frac{\dot{q}_{\max}r^{(\gamma+2)}z^{\delta}}{(\gamma+2)^2kR^{\gamma}L_1^{\delta}}-\frac{r^2}{4}\frac{\partial^2 T}{\partial z^2}+\frac{r^2}{4\alpha}\frac{\partial T}{\partial t}+C_1\ln(r)+C_2 \tag{13}$$

To solve the integration constants C_1 and C_2, it is necessary to use boundary conditions, where the first of the integration constants can be solved from the following condition:

$$\frac{\partial T}{\partial r}(0,z,t)=0\Rightarrow C_1=0 \tag{14}$$

The integration constant C_2 can be solved using the convective boundary condition:

$$\begin{gathered}-k\frac{\partial T}{\partial r}(R,z,t)=h_1(T(R,z,t)-T_{\infty 1})\\ \frac{\dot{q}_{\max}R^{(\gamma+1)}}{(\gamma+2)h_1R^{\gamma}}-\frac{\dot{q}_{\max}R^{(\gamma+1)}z^{\delta}}{(\gamma+2)h_1R^{\gamma}L_1^{\delta}}+\frac{kR}{2h_1}\frac{\partial^2 T}{\partial z^2}-\frac{kR}{2\alpha h_1}\frac{\partial T}{\partial t}=\\ =-\frac{\dot{q}_{\max}R^{(\gamma+2)}}{(\gamma+2)^2kR^{\gamma}}+\frac{\dot{q}_{\max}R^{(\gamma+2)}z^{\delta}}{(\gamma+2)^2kR^{\gamma}L_1^{\delta}}-\frac{R^2}{4}\frac{\partial^2 T}{\partial z^2}+\frac{R^2}{4\alpha}\frac{\partial T}{\partial t}+C_2-T_{\infty 1}\end{gathered} \tag{15}$$

$$C_2 = T_{\infty 1} + \frac{\dot{q}_{\max} R^{(\gamma+2)}}{(\gamma+2)^2 k R^\gamma} - \frac{\dot{q}_{\max} R^{(\gamma+2)} z^\delta}{(\gamma+2)^2 k R^\gamma L_1^\delta} + \frac{\dot{q}_{\max} R^{(\gamma+1)}}{(\gamma+2) h_1 R^\gamma} - \frac{\dot{q}_{\max} R^{(\gamma+1)} z^\delta}{(\gamma+2) h_1 R^\gamma L_1^\delta} + \\ + \frac{\partial^2 T}{\partial z^2}\left(\frac{R^2}{4} + \frac{kR}{2h_1}\right) - \frac{\partial T}{\partial t}\left(\frac{R^2}{4\alpha} + \frac{kR}{2\alpha h_1}\right) \tag{16}$$

Substituting the solved integration constants from (14) and (16) into Eq. (13) gives the form of the temperature equation with respect to radius r. It remains to solve the time domain and length dependence.

$$T = T_{\infty 1} + \frac{\dot{q}_{\max}\left(R^{(\gamma+2)} - r^{(\gamma+2)}\right)}{(\gamma+2)^2 k R^\gamma} - \frac{\dot{q}_{\max}\left(R^{(\gamma+2)} - r^{(\gamma+2)}\right)}{(\gamma+2)^2 k R^\gamma} \frac{z^\delta}{L_1^\delta} + \frac{\dot{q}_{\max} R^{(\gamma+1)}}{(\gamma+2) h_1 R^\gamma} - \\ - \frac{\dot{q}_{\max} R^{(\gamma+1)}}{(\gamma+2) h_1 R^\gamma} \frac{z^\delta}{L_1^\delta} + \frac{\partial^2 T}{\partial z^2}\left(\frac{R^2}{4} - \frac{r^2}{4} + \frac{kR}{2h_1}\right) - \frac{\partial T}{\partial t}\left(\frac{R^2}{4\alpha} - \frac{r^2}{4\alpha} + \frac{kR}{2\alpha h_1}\right) \tag{17}$$

To simplify the relations, the following substitutions can be used:

$$\begin{aligned} \frac{\dot{q}_{\max}\left(R^{(\gamma+2)} - r^{(\gamma+2)}\right)}{(\gamma+2)^2 k R^\gamma} &= A \\ \frac{\dot{q}_{\max} R^{(\gamma+1)}}{(\gamma+2) h_1 R^\gamma} &= B \\ \frac{R^2}{4} - \frac{r^2}{4} + \frac{kR}{2h_1} &= C \\ \frac{R^2}{4\alpha} - \frac{r^2}{4\alpha} + \frac{kR}{2\alpha h_1} &= D \end{aligned} \tag{18}$$

Therefore, Eq. (17) is simplified as follows:

$$T = T_{\infty 1} + A - A\frac{z^\delta}{L_1^\delta} + B - B\frac{z^\delta}{L_1^\delta} + C\frac{\partial^2 T}{\partial z^2} - D\frac{\partial T}{\partial t} \tag{19}$$

Now the time domain can be resolved. To do this, the following substitution can be used, assuming that the ambient temperature does not change over time.

$$\begin{gathered} T - T_{\infty 1} = \phi \\ \frac{\partial \varphi}{\partial t} = \frac{\partial T}{\partial t} - \frac{\partial T_{\infty 1}}{\partial t} \Rightarrow \frac{\partial T_{\infty 1}}{\partial t} = 0 \Rightarrow \frac{\partial \varphi}{\partial t} = \frac{\partial T}{\partial t} \end{gathered} \tag{20}$$

By substituting (20) into (19), the next expression has been obtained:

$$\varphi = A - A\frac{z^{\delta}}{L_1^{\delta}} + B - B\frac{z^{\delta}}{L_1^{\delta}} + C\frac{\partial^2 T}{\partial z^2} - D\frac{\partial \varphi}{\partial t} \tag{21}$$

After modifying Eq. (21), one can already integrate using a definite integral. In time $t = 0$, the temperature change with respect to the ambient temperature can be expressed, i.e. $\varphi = T - T_{\infty 1} = 0$. This means that at the beginning of the calculation we assume that the cylinder temperature is the same as the ambient temperature. On the other hand, the temperature, φ, can be expected to change at time, t, compared to the ambient temperature, $T_{\infty 1}$.

$$-\int_0^t dt = \int_0^{\varphi} \frac{D d\varphi}{\varphi - A + A\frac{z^{\delta}}{L_1^{\delta}} - B + B\frac{z^{\delta}}{L_1^{\delta}} - C\frac{\partial^2 T}{\partial z^2}}$$

$$-t = D\ln\left(\frac{\phi - A + A\frac{z^{\delta}}{L_1^{\delta}} - B + B\frac{z^{\delta}}{L_1^{\delta}} - C\frac{\partial^2 T}{\partial z^2}}{-A + A\frac{z^{\delta}}{L_1^{\delta}} - B + B\frac{z^{\delta}}{L_1^{\delta}} - C\frac{\partial^2 T}{\partial z^2}}\right)$$

$$= D\ln\left(\frac{\phi}{-A + A\frac{z^{\delta}}{L_1^{\delta}} - B + B\frac{z^{\delta}}{L_1^{\delta}} - C\frac{\partial^2 T}{\partial z^2}} + 1\right) \tag{22}$$

By modifying Eq. (22) and substituting it back into Eq. (20), the solved equation for the temperature along the radius, r, and at time, t, is obtained. Now it remains to solve the dependence on the length of the body.

$$T = T_{\infty 1} + (A + B)\left(1 - e^{-\frac{t}{D}}\right) - (A + B)\left(1 - e^{-\frac{t}{D}}\right)\frac{z^{\delta}}{L_1^{\delta}} + C\left(1 - e^{-\frac{t}{D}}\right)\frac{\partial^2 T}{\partial z^2} \tag{23}$$

To make the equation easier to solve, the following substitution can be used:

$$(A + B)\left(1 - e^{-\frac{t}{D}}\right) = E$$

$$C\left(1 - e^{-\frac{t}{D}}\right) = F \tag{24}$$

The remaining solution leads to a second order differential equation:

$$T = T_{\infty 1} + E - E\frac{z^{\delta}}{L_1^{\delta}} + F\frac{\partial^2 T}{\partial z^2} \Rightarrow F\frac{\partial^2 T}{\partial z^2} - T + E - E\frac{z^{\delta}}{L_1^{\delta}} + T_{\infty 1} = 0 \tag{25}$$

The solution of Eq. 25 is:

$$T = C_3 e^{\frac{z}{\sqrt{F}}} + C_4 e^{\frac{-z}{\sqrt{F}}} + E - E\frac{z^{\delta}}{L_1^{\delta}} + T_{\infty 1}$$

$$\frac{dT}{dz} = \frac{C_3}{\sqrt{F}} e^{\frac{z}{\sqrt{F}}} - \frac{C_4}{\sqrt{F}} e^{\frac{-z}{\sqrt{F}}} - \delta E \frac{z^{(\delta - 1)}}{L_1^{\delta}} \tag{26}$$

Now one of the length boundary conditions can be used to determine one of the integration constants:

$$\frac{dT}{dz}(r, 0, t) = 0 \Rightarrow \frac{C_3}{\sqrt{F}} e^{\frac{0}{\sqrt{F}}} - \frac{C_4}{\sqrt{F}} e^{\frac{-0}{\sqrt{F}}} - \delta E \frac{0^{(\delta - 1)}}{L_1^{\delta}} = 0 \Rightarrow C_3 - C_4 = 0$$

$$\Rightarrow C_3 = C_4 \tag{27}$$

In order to proceed with the solution, it is necessary to deal with the calculation for body 2 of the cylinder and therefore only the temporary solution of the body 1 can now be expressed in the form:

$$T = C_3 \left(e^{\frac{z}{\sqrt{F}}} + e^{\frac{-z}{\sqrt{F}}} \right) + E - E\frac{z^{\delta}}{L_1^{\delta}} + T_{\infty 1}$$

$$\frac{dT}{dz} = \frac{C_3}{\sqrt{F}} \left(e^{\frac{z}{\sqrt{F}}} - e^{\frac{-z}{\sqrt{F}}} \right) - \delta E \frac{z^{(\delta - 1)}}{L_1^{\delta}} \tag{28}$$

Since no internal heat generation load is assumed for body 2, Eq. (8) can be simplified to the form:

$$\frac{\partial^2 T}{\partial r^2} + \frac{1}{r}\frac{\partial T}{\partial r} + \frac{\partial^2 T}{\partial z^2} = \frac{1}{\alpha}\frac{\partial T}{\partial t} \tag{29}$$

Solving this equation in the same way as for the body 1, the following relations are obtained:

$$\frac{\partial T}{\partial r} = -\frac{r}{2}\frac{\partial^2 T}{\partial z^2} + \frac{r}{2\alpha}\frac{\partial T}{\partial t} + \frac{D_1}{r}$$

$$T = -\frac{r^2}{4}\frac{\partial^2 T}{\partial z^2} + \frac{r^2}{4\alpha}\frac{\partial T}{\partial t} + D_1 \ln(r) + D_2 \tag{30}$$

Using the first boundary condition, the form of the first integration constant D_1 can be obtained:

$$\frac{\partial T}{\partial r}(0,z,t)=0 \Rightarrow D_1=0 \tag{31}$$

And from the second boundary condition the second integration constant D_2 is expressed.

$$-k\frac{\partial T}{\partial r}(R,z,t)=h_2(T(R,z,t)-T_{\infty 2})$$

$$\frac{kR}{2h_2}\frac{\partial^2 T}{\partial z^2}-\frac{kR}{2\alpha h_2}\frac{\partial T}{\partial t}=-\frac{R^2}{4}\frac{\partial^2 T}{\partial z^2}+\frac{R^2}{4\alpha}\frac{\partial T}{\partial t}+D_2-T_{\infty 2}$$

$$D_2=T_{\infty 2}+\frac{kR}{2h_2}\frac{\partial^2 T}{\partial z^2}+\frac{R^2}{4}\frac{\partial^2 T}{\partial z^2}-\frac{kR}{2\alpha h_2}\frac{\partial T}{\partial t}-\frac{R^2}{4\alpha}\frac{\partial T}{\partial t} \tag{32}$$

Substituting the integration constants D_1 and D_2 back into Eq. (30), we obtain:

$$T=T_{\infty 2}+\frac{\partial^2 T}{\partial z^2}\left(\frac{R^2}{4}-\frac{r^2}{4}+\frac{kR}{2h_2}\right)-\frac{\partial T}{\partial t}\left(\frac{R^2}{4\alpha}-\frac{r^2}{4\alpha}+\frac{kR}{2\alpha h_2}\right) \tag{33}$$

Simplified substitutions have the form:

$$\frac{R^2}{4}-\frac{r^2}{4}+\frac{kR}{2h_2}=G$$

$$\frac{R^2}{4\alpha}-\frac{r^2}{4\alpha}+\frac{kR}{2\alpha h_2}=H \tag{34}$$

After simplification, the Eq. (33) has the following form:

$$T=T_{\infty 2}+G\frac{\partial^2 T}{\partial z^2}-H\frac{\partial T}{\partial t} \tag{35}$$

Now it is again possible to solve the time domain in the same way as for body 1 of the cylinder.

$$T=T_{\infty 2}+G\left(1-e^{-\frac{t}{H}}\right)\frac{d^2 T}{dz^2} \tag{36}$$

By introducing another simplifying substitution, the solution again leads to a second order differential equation.

$$G\left(1-e^{-\frac{t}{H}}\right)=I \tag{37}$$

$$T = T_{\infty 2} + I\frac{d^2T}{dz^2} \Rightarrow I\frac{d^2T}{dz^2} - T + T_{\infty 2} = 0 \tag{38}$$

The solution has the form:

$$\begin{aligned} T &= D_3 e^{\frac{z}{\sqrt{I}}} + D_4 e^{\frac{-z}{\sqrt{I}}} + T_{\infty 2} \\ \frac{dT}{dz} &= \frac{D_3}{\sqrt{I}} e^{\frac{z}{\sqrt{I}}} - \frac{D_4}{\sqrt{I}} e^{\frac{-z}{\sqrt{I}}} \end{aligned} \tag{39}$$

Using the convective boundary condition at the end of the cylinder length, one integration constant can be determined:

$$-k\frac{dT}{dz}(r, L_1 + L_2, t) = h_3(T(r, L_1 + L_2, t) - T_{\infty 3})$$

$$\frac{-kD_3}{\sqrt{I}h_3} e^{\frac{L_1+L_2}{\sqrt{I}}} + \frac{kD_4}{\sqrt{I}h_3} e^{\frac{-(L_1+L_2)}{\sqrt{I}}} = D_3 e^{\frac{L_1+L_2}{\sqrt{I}}} + D_4 e^{\frac{-(L_1+L_2)}{\sqrt{I}}} + T_{\infty 2} - T_{\infty 3}$$

$$D_3 = \frac{D_4\left(\frac{k}{\sqrt{I}h_3} - 1\right) e^{\frac{-(L_1+L_2)}{\sqrt{I}}} + T_{\infty 3} - T_{\infty 2}}{\left(\frac{k}{\sqrt{I}h_3} + 1\right) e^{\frac{L_1+L_2}{\sqrt{I}}}} \tag{40}$$

Using substitution, the integration constant D_3 can be simplified to the form:

$$\begin{aligned} \frac{\left(\frac{k}{\sqrt{I}h_3} - 1\right) e^{\frac{-(L_1+L_2)}{\sqrt{I}}}}{\left(\frac{k}{\sqrt{I}h_3} + 1\right) e^{\frac{L_1+L_2}{\sqrt{I}}}} &= J \\ \frac{T_{\infty 3} - T_{\infty 2}}{\left(\frac{k}{\sqrt{I}h_3} + 1\right) e^{\frac{L_1+L_2}{\sqrt{I}}}} &= K \\ D_3 &= JD_4 + K \end{aligned} \tag{41}$$

The temporary solution of the second body of the cylinder can be expressed as:

$$\begin{aligned} T &= D_4\left(Je^{\frac{z}{\sqrt{I}}} + e^{\frac{-z}{\sqrt{I}}}\right) + Ke^{\frac{z}{\sqrt{I}}} + T_{\infty 2} \\ \frac{dT}{dz} &= \frac{D_4}{\sqrt{I}}\left(Je^{\frac{z}{\sqrt{I}}} - e^{\frac{-z}{\sqrt{I}}}\right) + Ke^{\frac{z}{\sqrt{I}}} \end{aligned} \tag{42}$$

Solutions are now available for both bodies, expressed by Eqs. (28) for the first body and (42) for the second body, each containing one unknown integration constant. These constants can be determined using the two remaining boundary conditions connecting the solutions of the two bodies.

$$\frac{dT^{(1)}}{dz}(r, L_1, t) = \frac{dT^{(2)}}{dz}(r, L_1, t)$$

$$\frac{C_3}{\sqrt{F}}\left(e^{\frac{L_1}{\sqrt{F}}} - e^{\frac{-L_1}{\sqrt{F}}}\right) - \delta E \frac{L_1^{(\delta-1)}}{L_1^{\delta}} = \frac{D_4}{\sqrt{I}}\left(Je^{\frac{L_1}{\sqrt{I}}} - e^{\frac{-L_1}{\sqrt{I}}}\right) + Ke^{\frac{L_1}{\sqrt{I}}}$$

$$C_3 = \frac{D_4\sqrt{F}\left(Je^{\frac{L_1}{\sqrt{I}}} - e^{\frac{-L_1}{\sqrt{I}}}\right) + K\sqrt{F}\sqrt{I}e^{\frac{L_1}{\sqrt{I}}} + \frac{\delta E L_1^{(\delta-1)}\sqrt{F}\sqrt{I}}{L_1^{\delta}}}{\sqrt{I}\left(e^{\frac{L_1}{\sqrt{F}}} - e^{\frac{-L_1}{\sqrt{F}}}\right)} \tag{43}$$

The integration constant C_3 has the following form after simplifying substitutions:

$$\frac{\sqrt{F}\left(Je^{\frac{L_1}{\sqrt{I}}} - e^{\frac{-L_1}{\sqrt{I}}}\right)}{\sqrt{I}\left(e^{\frac{L_1}{\sqrt{F}}} - e^{\frac{-L_1}{\sqrt{F}}}\right)} = M$$

$$\frac{K\sqrt{F}\sqrt{I}e^{\frac{L_1}{\sqrt{I}}} + \frac{\delta E L_1^{(\delta-1)}\sqrt{F}\sqrt{I}}{L_1^{\delta}}}{\sqrt{I}\left(e^{\frac{L_1}{\sqrt{F}}} - e^{\frac{-L_1}{\sqrt{F}}}\right)} = N$$

$$C_3 = MD_4 + N \tag{44}$$

It remains to find the form of the last integration constant D_4, which is achieved using the last boundary condition:

$$T^{(1)}(r, L_1, t) = T^{(2)}(r, L_1, t)$$

$$C_3\left(e^{\frac{L_1}{\sqrt{F}}} + e^{\frac{-L_1}{\sqrt{F}}}\right) + E - E\frac{L_1^{\delta}}{L_1^{\delta}} + T_{\infty 1} = D_4\left(Je^{\frac{L_1}{\sqrt{I}}} + e^{\frac{-L_1}{\sqrt{I}}}\right) + Ke^{\frac{L_1}{\sqrt{I}}} + T_{\infty 2}$$

$$MD_4\left(e^{\frac{L_1}{\sqrt{F}}} + e^{\frac{-L_1}{\sqrt{F}}}\right) + N\left(e^{\frac{L_1}{\sqrt{F}}} + e^{\frac{-L_1}{\sqrt{F}}}\right) + T_{\infty 1} = D_4\left(Je^{\frac{L_1}{\sqrt{I}}} + e^{\frac{-L_1}{\sqrt{I}}}\right) + Ke^{\frac{L_1}{\sqrt{I}}} + T_{\infty 2}$$

$$D_4 = \frac{T_{\infty 1} - T_{\infty 2} + N\left(e^{\frac{L_1}{\sqrt{F}}} + e^{\frac{-L_1}{\sqrt{F}}}\right) - Ke^{\frac{L_1}{\sqrt{I}}}}{Je^{\frac{L_1}{\sqrt{I}}} + e^{\frac{-L_1}{\sqrt{I}}} - Me^{\frac{L_1}{\sqrt{F}}} - Me^{\frac{-L_1}{\sqrt{F}}}} \tag{45}$$

All the integration constants are already known, i.e. for cylinder body 1 using Eq. (28) and for cylinder body 2 using Eq. (42) it is possible to calculate the temperature in the cylinder at any position at any time.

2.3 FEM Approach

The transient thermal problem was solved with Ansys 2021R2 [21] FEM software. The computational model was a cylinder on which a finite elements mesh of linear hexahedral SOLID278 elements was created. A total of 14,025 nodes and 12,950 elements were used. Symmetry was not used in this case. The final time was set to 3,600 s with a total number of 1,000 steps.

The computational model included three types of convection, as in the analytical solution. An internal heat generator, $\dot{q}$, was used as the load, see Sect. 2.1.

3 Results and Discussion

In order to compare the results of both calculation methods, the following material parameters, boundary conditions (see Fig. 8), cylinder dimensions and fitting polynomials with the maximum value of the internal heat generator were chosen (Tables 1, 2, 3, and 4).

The temperature distribution along the cylinder is shown in Fig. 10. The results were evaluated at three different radii (see Fig. 9), i.e., at the outer radius R of the cylinder, at radius $R/2$, and at its center. From the steady-state temperature distribution curves (at time $t = 3{,}600$ s), it can be seen that the derived analytical equations provide the trend of the temperature distribution well and are in good agreement with the FEA results, see Fig. 10.

Further, the results were evaluated at four different distances from the center of the cylinder length (see Fig. 11), i.e. at the center of the cylinder length, $L_1/2$, L_1 and

Table 1 Thermal material parameters

Density, ρ [kg/mm^3]	Specific heat, C_p [J/kg K]	Thermal conductivity, k [W/mm K]
1e-6	1,600	1.5e-4

Table 2 Convection boundary conditions

Convection, h_1 [W/mm^2 K]	Convection, h_2 [W/mm^2 K]	Convection, h_3 [W/mm^2 K]	$T_{\infty 1}$, $T_{\infty 2}$, $T_{\infty 3}$ [°C]
3e-5	1e-2	3e-5	22

Table 3 Geometry dimensions

Outer radius, R [mm]	Length, L_1 [mm]	Length, L_2 [mm]
10	30	20

Table 4 Internal heat generator – analytical solution

Fit polynomial, γ [–]	Fit polynomial, δ [–]	Maximal internal heat generator $\dot{q}_{max}$ [W/mm^3]
2	2	2.1e-4

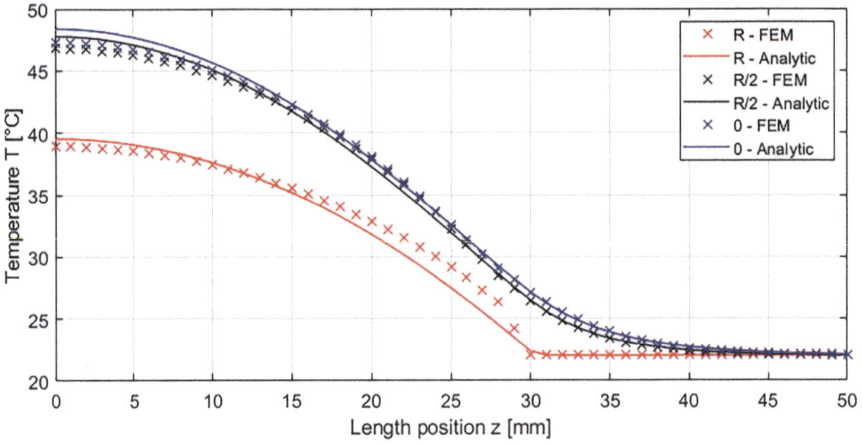

Fig. 10 Steady-state temperature distribution along the length of the cylinder (t = 3,600 s)

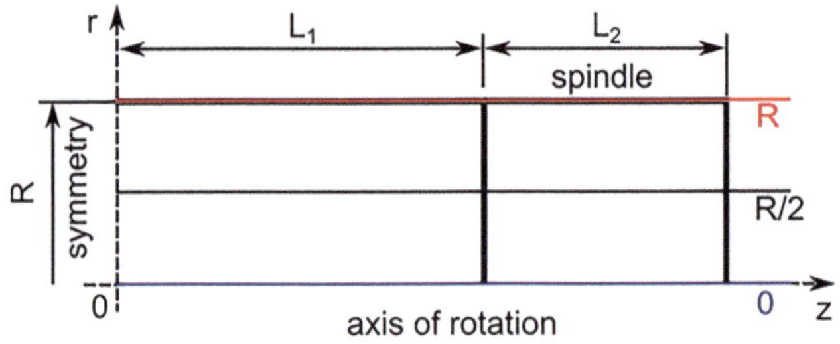

Fig. 9 Representation of the result paths – different radii

Fig. 11 Representation of the result paths – different lengths

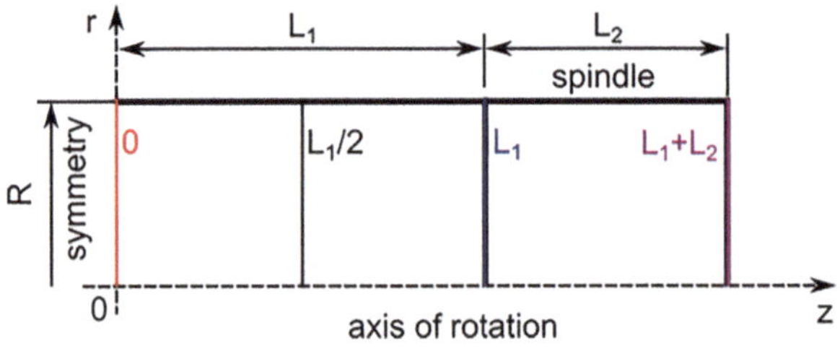

L_1 + L_2. The resulting temperature distribution along the cross-section at these distances is shown in Fig. 12. Again, a good agreement between the results obtained from the analytically derived relations and from FEM is evident. It can thus be said that these analytical equations can fully replace the thermal simulations using FEM, at least in terms of steady-state temperature prediction. The small differences between the results of the two methods are mainly due to the inaccuracy of the fit of the internal heat generator.

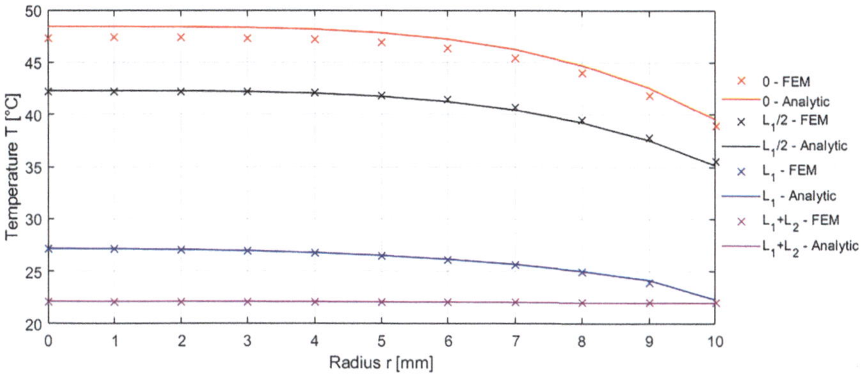

Fig. 12 Steady-state temperature distribution across the cylinder cross-section ($t = 3{,}600$ s)

Fig. 13 Representation of the result paths – selected points

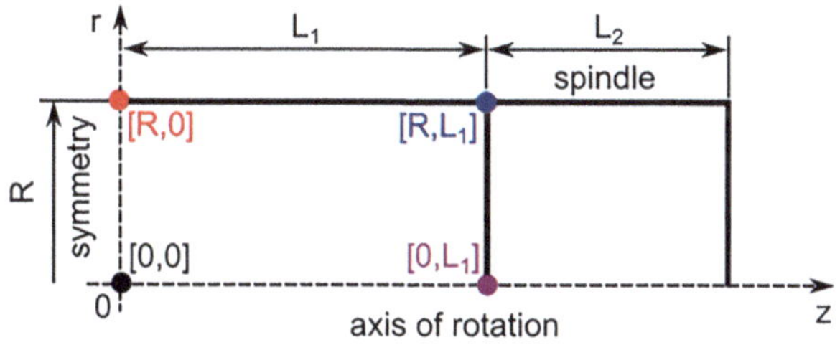

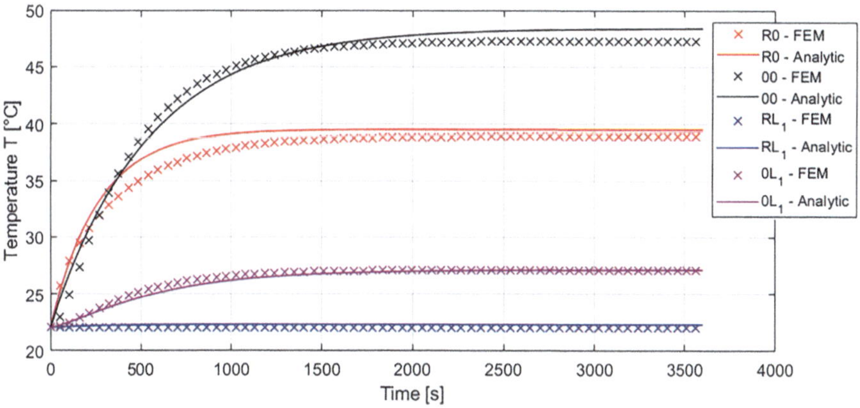

Fig. 14 Temperature distribution over time

As for the results in time domain, four points of the cylinder body 1 were selected for evaluation (see Fig. 13) and the results are shown in Fig. 14. As can be seen, the differences between both solutions are acceptable after the temperature has stabilized. Before that, the results differ considerably (see Fig. 15). The analytical solution does not correctly describe the trend of the temperature increase during the initial heating. Previous measurements reported in [17] show that the temperature

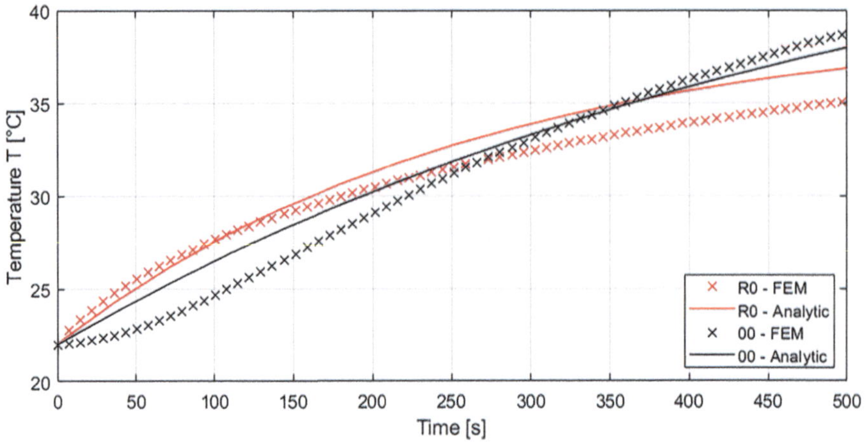

Fig. 15 Temperature distribution over time – detail

trend obtained by the FEA is in better agreement with reality than the analytical one. The maximum difference between both solutions is 1.85°C in the center of the cylinder at a time of approximately 100 s, which corresponds to an error of 7.4%.

4 Conclusion

This paper dealt with the 2D transient thermal analytical solution of the heat build-up in a rubber cylinder. The derivation of the analytical equations, based on the differential heat equation, was described in detail. The results of the analytical solution were compared with those of the FEA and their differences were discussed.

It has been shown that the presented analytical solution can predict the steady-state temperature at all locations of the cylinder very well compared to the FEM solution. The inaccuracy of the temperature results obtained analytically can only be due to the inaccuracy of the internal heat generator equations themselves. In the time domain, the analytical solution cannot correctly describe the temperature rise trend, especially at the beginning of the heating process.

The question remains whether and how the analytical solution can be modified to match the results of the FEA. Furthermore, whether there is a possibility to extend the analytical solution to handle material parameters that vary with temperature.

Due to the excellent accuracy of the steady-state temperature prediction between the FEA and the analytical results, the presented analytical solution can be used to fine-tune the convection values used in the calculation on the free surface and on the clamping surface of the cylinder in the spindle. For example, if a steady-state temperature distribution on the surface of the cylinder along its length is obtained by thermography and if the internal heat generator can be accurately determined, the magnitude of the surface convection can already be determined with knowledge of the thermal parameters of the material.

Declarations **Author contributions.** The manuscript was written through contributions of all authors. All authors have given approval to the final version of the manuscript.

Funding. This work was supported by the Ministry of Education, Youth and Sports of the Czech Republic—DKRVO (RP/CPS/2022/006) and within the project OP RDE Junior Grants of TBU in Zlín, Reg. No. CZ.02.2.69/0.0/0.0/19_073/0016941.

Data availability statement. The datasets generated and/or analyzed during the current study are available from the corresponding author on reasonable request.

Conflict of Interest. The authors declare that they have no conflict of interest.

References

1. Medalia AI (1991) Heat generation in elastomer compounds: causes and effects. Rubber Chem Technol 64(3):481–492. https://doi.org/10.5254/1.3538565
2. Gschwandl M, Kerschbaumer RC, Schrittesser B, Fuchs PF, Stieger S, Meinhart L (2019) Thermal conductivity measurement of industrial rubber compounds using laser flash analysis: applicability, comparison and evaluation. AIP Conf Proc 2065(1):030041. https://doi.org/10.1063/1.5088299
3. Gent AN, Hindi M (1988) Heat build-up and blowout of rubber blocks. Rubber Chem Technol 61(5):892–905. https://doi.org/10.5254/1.3536225
4. Saha T, Bhowmick AK (2021) Thermal and thermo-oxidative degradation of rubbers: some recent studies. In: Heinrich G, Kipscholl R, Stoček R (eds) Degradation of elastomers in practice, experiments and modeling. Springer, Cham, pp 209–238. https://doi.org/10.1007/12.2021108
5. Kaiser S, Rabbani R, Ahmed R, Kaiser S (2021) Temperature dependent mechanical properties of natural and synthetic rubber in practical structures. Acta Mechanica Slovaca 25(3):6–14. https://doi.org/10.21496/ams.2021.031
6. Ruellan B, Le Cam J-B, Jeanneau I, Canévet F, Mortier F, Robin E (2019) Fatigue of natural rubber under different temperatures. Int J Fatigue 124:544–557. https://doi.org/10.1016/j.ijfatigue.2018.10.009
7. Schieppati J, Schrittesser B, Wondracek A, Robin S, Holzner A, Pinter G (2019) Heat build-up of rubbers during cyclic loading
8. Luo W, Yin B, Hu X, Zhou Z, Deng Y, Song K (2018) Modeling of the heat build-up of carbon black filled rubber. Polym Test 69:116–124. https://doi.org/10.1016/j.polymertesting.2018.05.017
9. Li F, Liu J, Yang H, Lu Y, Zhang L (2016) Numerical simulation and experimental verification of heat build-up for rubber compounds. Polymer 101:199–207. https://doi.org/10.1016/j.polymer.2016.08.065
10. Ovalle Rodas C, Zaïri F, Naït-Abdelaziz M (2014) A finite strain thermoviscoelastic constitutive model to describe the self-heating in elastomeric materials during low-cycle fatigue. J Mech Phys Solids 64:396–410. https://doi.org/10.1016/j.jmps.2013.10.010
11. Ovalle Rodas C, Zaïri F, Naït-Abdelaziz M, Charrier P (2016) A thermovisco-hyperelastic model for the heat build-up during low-cycle fatigue of filled rubbers: formulation, implementation and experimental verification. Int J Plast 79:217–236. https://doi.org/10.1016/j.ijplas.2015.01.001
12. Tong X, Chen X, Xu J-S, Zheng Y, Zhi S-J (2018) The heat build-up of a polymer matrix composite under cyclic loading: experimental assessment and numerical simulation. Int J Fatigue 116:323–333. https://doi.org/10.1016/j.ijfatigue.2018.06.040

13. Le Saux V, Marco Y, Calloch S, Doudard C, Charrier P (2010) Fast evaluation of the fatigue lifetime of rubber-like materials based on a heat build-up protocol and micro-tomography measurements. Int J Fatigue 32(10):1582–1590. https://doi.org/10.1016/j.ijfatigue.2010.02.014
14. Stoček R, Stěnička M, Kipscholl R (2019) Heat build-up characterization under realistic load. In: Constitutive models for rubber XI – Proceedings of the 11th European conference on constitutive models for rubber, CRC Press/Balkema, Nantes, pp 157–162. https://doi.org/10.1201/9780429324710-28
15. Stoček R, Kratina O, Voldánová J, Kipscholl R, Heinrich G (2022) Realistische charakterisierung der wärmeentwicklung von gummi unter hochdynamischen, multiaxialen zyklischen belastungen. GAK Gummi-Fasern-Kunststoffe 4:14–21
16. Werner P, Baaser H (2021) Simulation of a heat-buildup process on rubber components. KGK Kautschuk-Gummi-Kunststoffe 3:43–47
17. Peter O, Stoček R, Kratina O (2023) Experimental and numerical description of the heat build-up in rubber under cyclic loading. In: Heinrich G, Kipscholl R, Stoček R (eds) Degradation of elastomers in practice, experiments and modeling. Springer, Cham, pp 121–141. https://doi.org/10.1007/12.2022.128
18. Akutagawa K, Hamatani S, Nashi T (2015) The new interpretation for the heat build-up phenomena of rubbery materials during deformation. Polymer 66:201–209. https://doi.org/10.1016/j.polymer.2015.04.040
19. Peter O, Stoček R (2023) Analytical calculation approach of heat-build up in rubber cylinder. In: Marano et al (eds) Constitutive models for rubber XII. CRC Press/Balkema, Boca Raton, pp 278–282
20. Holman JP (2010) Heat transfer.10th edn. The McGraw-Hill Book Company, New York
21. Ansys Mechanical Enterprise, Release 2021R2, ANSYS, Inc.

Adv Polym Sci (2024) 294: 53–70
https://doi.org/10.1007/12_2023_167

Published online: 14 November 2023

Kinetics of Self-Heat Build-Up in Carbon Black Filled Natural Rubber Caused by Cyclic Mechanical Loading

Ondřej Kratina, Radek Stoček, Petr Zádrapa, Reinhold Kipscholl, and Ondřej Peter

Contents

Ondřej Kratina, Radek Stoček, Petr Zádrapa, Reinhold Kipscholl, and Ondřej Peter contributed equally to this work.

O. Kratina and P. Zádrapa
Centre of Polymer Systems, Zlín, Czech Republic

R. Stoček (✉)
Centre of Polymer Systems, Zlín, Czech Republic

PRL Polymer Research Lab, Zlín, Czech Republic
e-mail: stocek@utb.cz

R. Kipscholl
Coesfeld GmbH & Co. KG, Dortmund, Germany

O. Peter
Centre of Polymer Systems, Zlín, Czech Republic

Bonatrans Group a.s., Bohumín, Czech Republic

Abstract The fact that heat is generated in viscoelastic materials like rubber progressed by energy dissipation while cyclically loaded is well known as self-heat build-up (HBU). In addition, the rubber components are reinforced with various types of fillers, the most used being carbon black (CB), which significantly affects the viscoelastic properties and therefore the development of heat. The exact course of heat generation depending on individual load cycles has not yet been sufficiently described in general terms, let alone defined with regard to the influence of different types of CB. More detailed knowledge would therefore be of great importance for the development of materials for many rubber applications.

For this reason, the aim of this study is to describe in detail the kinetics of HBU in natural rubber reinforced with 10 different types of CB under varied cyclic loading conditions.

In order to understand the effect of different filler reinforcements on the HBU behavior, the fundamental characterization of the studied materials, namely hardness, tensile properties as well as the determination of viscoelastic behavior with respect to the Payne effect, was first performed. The kinetic of HBU was characterized using a unique method of cyclic alternating tensile and compressive loading of a rotating bent rubber cylinder with in-situ temperature analysis inside as well as on the surface of the sample. Simple mathematical functions describing the heat evolution as a function of the type of CB and loading frequency were defined and these were discussed in relation to the fundamental properties of the materials studied.

An increasing temperature rise with increasing CB volume as well as aggregate surface area has been found which fully corresponds with increasing hardness and stiffness of the rubber. Moreover, it has been shown that the course of HBU kinetics as a function of CB type and amount is fully consistent with the G'values obtained in the Payne effect analyses.

Keywords Carbon black · Heat build-up · Kinetics · Multiaxial loading · Rubber · Testing

1 Introduction

Rubber materials are widely used in engineering applications due to their unique properties, which include high extensibility with almost no permanent deformation, high strength, high energy absorption, and high fatigue resistance as well [1]. In many applications, rubber products are commonly cyclically loaded, leading to self-heat development, thermal degradation, and reduction of mechanical properties and in consequence to failure.

The heat generation is an outcome of rubber's viscoelasticity, where part of the introduced mechanical energy is converted into heat during each loading cycle [2]. The kinetics of heat generation depends mainly on the type of rubber polymer and the compound's components, e.g., reinforcing fillers like CB [3–6] which is widely used in the rubber industry. It has been found that volume fraction and surface area of aggregates have a major influence on the heat development in rubber [7]. In particular, highly filled rubber compounds loaded with high strain rates and high frequencies cause significant temperature increases up to values leading to thermal degradation of the rubber. This can lead to a reduction in ultimate strength or even severe failures before the rubber components reach the end of their service life [8, 9]. Therefore, some studies have investigated the dependence of self-heat build-up (HBU) and rubber lifetime [10, 11], which can be clearly demonstrated. Currently, there are numerous studies discussing the influence of HBU on the fatigue process of rubber [12–16]. In the meantime, many scientific papers have been written to numerically simulate the HBU process in order to predict rubber lifetime [17–20].

To accurately determine the self-heating gradient of rubber samples with maximum regard to the simulation of real rubber loading in practice, a suitable experimental method as well as measuring equipment is essential. Furthermore, the loading conditions must be repeatable and should be measured and monitored just as the temperature inside and outside the sample must be measured in-situ and monitored simultaneously with the loading conditions. These prerequisites are fulfilled by the Heat Build-up Analyzer (HBuA) used in this study and presented, e.g., in [21, 22]. The temperature inside the rubber sample is measured by the contact method using a resistive flexible sensor. A non-contact and non-destructive optical imaging technique for the detection of invisible infrared radiation was used to measure the surface temperature. The applied measurement setup was also used to derive a numerical simulation of the HBU process [23–25].

In this study the effect of CB type and its volume fraction on the surface temperature development and its concentration inside the sample at different frequencies are investigated. The experimental part hereinafter defines the natural rubber-based compounds and the test procedures executed. The main part of the experiment is consecrated to the individual test results. In particular the gradient and the reached temperature values of HBU measurements at two deformation frequencies are presented. Additionally, it is shown how the dynamic moduli (G′, G′′) develop due to breakdown of the particle network formed by aggregates of carbon black depending on the degree of strain. In particular, the value of the loss moduli obtained by the method used is an indicator significantly correlating with the HBU process [26].

The Shore A hardness test and the tensile test were used to characterize the fundamental material properties.

2 Materials and Experimental

The rubber compounds used in this work were based on natural rubber (SVR CV60, Binh Phuoc, Vietnam), which was filled with various types (N115, N220, N330, N550, N772) and two quantities (20 phr, 60 phr) of CB (all from Cabot, Czech Republic) – see Table 1. All rubber compounds were prepared by a two-stage mixing procedure in an internal mixer (SYD-2L, Everplast, Taiwan) and a two-roll mill (Compuplast, Czech Republic). In the first step, which was carried out by the internal mixer, filler, zinc oxide (ZnO), stearic acid, and N-(1,3-dimethylbutyl)-N′-phenyl-p-phenylenediamine (6PPD) (all from Sigma-Aldrich Chemie, Germany) were added as additives to the elastomer to prepare masterbatches. The rotor speed was set at 40 rpm and the initial temperature was set at 80°C, while the dump temperature did not exceed 150°C. After the first step, the masterbatches were sheeted out on a two-roll mill and stored for 24 h at room temperature. In the second step, sulfur and the curing accelerator N-tert-butyl-2-benzothiazole sulfenamide (TBBS) were added to the masterbatch using a two-roll mill at 60°C for 5 min. The final rubber compounds were sheeted out and stored for 24 h at room temperature. The complete compositions of the prepared rubber compounds are listed in Table 1.

Basic data about the specific surface areas and structures of the CB types are illustrated in Fig. 1. These data were obtained from the CB manufacturer's data sheets. The Statistical Thickness Surface Area (STSA) is defined as the specific surface area that is accessible to rubber. STSA is determined from the amount of nitrogen adsorbed, at liquid nitrogen temperature, by a carbon black at several partial pressures of nitrogen according to ASTM D6556. The structure of CB was measured by oil absorption tests on pre-compressed CB (Compressed Oil Absorption Number; COAN) according to ASTM D3493. The level of COAN structure is assumed more indicative of the level of structure of CB after the aggregate attrition due to mixing with the rubber.

Table 1 Rubber compounds composition

	X_20		X_60	
	PHR	Volume fraction [−]	PHR	Volume fraction [−]
SVR CV 60	100	0.879	100	0.750
Carbon black	**20**	**0.086**	**60**	**0.220**
Zinc oxide	2	0.003	2	0.003
Stearic acid	1	0.010	1	0.008
6PPD[1]	1	0.008	1	0.007
TBBS[2]	1	0.008	1	0.007
Sulfur	2	0.006	2	0.005
Total	127	1.000	167	1.000

X according to CB designation – N115, N220, N330, N550, N772
1N-(1,3-Dimethylbutyl)-N′-phenyl-p-phenylenediamine
2N-tert. -butyl-2-benzothiazyl sulfenamide

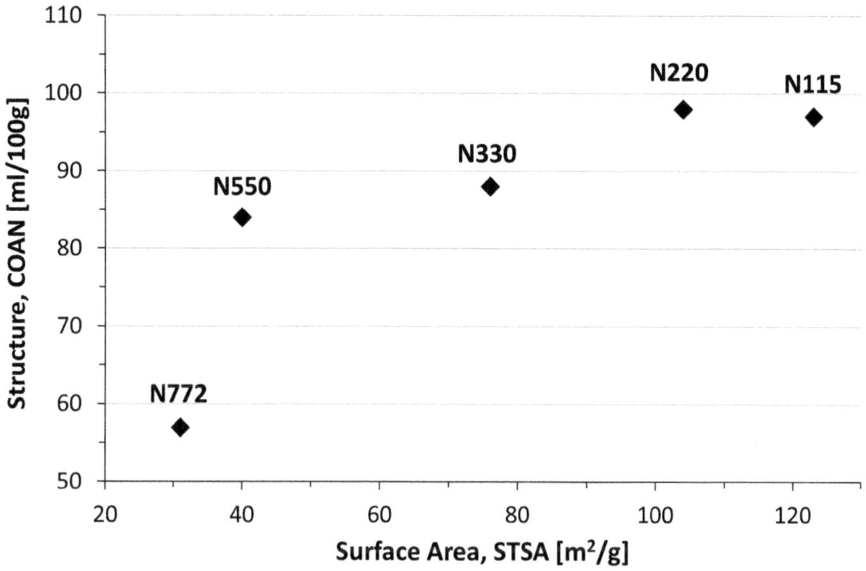

Fig. 1 The dependence of the structure (COAN) versus the surface area of used carbon black according to the manufacturer's data sheets

Table 2 Curing characteristics

Volume filler fraction [–]	Materials	S′ min [dN·m]	S′ max [dN·m]	TS2 [min]	t90 [min]
$\phi = 0.086$	N115_20	1.43 ± 0.09	11.18 ± 0.31	5.68 ± 0.05	9.31 ± 0.08
	N220_20	1.26 ± 0.19	11.38 ± 0.30	4.69 ± 0.01	8.36 ± 0.02
	N330_20	1.36 ± 0.05	10.54 ± 0.05	4.17 ± 0.42	7.38 ± 0.54
	N550_20	1.12 ± 0.21	10.50 ± 0.08	4.23 ± 0.38	7.15 ± 0.36
	N772_20	1.23 ± 0.05	10.04 ± 0.21	5.41 ± 0.07	8.60 ± 0.06
$\phi = 0.220$	N115_60	5.17 ± 0.01	23.86 ± 0.59	2.97 ± 0.21	10.49 ± 0.09
	N220_60	4.64 ± 0.06	22.35 ± 0.38	2.18 ± 0.09	6.55 ± 0.34
	N330_60	3.64 ± 0.14	24.21 ± 0.17	1.98 ± 0.01	5.63 ± 0.07
	N550_60	3.04 ± 0.03	24.12 ± 0.01	2.05 ± 0.02	5.64 ± 0.02
	N772_60	1.86 ± 0.02	17.62 ± 0.13	3.17 ± 0.01	7.02 ± 0.03

The curing properties were determined using a moving die rheometer (MDR 3000 Basic, MonTech, Germany) according to ASTM 5289 at 150°C for 0.50 arc at 1.7 Hz.

The curing of samples of different shapes required for the experimental mechanical tests was carried out in a heat press (LabEcon 300, Fontijne Presses, The Netherlands) at 150°C according to the optimum curing time t90 (Table 2), which was extended by surcharge of heating time according to the geometry of the sample.

2.1 Tensile Properties

The effect of rubber composition on the mechanical properties in tension was determined by quasi-static tensile testing, which was carried out on an M350-5CT universal testing machine (Testometric, UK) according to ISO 37 using Type 2 specimens. The maximum stress and elongation at break values were determined as results characterizing the mechanical properties. These values were supplemented with stress values at 50 and 100% elongation.

2.2 Hardness Shore A

The Shore A hardness test was carried out in accordance with ISO 868 using a Shore A hardness tester (Wallace, Kingston, UK). The result is the average hardness values in Shore A units calculated from five measurements taken for each specimen.

2.3 Payne Effect

The Payne effect is generally a typical nonlinear softening response of rubber materials under dynamic increasing strain amplitude. This mechanical behavior is influenced by several mechanisms, in particular the breakdown of the filler–filler or polymer–filler network [27]. These mechanisms affect the degree of viscoelasticity of the material, which is of great practical importance and is particularly crucial in tire applications where it affects traction, rolling resistance, and ultimately heating. The Payne effect value, which is given by the difference between the upper and lower values of storage modulus (G'), is an important parameter used to describe filler–filler and filler–matrix interactions. A low value of the Payne effect indicates a better interaction between filler and matrix, while a high value indicates a better filler–filler interaction. The loss modulus (G') characterizes the part of energy that is dissipated to heat during cyclic deformation. The measurement of the Payne effect was performed using a Rubber Process Analyzer (RPA2000, Alpha Technologies, USA), which is a dynamic mechanical rheometer. The strain amplitude was varied from 0.07 to 18.75% at a frequency of 10 Hz and a temperature of 60°C which is the temperature of the sample after several hundred cycles. As a result, a course of G' and G' values was obtained for all characterized materials.

2.4 Heat Build-Up Analyses

The HBU behavior of the listed materials was determined by recording surface and core temperatures of samples taken in every loading cycle. All HBU experiments were performed on HBuA™ (Coesfeld GmbH & Co. KG, Germany) [21, 22]. Figure 2 shows a schematic of the loading conditions of the specimen. The specimens were cylinders of diameter 20 mm and length 100 mm. During the molding process, a cylindrical hole with diameter 5 mm and length 50 mm from the end to the center of the specimens was made on one end using an insert. The individual principle of measurement is described in several steps. Firstly, the sample is clamped. A flexible temperature sensor, which is part of the equipment, is inserted into the cylindrical hole of the sample. The spherical tip of the temperature sensor touches the surface in the middle of the sample. In the second step, clamped sample is bent to the given angle, β. The bending of the test specimen results in tensile stress on one side of the neutral axis and compressive stress on the opposite side. Finally, the bent specimen starts to rotate about its neutral axis. Rotation results in a dynamic change of torque and multi-planar bending of the specimen as a combined load under fatigue conditions. From the schematic visualization of the deformed specimen in Fig. 2, it can be seen that the maximum deformation and hence the maximum stress occurs at the surface of the specimen, while toward the center of the specimen it decreases to zero in the neutral axis. Therefore, it is expected that in time the temperature first increases at the surface of the sample and further concentrates inside the sample.

Specifications and technical parameters of the test equipment are given in [21].

Although many combinations of test parameters can be chosen, this research was conducted with only one set of parameters for all compounds. These parameters were chosen to be close to the mean operating conditions of tire loading in practice and correspond to the following values:

- Two sample loading frequencies of 8.33 Hz (corresponds to 500 rpm) and 16.67 Hz (corresponds to 1,000 rpm) were applied. The applied frequencies

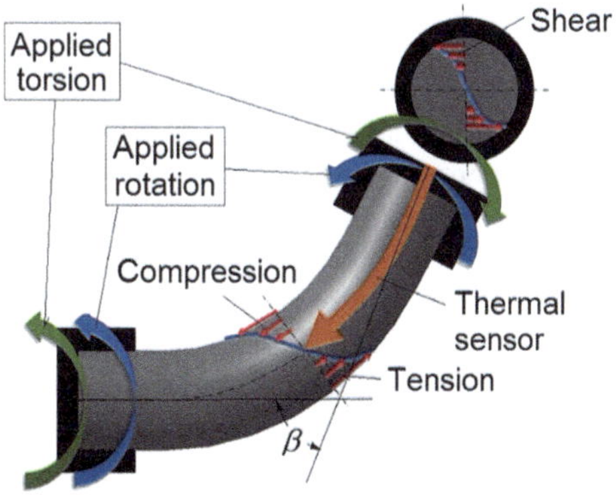

Fig. 2 Scheme of the measuring principle

simulate approximately at the tire situations at speeds of 50 Miles/Hour, respectively, 25 Miles/h)
- Bending angle, β: 50°
- Torque regulation: 0 Nm
- Chamber temperature: 23°C

Three replicates have been analyzed per compound.

3 Results and Discussion

3.1 Tensile Properties and Hardness Shore A

The test results describing fundamental mechanical properties are shown in Table 3. The tensile test confirms the expectation that materials listed in order of decreasing volume filler fraction achieve higher strain and stress at break. Conversely, the moduli at strains of 50% and 100% decrease with decreasing volume filler fraction. Additionally, the Shore A test shows that the hardness of the materials increases with the increasing volume filler fraction as well as with the increasing specific surface area of CB. This confirms the well-known experience, which also applies to the materials studied, that the stiffness of the material increases with increasing volume fraction and specific surface area of the reinforcing filler in general.

3.1.1 Payne Effect

Figure 3 shows the G'-value dependencies on strain of rubber types filled with $\phi = 0.086$ CB in comparison with those filled with $\phi = 0.220$ CB. The results show that the degree of network disintegration increases in the same order with lower numerical designation of CB, and thus with higher stiffening ability, which is effectuated to the larger surface area of CB. The differences of G' are roughly one power of ten (Fig. 3 left/ right) resulting from the CB volume fraction. These results confirm the reported properties of the used CB, which are shown in Fig. 1. In fact, it can be seen that the rubber with the presence of CB with lower numerical designation and therefore larger specific surface area show more pronounced differences between the values of the storage modulus (G') at the beginning and at the end of the measurements. It can therefore be concluded that these results are consistent with the hardness test results and clearly indicate a higher stiffening ability of CB with a larger specific surface area and therefore a lower numerical designation.

Furthermore, the values of the loss modulus (G') over strain were obtained. From the results, presented in Fig. 4, it obvious that the values of G' increase with both higher volume filler fraction and increasing specific surface area. It can be concluded that the dissipated energy, which is generally influenced by the disintegration of the

Table 3 Mechanical properties of materials studied

Volume filler fraction [−]	Materials	Stress at break [MPa]	Strainat break [%]	M50 [MPa]	M100 [MPa]	Shore A [−]
$\phi = 0.086$	N115_20	31.53 ± 1.09	653.24 ± 19.17	0.95 ± 0.06	1.47 ± 0.08	46.1 ± 0.4
	N220_20	32.30 ± 0.92	642.24 ± 3.24	0.97 ± 0.01	1.53 ± 0.05	45.5 ± 0.7
	N330_20	32.44 ± 0.87	647.92 ± 14.61	0.93 ± 0.03	1.50 ± 0.06	45.2 ± 0.3
	N550_20	30.82 ± 0.54	516.24 ± 19.51	1.11 ± 0.09	1.96 ± 0.18	45.0 ± 0.5
	N772_20	29.74 ± 2.39	624.52 ± 29.88	0.88 ± 0.04	1.40 ± 0.06	43.6 ± 0.4
$\phi = 0.220$	N115_60	27.01 ± 0.70	451.36 ± 9.45	2.18 ± 0.08	4.06 ± 0.19	68.5 ± 0.5
	N220_60	27.60 ± 1.33	437.01 ± 14.62	2.25 ± 0.20	4.79 ± 0.27	67.9 ± 0.2
	N330_60	27.95 ± 0.94	435.49 ± 10.66	2.31 ± 0.12	4.90 ± 0.26	64.7 ± 0.4
	N550_60	25.88 ± 0.35	355.68 ± 13.44	2.88 ± 0.22	6.95 ± 0.50	63.8 ± 0.6
	N772_60	24.61 ± 0.68	458.55 ± 13.68	1.73 ± 0.06	3.46 ± 0.16	56.8 ± 0.6

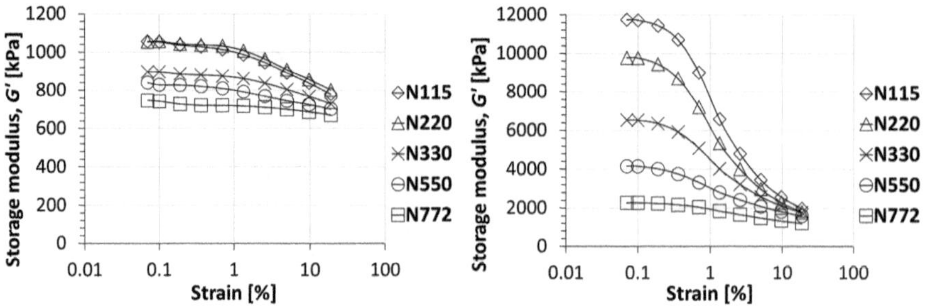

Fig. 3 Payne effect results of storage modulus G' of ϕ = 0.086 filled series (left), ϕ = 0.220 filled series (right)

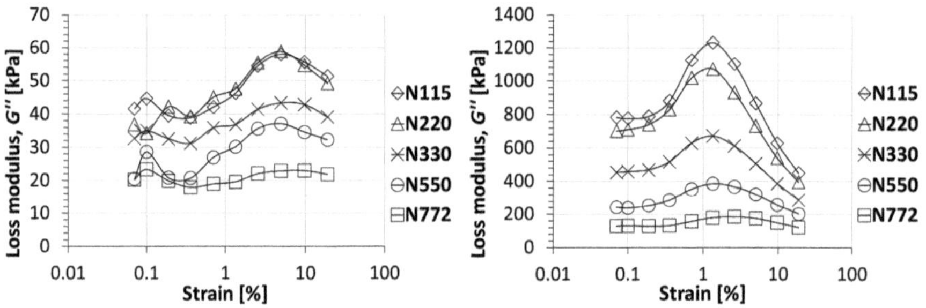

Fig. 4 Payne effect results of loss modulus G' of ϕ = 0,086 filled series (left), ϕ = 0,220 filled series (right)

filler network and the slip of the filler particles and polymer molecules to each other during deformation, is highest for the material with the highest reinforcement.

It can be concluded that the loss energy, which is generally influenced by the disintegration of the filler network and the mutual slip between filler particles and polymer molecules during deformation, is highest for the materials with the higher reinforcement.

3.2 *Heat Build-Up*

Two graphs in Fig. 5 are exemplary presenting the HBU development of one tested rubber sample. The graph, left presents curves temperature developments on the surface and inside the sample in dependence on loading cycle executed by HBuA. It is evident that the both temperatures start to rise from the first loading cycle and, as expected, a higher rate of temperature rise at the beginning of loading is recorded for the surface temperature compared to the internal temperature. While the surface temperature gradient slows down shortly after the start, the internal temperature rise

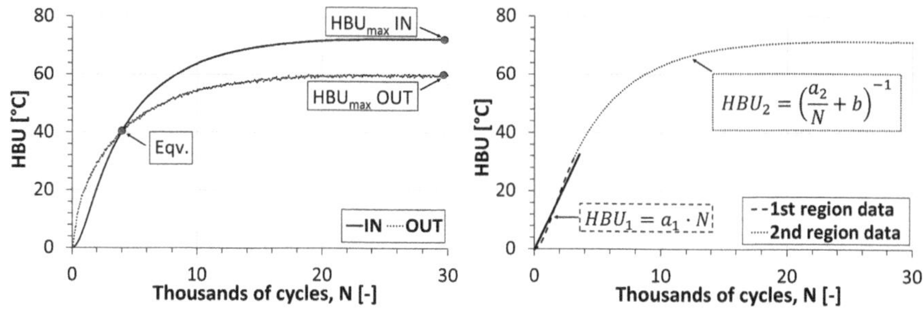

Fig. 5 General course of HBU curves obtained by HBuA for material N550_0.220 at 16.67 Hz (left) and definition of HBU periods (right)

remains constant until both temperatures reach equilibrium. Once equilibrium is reached, the internal temperature begins to exceed the surface temperature in value and the rate of temperature increase inside the sample begins to decrease rapidly. Both curves are downward sloping until a temperature maximum is reached, followed by a temperature plateau or even a temperature drop in some specific cases. Heat build-up stops when self-heat generation and heat transfer to the ambient air are balanced. The observations made confirm earlier studies [15].

Mathematically, each obtained HBU curve can be divided into two regions, as shown in Fig. 5, right. The first region, corresponding to the beginning of loading, which is in the order of hundreds of cycles for the surface temperature and in the order of thousands for the internal temperature, provides information about the HBU kinetics at the beginning of loading, which is essential information describing the initial evolution of the HBU kinetics of the rubber when the initial loading occurs and the material exhibits an initial temperature equal to the ambient temperature. The temperature in this region increases almost linearly and can be mathematically described by the slope of the linear function:

$$\mathrm{HBU}_1 = a_1.N \tag{1}$$

where HBU_1 = Heat build-up; N = cycle number; a_1 = slope.

The second region of the HBU curve presented in Fig. 5, right describes the course of the HBU rate at which the maximum temperature is reached. In order to describe this course with a continuous analytical function, moreover, with the highest possible accuracy, the specific function was required to be found, whereas the complete searching process has firstly been introduced in Ref. [28] and identically applied here. Practically, the used procedure is based on searching from a set of given functions, $y = f(x)$, with two coefficients, which one has the smallest sum of squares of deviations respective regression, R^2, with respect to the given set of data (x_i, y_i).

The final mathematical description of the course of the second region is therefore finally characterized by the following equation:

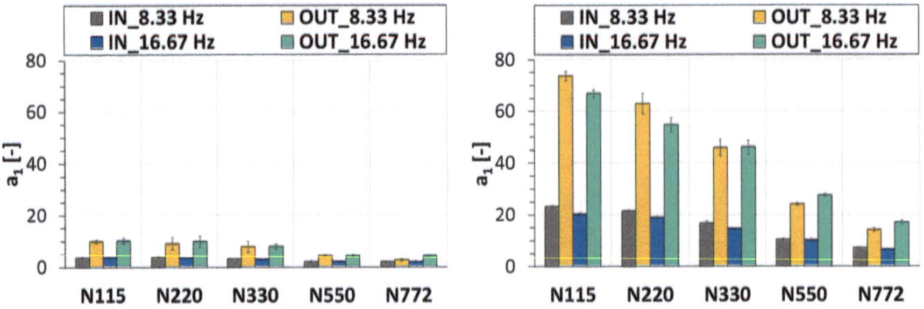

Fig. 6 Results of a_1 of initial HBU of $\phi = 0.086$ filled series (left), $\phi = 0.220$ filled series (right)

$$\mathrm{HBU}_2 = \left(\frac{a_2}{N} + b\right)^{-1} \quad (2)$$

where HBU_2 = Heat build-up; N = thousands of cycles; a_2 = coefficient, b = coefficient.

Figure 6 shows the results of the values of the a1 coefficient, which mathematically evaluates the HBU rate in the first cyclic loading region. From the results seen in Fig. 8, left, which compares rubber materials with filler volume fraction $\phi = 0.086$ and different CB grades, it is clear that a higher degree of HBU occurs for CB-filled rubber with a larger specific surface area. This trend was also confirmed for materials with higher filler volume fraction, the dependence of which is shown in Fig. 8, right.

The determinations confirmed the trend of higher mechanical energy dissipation with both increasing volume fraction and specific surface area of CB, which correlates closely with the *G*' values obtained in the Payne effect analysis and presented in Fig. 4. In addition, it is also numerically confirmed that in the range of validity of the first region, the kinetics of temperature rise at the surface of the specimen increases significantly in comparison with the central part of rubber sample. Regarding the effect of frequency, it is evident that samples loaded at lower strain frequency showed a higher rate of HBU. This behavior is due to the lower number of cycles required to reach the end of the first HBU region for the lower frequency value. A comparison of the number of cycles in the first period for both used frequencies is shown in Fig. 7.

The dependence of the a_2 values describing the HBU kinetics after the initial region up to the maximum value of the self-heat increase is shown in Fig. 8. For both series of fillings, it can be seen that the increase in a_2 values occurs with decreasing stiffness or specific surface area or volume fraction of CB. The effect of strain frequency on the a_2 value was not significantly evident from the determinations made.

The combination of the two mathematical equations can be used to describe the kinetics of temperature evolution along the entire length of the curve up to the value of the maximum HBU temperature, which is an important value for predicting the behaviour and service life of the material under cyclic mechanical loading. The mathematical equations presented were chosen to be as close as possible to the

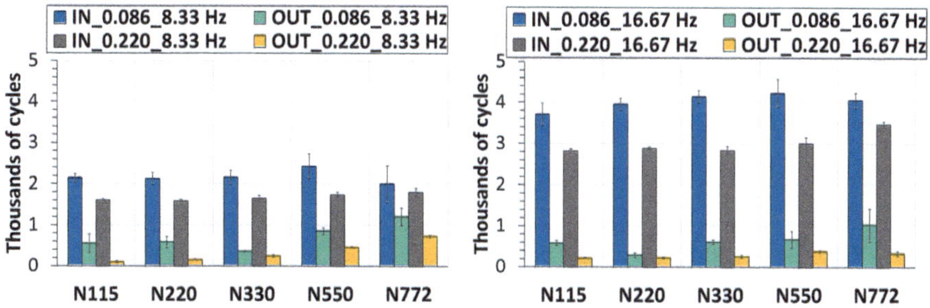

Fig. 7 Number of cycles characterizing first region of HBU for inside (left), and outside (right)

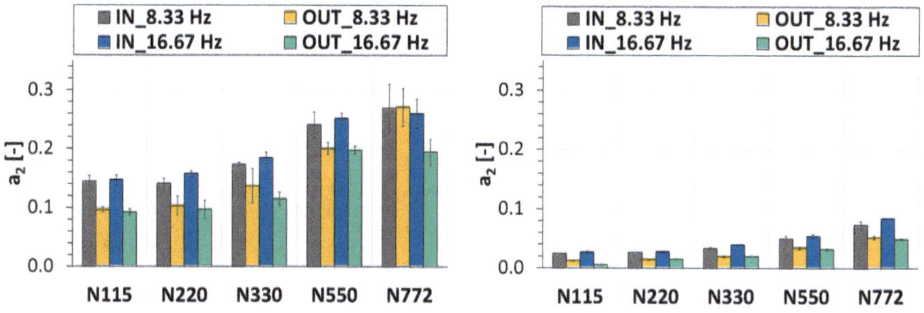

Fig. 8 Results of a_2 for $\phi = 0.086$ series (left), $\phi = 0.220$ series (right)

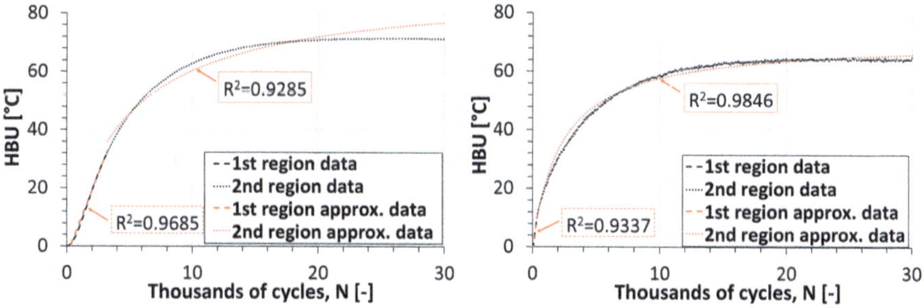

Fig. 9 Comparison of real data obtained for material N550_0.220 at 16.67 Hz versus approximated data on the surface (left) and inside (right) the sample

experimentally obtained data. Figure 9 therefore shows sample actual data obtained for material N550_0.220 at a strain rate of 16.67 Hz compared with approximated data representing the proposed equations expressed inclusive of the numerical reliability values.

The diagram, left shows the data for surface temperature, while the diagram, right presents the data for temperature development inside the sample.

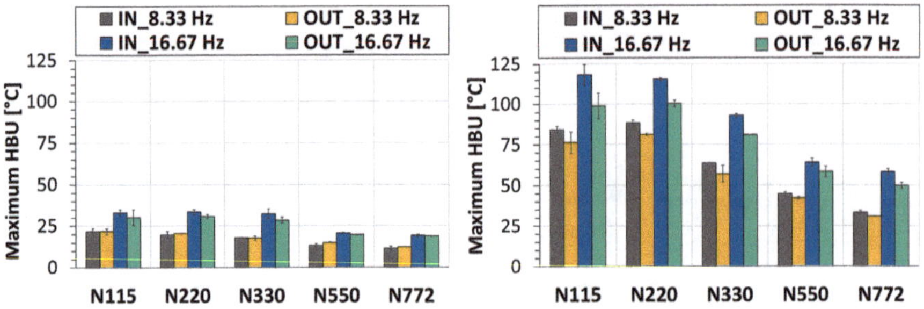

Fig. 10 Results of maximum HBU temperature for $\phi = 0.086$ series (left), $\phi = 0.220$ series (right)

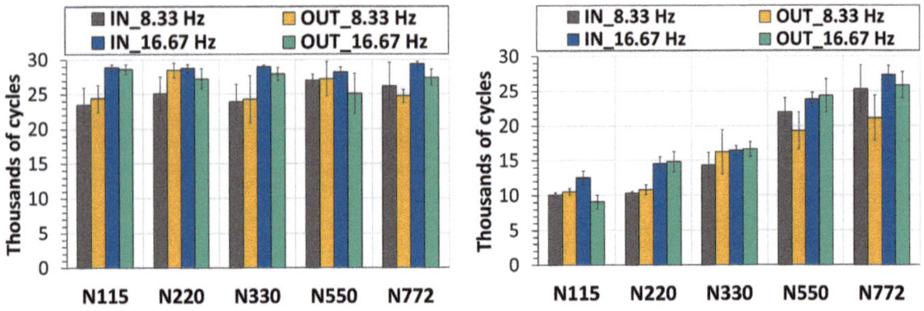

Fig. 11 Number of cycles necessary to reach maximum HBU temperature for $\phi = 0.086$ series (left), $\phi = 0.220$ series (right)

The maximum developed temperature values give an idea about the self-heating of the material in response to mechanical loading. These values show the extent to which the lifetime of cyclically deformed materials can be affected by thermal degradation. A comparison of the maximum temperatures of the rubbers is shown in Fig. 10, where the results for rubber filled with $\phi = 0.086$ of CB are shown on the left and rubber filled with $\phi = 0.220$ of CB on the right. It is evident from both series of materials that the type and amount of CB have an effect on the temperature development in the rubber, which is consistent with the G' values obtained using the Payne effect test. From the dependence of loading frequency on the maximum developed temperature, it was found that higher loading frequency leads to higher heating. It was also found that a higher loading frequency leads to an acceleration of the self-heating process, as shown in Fig. 11, where a comparison of the total number of cycles till reaching the maximum temperature inside and on the surface of the samples is shown. After reaching the maximum HBU value, the temperature development changes with further loading. A drop in temperature after reaching the maximum was observed in particular for materials that were considerably stiffened. A temperature plateau was evident for less stiffened materials. In order to describe the temperature behavior after reaching the maximum, it would be required to extend the existing experiment by a much larger number of load cycles.

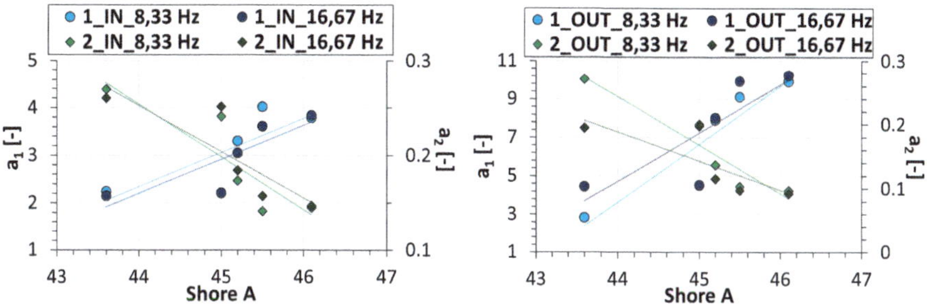

Fig. 12 Effect of hardness on the temperature development coefficients inside (left) and on the surface (right) of a test specimen made of rubber material filled with $\phi = 0.086$ CB

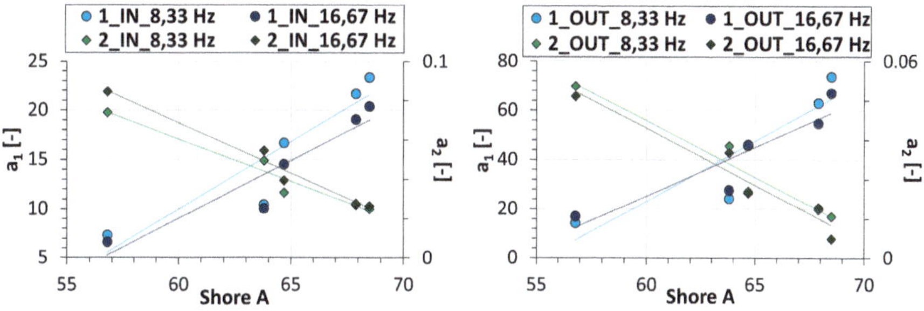

Fig. 13 Effect of hardness on the temperature development coefficients inside (left) and on the surface (right) of a test specimen made of rubber material filled with $\phi = 0.220$ CB

In addition, the results also show that the value of the maximum temperature at the surface is lower than inside the specimen, due to heat transfer to the surrounding environment, which has a lower temperature than the specimen surface, while the temperature inside increases due to conduction from the surface to the center, where it accumulates.

From the results presented above, it is quite clear that the degree of material stiffening has a significant effect on the development of mechanically initiated heat in viscoelastic rubber materials. Figures 12 and 13 illustrate the dependence of the material hardness, which is a very effective and fast test to characterize the changes in the degree of stiffening of the materials, on the values of the coefficients describing kinetics of temperature development for the both investigated regions of the HBU curve. The dependence of the increasing value of the temperature development coefficient both inside and on the surface of the sample on the increasing hardness of the material is clearly visible from the figures that give information about the two series of prepared materials.

4 Conclusion

The aim of this study was to experimentally describe the process of self-heat build-up of carbon black reinforced rubber under cyclic mechanical loading at different frequencies and to complement this with a mathematical expression of the kinetics of temperature rise up to the maximum values of developed temperature. The self-heat build-up was described for 10 natural rubber compounds filled with different types and volume fractions of carbon black. The investigated rubber materials were subjected to fundamental mechanical tests, which showed an increasing rate of stiffening of the material as a function of increasing carbon black volume and specific surface area. In addition, the trend of increasing values of the loss modulus under cyclic mechanical loading with increasing stiffening was also confirmed.

To determine the temperature development, a unique device Heat build-up analyzer (HBuA) capable of characterizing in-situ temperature rise on the surface and inside the rubber sample was used. From the temperature development values obtained with the HBuA device, it was clearly found that materials with higher stiffness and therefore higher loss modulus values showed faster temperature development and reached higher maximum temperature values. The effect of loss modulus on the temperature rise is consistent with the theoretical assumption where increasing values of the loss modulus lead to a higher rate of dissipation of mechanical energy.

Furthermore, the difference in kinetics of HBU inside and on the surface of the sample was clearly demonstrated from the experimentally obtained data, where a delayed temperature evolution inside the material was evident due to heat accumulation directed away from the sample surface, where the largest dissipation of mechanical energy to heat occurs. Based on the experimental data a mathematical description of the HBU kinetic has been determined.

The presented methodology appears to be a useful tool for obtaining accurate data, which are necessary for the development of computational models capable of characterizing the temperature evolution for the rubber products under realistic loading conditions, where high heating capacity is limiting for both mechanical properties and durability.

Acknowledgments This work was supported by the Ministry of Education, Youth and Sports of the Czech Republic – DKRVO (RP/CPS/2022/006) and within the project OP RDE Junior Grants of TBU in Zlín, Reg. No. CZ.02.2.69/0.0/0.0/19_073/0016941.

References

1. (2009) Engineered rubber products: introduction to design, manufacture and testing by John Sommer. Hanser
2. Luo W, Yin B, Hu X, Zhou Z, Deng Y, Song K (2018) Modeling of the heat build-up of carbon black filled rubber. Polym Test 69:116–124. https://doi.org/10.1016/j.polymertesting.2018.05.017

3. Park DM, Hong WH, Kim SG, Kim HJ (2000) Heat generation of filled rubber vulcanizates and its relationship with vulcanizate network structures. Eur Polym J 36(11):2429–2436. https://doi.org/10.1016/s0014-3057(00)00020-3
4. Ferry SD (1980) Viscoelastic properties of polymers.3rd edn. Wiley, New York, pp 356–359
5. Isono Y, Ferry JD (1984) Stress relaxation and differential dynamic modulus of carbon black filled styrene-butadiene rubber in large shearing deformations. Rub Chem Tech 57:225–943
6. Bokobza L (2004) The reinforcement of elastomeric networks by fillers. Macromol Mater Eng 289(7):607–621. https://doi.org/10.1002/mame.200400034
7. Soltani S, Abbassi SF (2005) Effect of carbon black type on viscous heating, heat build up, and relaxation behavior of SBR compounds. Iran Polym J 14(8):745–751
8. Gent AN, Hindi M (1988) Heat build-up and blowout of rubber blocks. Rubber Chem Technol 61(5):892–905. https://doi.org/10.5254/1.3536225
9. Kerchman V, Shaw C (2003) Experimental study and finite element simulation of heat build-up in rubber compounds with application to fracture. Rubber Chem Technol 76(2):386–405. https://doi.org/10.5254/1.3547750
10. Luo W, Huang Y, Yin B, Jiang X, Hu X (2020) Fatigue life assessment of filled rubber by hysteresis induced self-heating temperature. Polymers 12(4):846. https://doi.org/10.3390/polym12040846
11. Masquelier I, Marco Y, Le Saux V, Calloch S, Charrier P (2013) Thermal measurements on elastomeric materials: from the characterization of the dissipation gradients to the prediction of the fatigue properties. Procedia Eng 66:661–668. https://doi.org/10.1016/j.proeng.2013.12.118
12. Rosa GL, Risitano A (2000) Thermographic methodology for rapid determination of the fatigue limit of materials and mechanical components. Int J Fatigue 22:65–73. https://doi.org/10.1016/S0142-1123(99)00088-2
13. Jegou L, Marco Y, Le Saux V, Calloch S (2013) Fast prediction of the Wöhler curve from heat build-up measurements on short fiber reinforced plastic. Int J Fatigue 47:259–267. https://doi.org/10.1016/j.ijfatigue.2012.09.007
14. Serrano L, Marco Y, Le Saux V, Robert G, Charrier P (2013) Fast prediction of the fatigue behavior of short fiber reinforced thermoplastics from heat build-up measurements. Procedia Eng 66:737–745. https://doi.org/10.1016/j.proeng.2013.12.127
15. Le Saux V, Marco Y, Calloch S, Doudard C, Charrier P (2010) Fast evaluation of the fatigue lifetime of rubber-like materials based on a heat build-up protocol and micro-tomography measurements. Int J Fatigue 32:1582–1590. https://doi.org/10.1016/j.ijfatigue.2010.02.014
16. Marco Y, Masquelier I, le Saux V, Charrier P (2017) Fast prediction of the Wöhler curve from thermal measurements for a wide range of Nr and Sbr compounds. Rubber Chem Technol 90(3):487–507. https://doi.org/10.5254/rct.16.83755
17. Huang J, Pastor M, Garnier C, Gong X (2019) A new model for fatigue life prediction based on infrared thermography and degradation process for CFRP composite laminates. Int J Fatigue 120:87–95. https://doi.org/10.1016/j.ijfatigue.2018.11.002
18. Behnke R, Kaliske M (2018) Finite element based analysis of reinforcing cords in rolling tires: influence of mechanical and thermal cord properties on Tire response. Tire Sci Technol 46(4): 294–327. https://doi.org/10.2346/tire.18.4604010
19. Behnke R, Berger T, Kaliske M (2018) Numerical modeling of time- and temperature-dependent strain-induced crystallization in rubber. Int J Solids Struct 141–142:15–34. https://doi.org/10.1016/j.ijsolstr.2018.01.034
20. Behnke R, Kaliske M (2016) The extended non-affine tube model for crosslinked polymer networks: physical basics, implementation, and application to thermomechanical finite element analyses. In: Stöckelhuber K, Das A, Klüppel M (eds) Designing of elastomer nanocomposites: from theory to applications. Advances in polymer science, vol 275. Springer, Cham. https://doi.org/10.1007/12_2016_9
21. Stoček R, Stěnička M, Kipscholl R (2019) Heat build-up characterization under realistic load. Constitutive models for rubber XI, pp 157–162. https://doi.org/10.1201/9780429324710-28

22. Kratina O, Stoček R (2023) How the rubber compounds of different tire's components heat up? In: Marano et al (eds) Constitutive models for rubber, vol XII. CRC Press/Balkema, Boca Raton, pp 199–204. https://doi.org/10.1201/9781003310266-34
23. Peter O, Stoček R, Kratina O (2022) Experimental and numerical description of the heat build-up in rubber under cyclic loading. In: Heinrich G, Kipscholl R, Stoček R (eds) Degradation of elastomers in practice, experiments and modeling. Advances in polymer science, vol 289. Springer, Cham. https://doi.org/10.1007/12_2022_128
24. Peter O, Stoček R (2023) Analytical calculation approach of heat-build up in rubber cylinder. In: Marano et al (eds) Constitutive models for rubber, vol XII. CRC Press/Balkema, Boca Raton, pp 278–282. https://doi.org/10.1201/9781003310266-46
25. Werner P, Baaser H (2021) Simulation of a heat-buildup process on rubber components. Kautschuk und Gummi Kunststoffe 3:43–47
26. He H, Liu J, Zhang Y, Han X, Mars WV, Zhang L, Li F (2022) Heat build-up and rolling resistance analysis of a solid Tire: experimental observation and numerical simulation with thermo-mechanical coupling method. Polymers 14(11):2210. https://doi.org/10.3390/polym14112210
27. Shi X, Sun S, Zhao A, Zhang H, Zuo M, Song Y, Zheng Q (2021) Influence of carbon black on the Payne effect of filled natural rubber compounds. Compos Sci Technol 203:108586. https://doi.org/10.1016/j.compscitech.2020.108586
28. Stoček R (2020) Some revisions of fatigue crack growth characteristics of rubber. In: Heinrich G, Kipscholl R, Stoček R (eds) Fatigue crack growth in rubber materials. Advances in polymer science, vol 286. Springer, Cham. https://doi.org/10.1007/12_2020_72

Adv Polym Sci (2024) 294: 71–94
https://doi.org/10.1007/12_2024_169

Published online: 13 February 2024

Thermo-Mechanical Behavior of Tread Rubber During High-Speed Friction

J. Wu, S. An, F. Teng, B. L. Su, and Y. S. Wang

Contents

Abstract During the landing process, an aircraft tire is subjected to alternating sliding friction. High-speed friction occurs between the tire and the runway, causing a rapid increase in the tire's surface temperature. This constant alternating stress causes the aircraft tire rubber to suffer hysteresis energy loss and generate heat. In this paper, experimental and simulation analyses are combined to investigate the heat generation behavior of aircraft tire rubber materials under different sliding conditions. A high-speed sliding friction test was conducted to investigate the temperature distribution law of the rubber, which proved that the heat was mainly generated by sliding friction. The rubber tire heat generation test was conducted in such a way that hysteresis and frictional losses interact at different sliding speeds. The simulation model for thermo-mechanical coupling was established for the combined effect of

J. Wu, S. An (✉), F. Teng, B. L. Su, and Y. S. Wang
Harbin Institute of Technology, Center for Rubber Composite Materials and Structures, Weihai, China
e-mail: wujian@hitwh.edu.cn

heat generation by hysteresis and friction losses. The finite element model for heat generation prediction was created and verified based on the test results. The validity and accuracy of the simulation method for thermo-mechanical coupling was systematically analyzed, and the maximum error between the simulation results and experimental data did not exceed 7%.

Keyword High-speed friction · Rubber · Simulation and experiment · Slip rate · Slip rate · Temperature filed · Thermo-mechanical coupling

1 Introduction

The aircraft tire is an important component to ensure the safety of the aircraft, and landing is a critical phase in the flight process. During the landing process, the aircraft tire is exposed to various sliding conditions with high speed and high pressure [1]. When the tire first touches the ground, it starts to rotate rapidly from a standstill under the influence of friction. The tire speed quickly increases until it is as same as the forward speed of the aircraft. The tire changes from completely slipping to free rolling, and then enters the braking stage. When braking an aircraft, the tire begins to slip relative to the runway surface due to the braking torque, and the slip rate has a significant impact on braking performance. During the entire landing process, the tires are on the ground at high speed. The rapid temperature rise of the tire tread was caused by the friction energy loss [2]. At the same time, cyclic contact occurs between different positions of the tire circumferentially and the road surface, the stress-strain cycle produces hysteresis energy loss to generate a large amount of heat [3]. The combination of these two energy losses results in combustion and wear of the tread rubber, with some adhering and depositing on the track surface, and the other exhibiting a large amount of white smoke [4], resulting in a loss of tire quality and a reduction in service life and wear resistance. Temperature rise causes changes in the mechanical properties of aircraft tire rubber, such as friction and wear [5, 6], which in turn affects the temperature field of the tire again. Therefore, the tire temperature field is a complex thermo-mechanical coupling problem, and many scholars investigate the temperature field distribution of tire rubber through experiments and finite element simulations.

Rosu conducted tests on aircraft tires on steel cylinders to investigate the thermal evolution of aircraft tires during high-speed rolling, using rolling conditions, vertical loads, speeds, and turning conditions to correspond to the actual situation of aircraft tires skidding or skidding on a flat runway. The comparison between numerical results and test data showed that the tire tread temperature distributions have the same general trend [7]. Korunovic et al. investigated the braking and steering characteristics of tires, as well as the tread friction coefficient, using a drum testing machine, and compared the results with finite element simulation results [8]. Behnke used a roller test bench with a measuring device to test tires under different rolling loads and speeds, the temperature of tires was measured by an infrared temperature measurement system, and the simulation results of rolling resistance and surface temperature changes were verified [9]. The indoor whole tire testing platform has good repeatability and can simulate various working conditions such as rolling, braking, and turning. However, during the landing process, the aircraft tires are in the harsh conditions of high-speed, high-pressure, and heavy-duty, which requires high requirements for the whole tire testing platform. As a result, it is difficult and expensive to implement. Therefore, specimen-level testing is more used to study the properties of rubber friction, temperature, and mechanics. Wallaschek et al. used a linear friction tester to investigate the friction behavior of the tire blocks made from different viscoelastic materials, which revealed the relationships between the friction coefficient and the sliding speed at different temperatures [10]. Rosu et al. conducted a high-speed sliding friction test using rubber block specimens to discuss the friction and wear characteristics and temperature evolution process of rubber blocks under sliding conditions. The test parameters such as load and speed were chosen to be similar to the real parameters of an aircraft tire under rolling and skidding conditions. The results of two-dimensional and three-dimensional numerical simulations were obtained [11]. Tong Xin et al. conducted cyclic uniaxial strain control fatigue tests on rubber composite materials under various load conditions, real-time temperature monitoring, and measured that the surface temperature of the specimen increased by tens of degrees Celsius during the fatigue test. At the same time, the mechanical and thermal responses of the tested materials were simulated through thermo-mechanical coupling simulation, and the simulation results were in good agreement with the experimental results, which helps to verify the accuracy of the proposed model [12]. Peter applied cyclic loads to cylindrical rubber specimens and recorded surface and internal temperatures, which explored the effects of different rotational speeds and bending angles and simulated by ANSYS [13]. Linke et al. tested tread block specimens at a constant slip rate of 100% on a high-speed linear tester and found that the effect of convective heat dissipation was less than that of contact heat generation. Subsequently, a test bench was used to study the thermal behavior of aircraft tires. The results showed that the slip angle had the greatest impact on temperature, followed by speed and load [14].

Thermo-mechanical coupling simulation can be classified into unidirectional decoupling and bidirectional decoupling based on whether the influence of the temperature field on the stress-strain field is considered. Unidirectional coupling only considers the influence of mechanical properties on the temperature field, without studying the influence of the temperature field on structural deformation and mechanical properties. Whicker et al. proposed the method of bidirectional coupling for the first time, taking into account the mutual influence between the stress-strain field and the temperature field, and used this method to carry out thermo-mechanical coupling analysis on the rolling tire model [15]. However, a large amount of simplification was made to the tire structure, so the results were not very accurate. For the implementation of bidirectional coupling, the main commonly used methods currently include fully coupled analysis and modular simulation. The fully coupled method directly adopts the coupled elements with temperature and displacement degrees of freedom, and calculates the mechanical and heat transfer responses in each iteration step of the tire rolling simulation analysis, so as to realize the real-time prediction of stress and heat generation in the rolling process. Alroqi et al. used the structural-thermal coupling transient simulation in ANSYS. Through the fully coupling method, the tread temperature of the landing aircraft tire was simulated under different conditions before rotation. The effect of reducing the tread temperature by pre-rotating the tire before landing was verified [16]. The modular method separates mechanics and thermodynamics, independently solving mechanical analysis and heat transfer analysis in sequence. The interaction between force and temperature through iteration was introduced. Li used the finite element method to study the transient temperature of the solid rubber tire. This method is divided into deformation module, heat dissipation module, and heat transfer module. The three-dimensional static contact analysis between the tire and the road surface was carried out in the deformation module. In the heat dissipation module, the approximate strain amplitude and heat generation rate were calculated. In the heat transfer module, the thermal parameters and thermal boundary conditions were set, and finally the tire transient temperature distribution was obtained [17]. Konde et al. used experimental equipment to predict the thermal-mechanical coupling of aircraft tires during cornering at medium to low speeds. Based on the frictional heat generated by tire slip, it was assumed that the internal hysteresis heat generation can be ignored [18]. Song calculated the heat using the internal hysteresis loss due to rubber deformation and the energy dissipated by frictional heat generation. The results were applied to the simulation model through a user subroutine for thermo-mechanical properties prediction. The reliability of the simulation method was verified by comparing the simulation results of stress and experimental results [19].

The fully coupled method in bidirectional decoupling is simple. However, it takes a long time to calculate once, which makes it difficult to adjust parameters. The modular solution method greatly improves the computational efficiency of thermo-mechanical coupling simulation, but the iterative steps are complex. In this paper, a new thermo-mechanical coupling method is proposed by combining the fully coupled method with the modular method. Firstly, experimental research on the thermodynamic characteristics of tread rubber under service conditions is conducted. The tread rubber material of an aircraft tire is prepared into the shapes of block and wheel specimens. The high-speed friction tests of the rubber block and high-speed rolling tests of the rubber wheel under different slip conditions are carried out using a high-speed friction test platform. The temperature field distribution and temperature rise curve on the rubber surface are recorded using an infrared thermal imager. Afterward, the thermo-mechanical coupling process of rubber under the combined action of hysteresis heat generation and friction heat generation is established, and the corresponding finite element model is established using ABAQUS based on experimental conditions. The simulation results are compared with the test data for verification.

2 Experimental

2.1 *Material*

Solid rubber wheels and rubber blocks need to be prepared for testing. The rubber raw materials are provided by the cooperating unit and taken from the tread of an aircraft tire. Firstly, the raw rubber is vulcanized with a temperature of 151°C for 60 min. In the initial stage of vulcanization, pressurize for 10 s, then unload for 5 s, and repeat twice to remove bubbles in the material to ensure vulcanization quality. Using a wheel-shaped mold and a block-shaped mold for vulcanization, the rubber wheel and rubber block specimens are obtained as shown in Fig. 1. The outer diameter of the rubber wheel is 80 mm, the thickness is 18 mm, and the inner diameter is 35 mm. The rubber block specimen is approximate to sector shape with an outer arc radius of 85 mm, the short side length of isosceles trapezoid is 22 mm, and the thickness is 14 mm.

2.2 *Testing Equipment*

Tests are conducted by the high-speed friction testing machine as shown in Fig. 2. During the high-speed friction machine test, the grinding wheel is rotated by a high-speed motor, and the size of the grinding wheel is 300 mm × 32 mm, with particle sizes on the surface of the grinding wheel ranging from 160 μm to 200 μm, simulating the airport road surface. The rotational speed of the grinding wheel is in the range of 0–1,500 rpm. The clamping device moves along the guide rail driven by the motor, which can replace different fixtures for different rubber specimens. The load is applied to the rubber specimen through a steel wire rope connected to the loading device, and the magnitude of the loading force is determined by weight adjustment. After the specimen contacts the grinding wheel, it is pressed against the

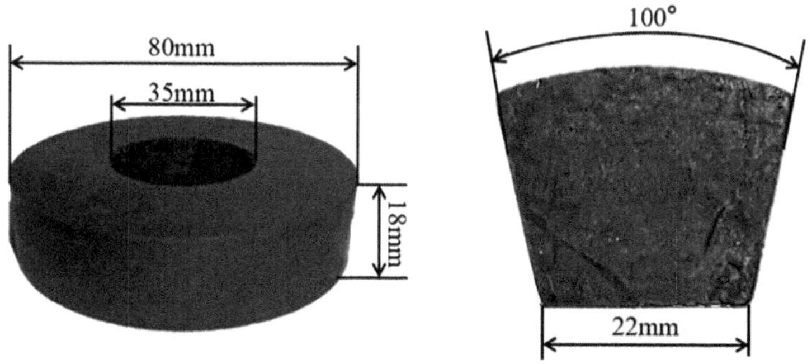

Fig. 1 Rubber wheel specimen (left) and block specimen (right)

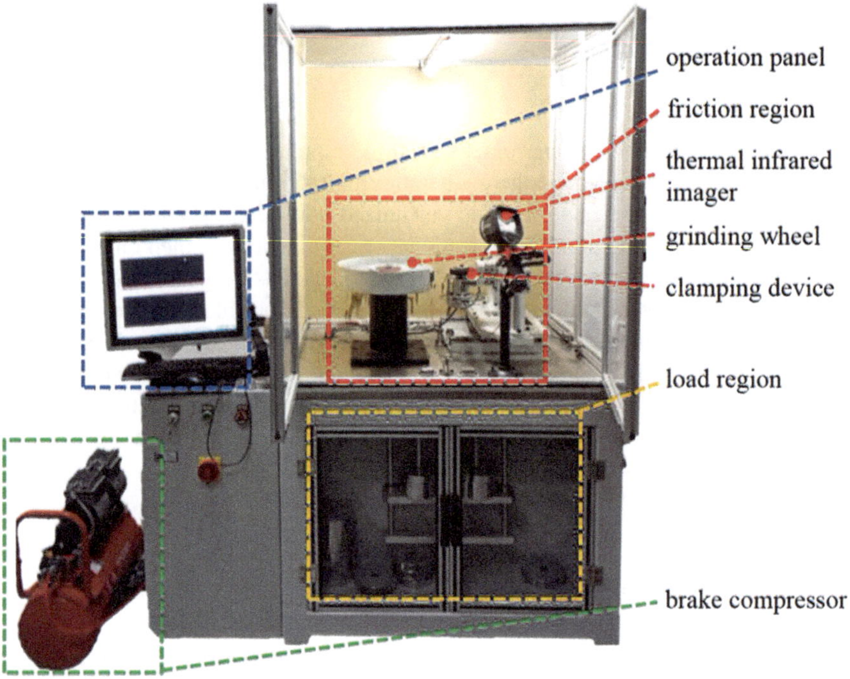

Fig. 2 High-speed friction testing machine

cylindrical surface of the grinding wheel with a constant loading force, and the grinding wheel spindle rotates at a set constant speed.

The three-dimensional force sensor collects force values in the X, Y, and Z directions, and the speed sensor records the rotational speed of the rubber specimen in real time. The distribution of the temperature field on the surface of the rubber specimen and the maximum temperature change process in real time are collected using the infrared thermal imager FLIR E4 WiFi. The temperature collection range of the thermal imager is −20 to 280°C with a collection frequency of 9 fps. This FLIR thermal imager automatically records the highest temperature on the specimen surface. During the test, the FLIR is connected to the computer via a USB cable, the computer has support software FLIR Tools installed, which is used to synchronize the measurement images from the camera. In addition, the software processes the video surveillance images and reads the temperature data at any point on the screen. According to the technical parameters provided by the thermal imager company, the temperature measurement accuracy of this type of imager is ±2°C for test targets with temperature higher than 0°C in ambient temperatures of 10–35°C. After the test time reaches the set value, the grinding wheel stops rotating, and the specimen is driven by the motor to swing the crank to detach from the grinding wheel, and the test is completed.

2.3 Rubber Block Sliding Friction Test

We used a special fixture to clamp the rubber block as shown in Fig. 3. The spindle speed of grinding wheel, loading time, and other parameters are set in the control panel before the test. The crank swing drives clamp to make the rubber block contact with the surface of the grinding wheel; the rubber block remains stationary during the test. Spindle motor drives the grinding wheel to rotate at a high speed, and high-speed sliding friction occurs between the grinding wheel and the rubber block. The FLIR infrared camera records the distribution of the temperature field on the surface of the rubber block and the maximum temperature in real time.

The frictional heat generation characteristics of rubber blocks under pure sliding conditions are obtained through this experiment, which provides data support for subsequent thermo-mechanical coupling simulation. The horizontal speed of the aircraft tire at the time of landing is 220 km/h, corresponding to a rolling angular velocity of 120 rad/s. Therefore, the grinding wheel angular velocity is taken as 120 rad/s, corresponding to a rotational speed of 305.6 r/min. The basic parameters of this experiment are shown in Table 1.

2.4 Rubber Wheel Test Under Different Slip Conditions

Rolling tests of the rubber wheels under different slip conditions are carried out on the high-speed friction testing machine. As shown in Fig. 4, the rubber wheel is fastened to the brake disc, and the braking force on the disc is adjusted by the

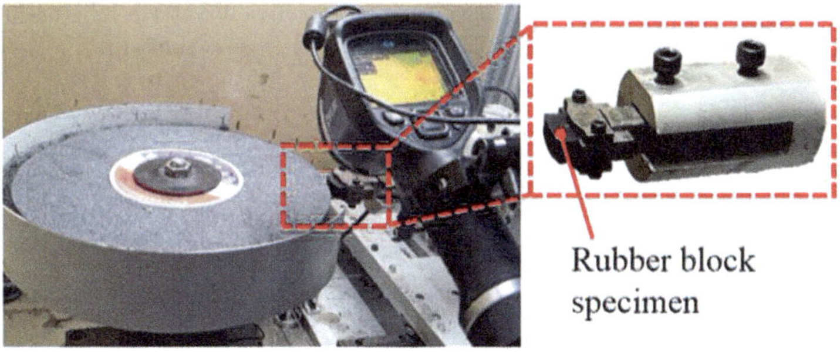

Fig. 3 Rubber block sliding friction test device

Table 1 Basic parameters of rubber block sliding test

Experimental condition	Parameter setting
Grinding wheel speed (r/min)	305.6
Load(N)	30
Time(s)	4
Indoor temperature (°C)	20

Fig. 4 Rubber wheel high speed test device

pneumatic brake cylinder to achieve different slip rates between the rubber wheel and the grinding wheel.

After the rubber wheel contacts the grinding wheel, the grinding wheel spindle rotates at a set constant speed. The relationship between the grinding wheel speed and linear speed is as shown in Eq. (1):

$$n = \frac{60v}{2\pi \cdot r} \tag{1}$$

where n is the grinding wheel speed (r/min), v is the grinding wheel line speed (m/s), and r is the grinding wheel radius (m). The speed sensor records the speed of the rubber wheel in real time, and the relationship between the speed of the rubber wheel and the linear speed is:

$$v_0 = 2\pi n_0 r_0 \tag{2}$$

where v_0 is the rubber wheel line speed (m/s), n_0 is the rubber wheel rotating speed (r/min), and r_0 is the rubber wheel radius (m). The rubber wheel slip rate is given as follows:

$$\lambda = \frac{v - v_0}{v} \tag{3}$$

When an aircraft lands, the instantaneous angular velocity of the aircraft tires is zero. The tire rotates rapidly under the action of friction after contact with the ground. When the wheel speed matches the aircraft speed, most of the weight of this aircraft falls on the ground, and the braking pressure is applied to slow down the aircraft. The aircraft tires are under different slip conditions throughout the landing process. The slip rate-controlled anti-slip braking system is used to brake the aircraft by controlling the slip rate, the desired slip rate is set in advance. The slip rate is continuously controlled according to the actual calculated slip rate. It is known that the optimal slip rate is between 0 and 0.2, so the range of the test slip rate is set to be between 0 and 0.2. The basic parameters set for the test are shown in Table 2, the contact load is set to 120N and all tests are conducted for 30 s. The pure rolling

Table 2 Basic parameters of rubber wheel test under different sliding conditions

Testing program	Velocity (km/h)	Load (N)	Slip rate
1	20	120	0
2	20	120	0.07
3	20	120	0.12
4	20	120	0.17

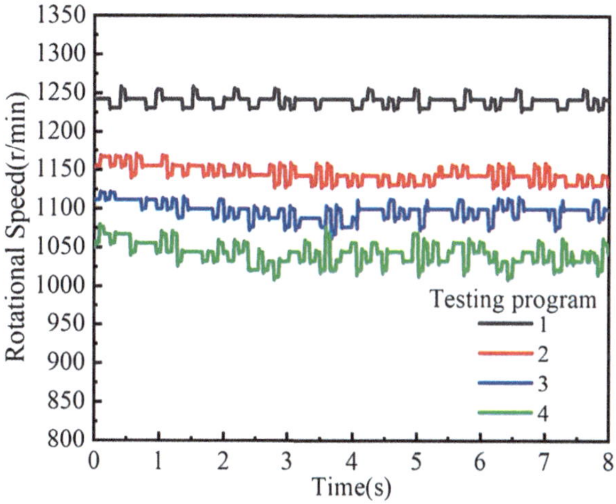

Fig. 5 Rotation speed curve with time

condition of the rubber wheel with a slip rate of 0 is equivalent to the free rolling of the tire. If the slip rate is not 0, it indicates that the rubber wheel is in a sliding process, with both rolling and sliding.

During the experiment, the measured speed variation curve of the rubber wheel is derived as shown in Fig. 5. It can be seen that the speeds of the rubber wheel are relatively stable and the fluctuations are small with different slip rates, which indicates that a relatively stable braking process can be provided by the brake disc through the air pump.

Based on the rotational speed data of the rubber wheel, the slip rate curve of the rubber wheel after applying different braking forces is calculated as shown in Fig. 6, which can be seen that the obtained slip rate curve is relatively stable due to the stable rotational speed. It is indicated that the rubber wheel can move under a fixed slip rate during the test. After applying the braking force, the rotational speed decreases compared to pure rolling. The slip rate during pure rolling is basically 0, and the slip rate curves obtained from experiments 1, 2, and 3 fluctuate around the set slip rate values of 0.07, 0.12, and 0.17, within an error of 0.01.

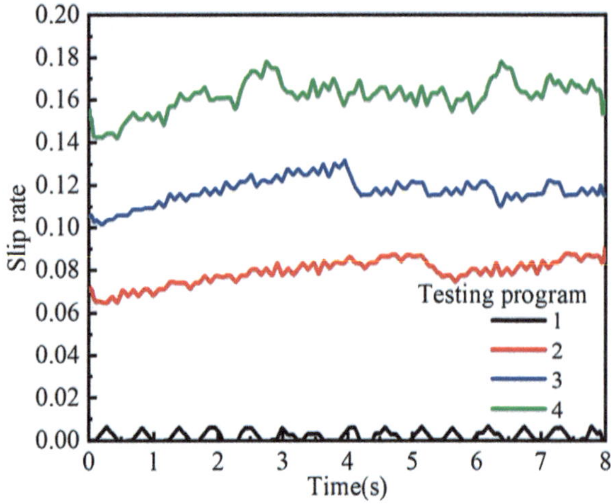

Fig. 6 Slip rate curves with time

3 Thermo-Mechanical Coupling Simulation Analysis

3.1 Theory of Rubber Hysteresis and Friction Heat Generation

The viscoelasticity of the rubber material is one of the important factors for the periodic stress and strain inside the tire during the rolling process. Under the action of alternating stress, the changes in strain always lag behind the changes in stress, leading to delayed heat generation. When harmonic tensile strain is applied to viscoelastic material, the stress also changes in harmonic form. Energy loss of harmonic loading in one cycle is expressed as Eq. (4):

$$
\begin{aligned}
U_{loss} &= \int_0^{\frac{2\pi}{w}} \sigma(t)d\varepsilon(t) = \int_0^{\frac{2\pi}{w}} \sigma_0\varepsilon_0 \sin(wt+\delta)d\sin(wt) \\
&= \sigma_0\varepsilon_0 \sin\delta \int_0^{\frac{2\pi}{w}} \cos^2(wt)d(wt) \\
&= \pi\sigma_0\varepsilon_0 \sin\delta
\end{aligned} \tag{4}
$$

where σ_0 is stress amplitude, ε_0 is strain amplitude, δ is the phase difference between stress and strain, and w is the angular velocity of a harmonic response. The stress formula is developed as follows:

$$\sigma(t) = \sigma_0 \sin(wt)\cos\delta + \sigma_0 \cos(wt)\sin\delta \tag{5}$$

Then the relationship between stress and strain can be expressed as:

$$\sigma(t) = \varepsilon_0 E_1 \sin(wt) + \varepsilon_0 E_2 \cos(wt) \tag{6}$$

$$E_1 = \frac{\sigma_0 \cos\delta}{\varepsilon_0} \tag{7}$$

$$E_2 = \frac{\sigma_0 \sin\delta}{\varepsilon_0} \tag{8}$$

where E_1 is the storage elastic modulus, and is equivalent to the elastic modulus of static deformation, which is used to overcome the contraction force of ideal rubber. The strain energy is alternately stored and released in the process of reciprocating deformation. E_2 is a loss elastic modulus, which is used to overcome the viscosity of rubber. It is completely lossy and will all be converted into heat. If the overall modulus is represented by E^*, then:

$$E^* = E_1 + iE_2 \tag{9}$$

$$\tan\delta = \frac{E_2}{E_1} \tag{10}$$

where E^* is the apparent elastic modulus, tanδ is loss factor which is an important parameter to characterize the energy loss of the tire. The smaller the value, the lower the heat generation of the tire. Substitute E_2 into U_{loss} to get:

$$U_{loss} = \pi\sigma_0\varepsilon_0 \sin\delta = \pi\varepsilon_0^2 E_2 = \pi\varepsilon_0^2 E_1 \tan\delta = \pi\sigma_0^2 \frac{E_2}{|E^*|} \tag{11}$$

The stress-strain course of any material point of the rolling rubber wheel is often non-harmonic, so it is advisable to use the Fourier series to decompose it into several superposition values of harmonic responses. Fourier series expression is:

$$\sigma(\theta) = A_0^\sigma + \sum_{n=1}^{N} A_n^\sigma \sin\left(n\theta + \phi_n^\sigma + \delta_n\right) \tag{12}$$

$$\varepsilon(\theta) = A_0^\varepsilon + \sum_{n=1}^{N} A_n^\varepsilon \sin\left(n\theta + \phi_n^\varepsilon\right) \tag{13}$$

where A_n^ε is the coefficients of simple harmonic components, ϕ_n^ε is the phase angle of simple harmonic components, and n is the Fourier series. During the wheel rolling, the wheel endured forces from three forces and three corresponding moments along three axes. All of them can cause the rubber wheel deformation and the hysteresis energy loss. Thus, the total hysteresis energy loss of any material point on the rubber wheel within one rolling circle is the sum of the energy losses of the six-axis wheel force.

When the loading frequency is f, the heat generation rate is:

$$Q = U_{\text{loss}} \cdot f \tag{14}$$

During the aircraft tire skidding process, high-speed friction occurs between the tire tread and the runway surface, the tread heats up sharply. The friction heat generation rate at any node on the rubber surface is:

$$\dot{Q}_f = \eta \sigma_t v_t \tag{15}$$

where η is the heat flow distribution coefficient; σ_t is tangential stress at contact zone joints; and v_t is the relative velocity of the rubber on the contact surface.

3.2 Thermo-Mechanical Coupling Method

In general, when a tire is rolling in a steady state, only the temperature rise caused by the hysteresis heat generation of the rubber material needs to be considered. When the tire is locked and sliding, only the frictional heat generation needs to be considered. In situations such as rapid speed changes, the relative sliding between the tire and the road surface occurs. The distribution of the rubber temperature field is affected by the generated hysteresis heat and friction heat. Aiming at the above three kinds of heat generation, a thermo-mechanical coupling simulation method for rubber is established as shown in Fig. 7. This method combines the fully coupled method with the modular method, and considers the mutual influence between the stress-strain field and the temperature field.

The blue, red, and black parts represent the mechanics module, the thermal coupling module, and the iteration module, respectively. First, in the mechanics module, the boundary conditions were set according to the test conditions. The finite

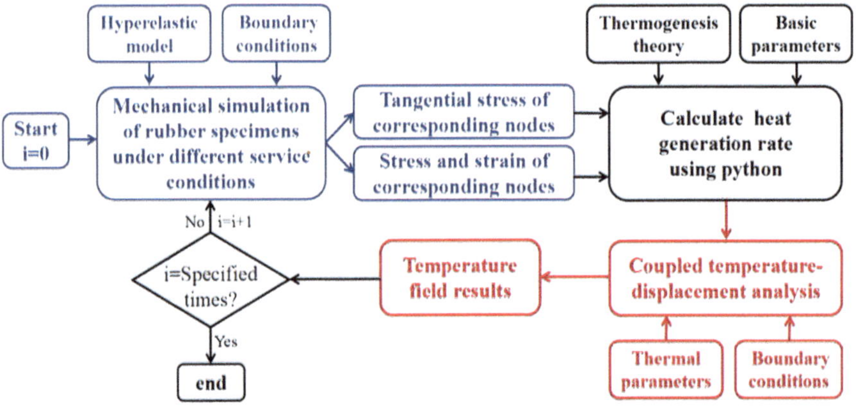

Fig. 7 Thermo-mechanical coupling flowchart

element modeling and three-dimensional mechanical simulation were conducted for the rubber block and the solid rubber wheel. The Python script is used to read the values of stress-strain fields and tangential stresses at the required nodes. The Ogden model was used to define the hyperelastic model according to the tensile test data. Then, in the thermal coupling module, based on the heat generation theory mentioned in Sect. 3.1, the heat generation rate is calculated using Python. The heat generation rate values of each node are defined as heat sources in the ABAQUS thermal simulation model. Combined with material thermal parameters, the coupled temperature-displacement element is used for thermo-mechanical coupling simulation to obtain temperature and force field results. In the iteration module, the obtained temperature field and force field results are used as the initial conditions for the next thermal coupling, and iterated for several times until the whole heating process is completed.

The thermo-mechanical coupling simulation of pure sliding friction of the rubber block only calculates friction heat generation, and the pure rolling of the rubber wheel only calculates hysteresis heat generation. When the rubber wheel is in different slip rate conditions, the required rolling period for the hysteresis heat generation is calculated from the equivalent angular velocity, which is determined by the rolling ratio and the angular velocity during pure rolling. The relative velocity required for friction heat generation calculation is equivalent to the proportion of sliding and the linear velocity during pure sliding. The calculated hysteresis loss heat source and friction loss heat source are simultaneously assigned to each node of the rubber specimens.

3.3 Finite Element Modeling

The sliding friction model of the rubber block is established according to the test for mechanical analysis, which is shown in Fig. 8. The simulation model includes grinding wheel, rubber block, and fixture, with a total of 3,264 units. The rubber

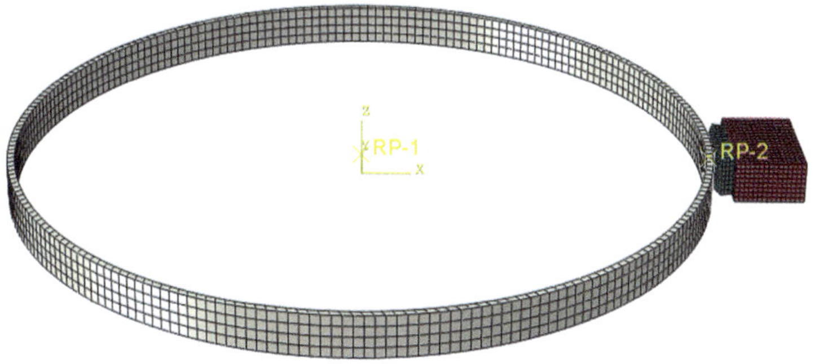

Fig. 8 Rubber block sliding friction mechanics simulation model

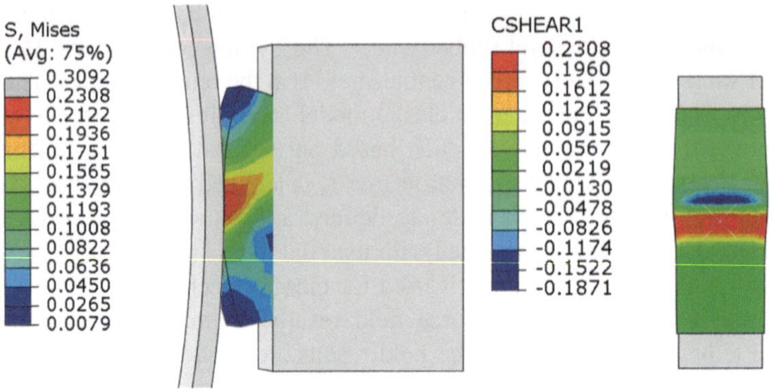

Fig. 9 Stress result (left) and tangential shear result (right)

block, fixture, and grinding wheel are respectively assigned with corresponding strength of material parameters. The C3D8RHT unit type is used for rubber. The grinding wheel and the fixture are set as rigid bodies and bound to reference points respectively. The rotational movement is adopted on the grinding wheel through the reference point by certain constrain. The rubber block and fixture are applied with a load of 30N in the x direction to make them contact with the grinding wheel. The friction coefficient of the contact attribute between the rubber block and the grinding wheel is set to be 0.65. Finally, the mechanical simulation results of sliding friction of the rubber block are obtained as shown in Fig. 9.

The model of pure rolling solid rubber wheel was also established. In the mechanical analysis, a two-dimensional model was firstly established to simulate the clamping process. The model contains 1,620 units in total with the unit type of CAX4HT. The material parameters of mechanical performance, heat transfer coefficient, specific heat capacity, and other thermal parameters were determined to simulate the clamping process of the two-dimensional model. The 3D rubber tire model was obtained by rotating the established 2D model for further analysis. A rigid cylindrical surface was also built to simulate the grinding wheel, with unit type C3D8RHT. The assembled tire-wheel rolling model is shown in Fig. 10. The rigid cylindrical surface is bound to the reference point and the Z-direction load is applied to simulate the contact loading. As there is no relative slip between the rubber wheel and the grinding wheel during pure rolling, static loading is used to simulate the pure rolling condition, with a load of 120N. The simulation stress results are shown in Fig. 11.

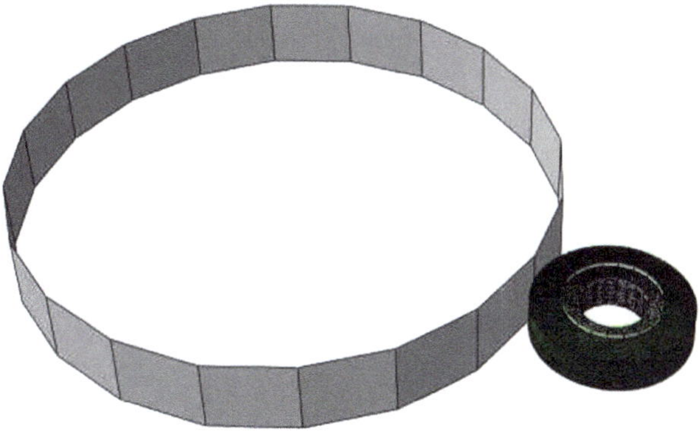

Fig. 10 Three-dimensional rubber wheel finite element model

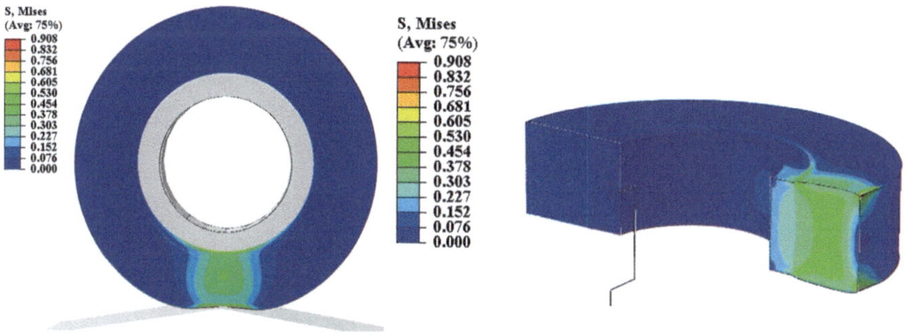

Fig. 11 Stress field simulation results

4 Results and Discussion

The result of the high-speed friction temperature of the rubber block is shown in Fig. 12. From the temperature field distribution screen, it can be seen that the rubber block shows a significant temperature rise, with the highest temperature occurring in the contact area between the outer surface and the grinding wheel. High-speed sliding causes friction and wear on the rubber block surface, resulting in irregular surface deformation. The temperature value gradually decreases away from the center of the contact area.

Figure 13 shows the curve of the maximum temperature change on the surface of the rubber block. It can be seen that the temperature rises slowly in the first 4 s, then rises abruptly after 5 s, rapidly increasing from 30 to 109°C, and then gradually decreasing from 109°C. This is because the rubber block is in the process of high-speed sliding friction in the first 4 s, the surface of the rubber block is in close contact with the grinding wheel, and infrared thermal imager cannot monitor temperature of

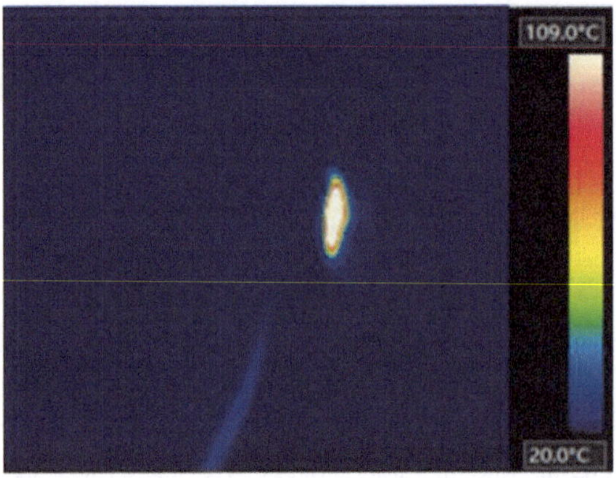

Fig. 12 Temperature field distribution measured by infrared thermal imager

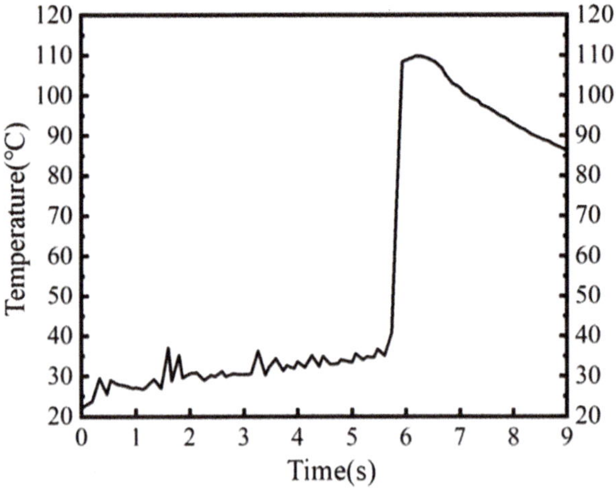

Fig. 13 History of maximum temperature change

the contact area. After 4 s, the rubber block leaves the grinding wheel, and the temperature field of the contact area appears in the monitoring screen. The maximum temperature reached 109°C. Then the rubber block dissipated heat into the air, which caused the gradual decrease in the temperature. Heat generated by high-speed sliding friction is rapid, with the temperature rapidly rising from 20 to 109°C within 4 s.

According to the above thermo-mechanical coupling simulation process, since calculating the heat generation rate requires the value of tangential stress, and frictional heat generation occurs on the contact surface, it is necessary to extract the tangential stress of all elements on the surface of the rubber block. In fact, there are tangential stresses in both the circumferential (X direction) and radial (Y direction) directions of the rubber block. Only the tangential stress in the circumferential direction needs to be extracted since the lateral deflection condition is not considered. The tangential stress values of the four nodes corresponding to each unit on the outer surface were extracted. These four tangential stress values

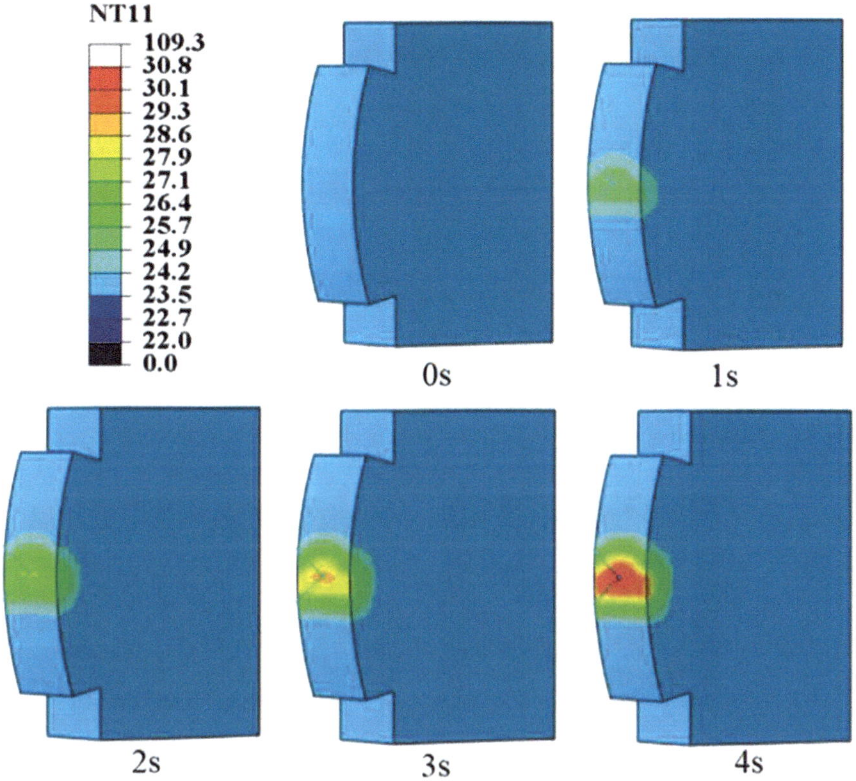

Fig. 14 Simulation results of sliding temperature field of rubber block

were averaged through Python programming and then assigned to the corresponding units. Based on the friction heat generation rate calculation formula, the heat generation rates of each unit were calculated and then written into the inp file by the keyword Dflux. As a result, the heat source can be attached to each corresponding unit. Using the Coupled Temperature-displacement analysis step in the inp file for thermo-mechanical coupling analysis, and time is set to 4 s. The results of temperature field distribution are obtained as shown in Fig. 14.

Comparing the simulation and experimental results, it can be concluded that the simulated and measured data of the highest temperature of rubber block are both 109°C, indicating good consistency. The distribution of the simulated temperature field is in good agreement with the measured temperature field, in which the highest temperature occurs at the center of the contact area on the outer surface, and gradually decreases away from the contact area. Heat is gradually transferred from the outer surface to the inside of the rubber block considering the relatively low thermal conductivity. The temperature field distribution is related to the distribution of tangential stress results, which is consistent with the calculation formula of heat generation rate. The heat generation rate is proportional to the tangential stress for

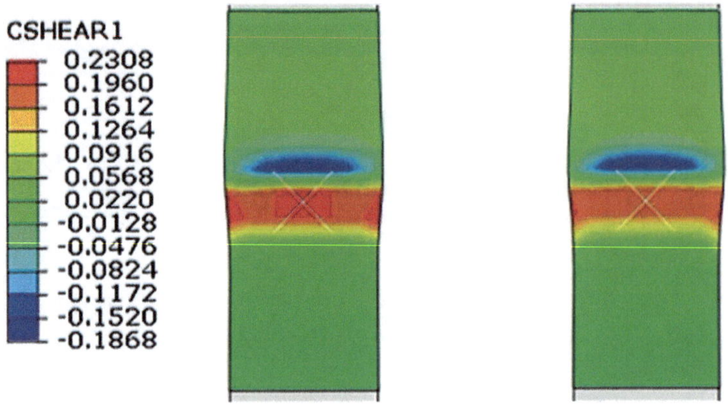

Fig. 15 Comparison of tangential stress before and after thermo-mechanical coupling

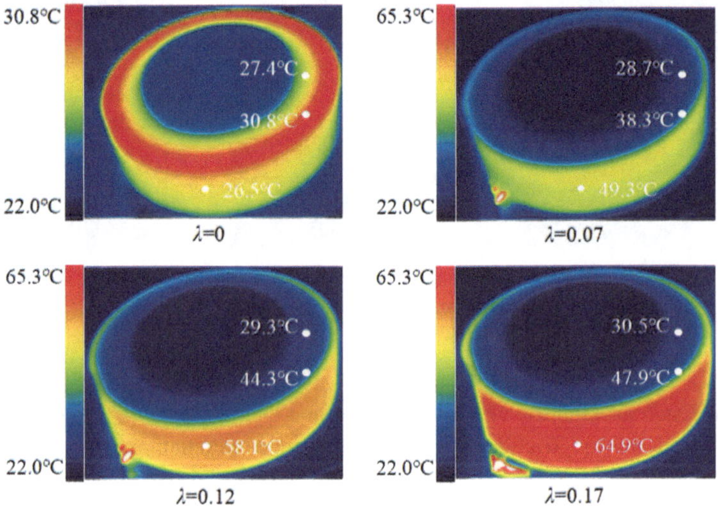

Fig. 16 Temperature field of rubber wheels under different slip rates measured in tests

the same heat flow distribution coefficient and relative velocity. The comparison between the updated tangential force field after thermo-mechanical coupling and the initial tangential force field is shown in Fig. 15. As shown in Fig. 15, after thermo-mechanical coupling simulation, the temperature field has an impact on the force field, and the tangential stress decreases gradually with the increase of temperature.

Temperature fields of rubber wheels under different slip rates captured by an infrared thermal imager are shown in Fig. 16. It indicates a significant difference in the temperature field distribution between the pure rolling condition and the sliding condition. Figure 16a shows the temperature field distribution under pure rolling condition. The heat generation in pure rolling is mainly generated by hysteresis loss inside the rubber, and the heat is diffused from inside to outside. Therefore, the highest temperature of the surface occurs in the region near the outer side of the upper surface of the rubber wheel, and it is distributed in a circular shape, with a small temperature rise on the side surface. Figure 16b, c, and d shows the temperature field distributions under sliding conditions. When there is braking, the slip rate $\neq 0$, and the heat generation is generated by the friction loss on the contact surface and internal hysteresis loss. The highest temperature appears on the side of the rubber wheel in contact with the grinding wheel, which indicates that most of the detected heat is generated from the friction loss between the side of the rubber wheel and the grinding wheel. The braking disc also generates a large amount of heat under severe friction, and the rubber wheel temperature data should be extracted in a way to avoid this interference.

When the slip rates are 0.07, 0.12, and 0.17, the temperature distribution patterns of the rubber wheel are similar. The upper and side surfaces of the rubber wheel both have temperature rises. Relatively low temperature is observed around the inner part of the rubber wheel, which is away from the friction area. The highest temperatures on the upper surface are 38, 44, and 48°C, which correspond to slip rates of 0.07, 0.12, and 0.17, respectively. The severity of temperature rise on the side surface is significantly higher than that on the upper surface. The highest temperatures on the side surface are 49, 58, and 65°C. It can be seen that at the same speed, the final temperature reached at each place of the rubber wheel surface increases with the increase of slip rate, and the highest temperature occurs on the outer side.

Figure 17 shows the variation process of the highest temperature on the surface of the rubber wheel under different slip rates obtained from experiments. According to the temperature rise curves of experiments 2, 3, and 4, when the speed is the same, the slip rates of 0.07, 0.12, and 0.17 ultimately reach temperatures of 49°C, 59°C, and 65°C, respectively. Compared with the 30°C achieved by the pure rolling

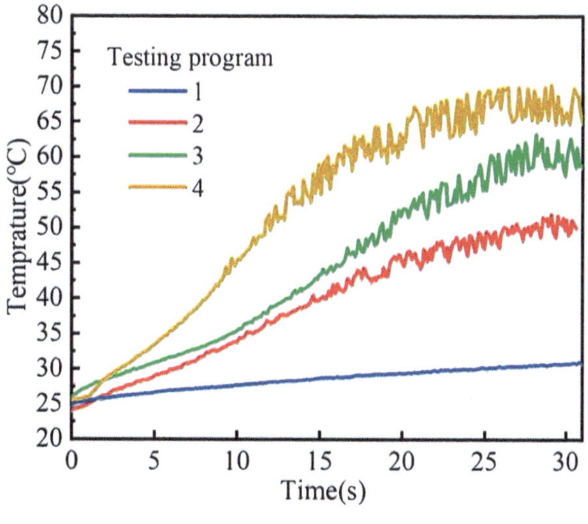

Fig. 17 Variation curve of the maximum temperature of rubber wheel surface with time under different slip rates

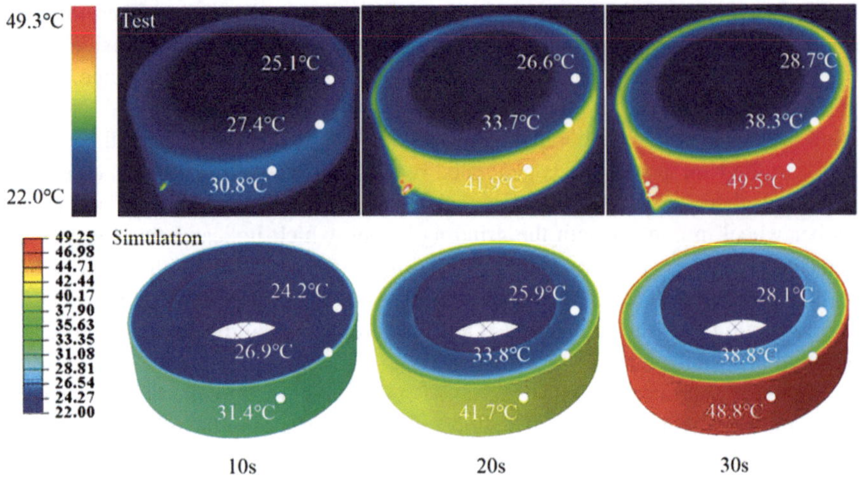

Fig. 18 Comparison of simulated and experimental temperature field at slip rate 0.07

conditions, there is a significant increase in the temperature with slip. The larger the slip rate, the faster the temperature rises, and the higher the final temperature reached. The temperature fluctuation amplitude increases when there is a slip, which is caused by the oscillation of the brake disc when applying braking force.

According to the previous thermo-mechanical coupling process considering two kinds of heat generation at the same time, the comparison between the simulation and experimental temperature fields at a slip rate of 0.07 is shown in Fig. 18. The temperatures of three positions on the rubber surface at three time points are selected for comparison between experiments and simulations. Three selected time points are 10, 20, and 30 s. The lower limit of temperature display for both experiments and simulations is taken to be 22°C, and the upper limit of temperature display is 49.3°C. As can be seen from the previous text, the generated heat is mainly distributed on the upper surface of the rubber wheel during pure rolling. The highest temperature is distributed in a circular shape in the outer circumference area of the upper surface. The upper surface of the rubber wheel and the side of the wheel both have obvious temperature rises when the slip condition is applied. From Fig. 18, it can be seen that the temperature at two selected points on the upper surface of the rubber wheel obtained from the test is 28.7 and 38.3°C, and the temperature at the selected point on side surface is 48.8°C. It can be indicated that the effects of hysteresis heat generation and friction heat generation cannot be ignored. The side surface temperature is higher than the upper surface temperature. As time increases, the difference between the side temperature and the upper surface temperature gradually increases, indicating that frictional heat generation is more severe than hysteresis heat generation under slip condition.

With the increase of time, the temperature of the rubber wheel surface increases at all points. The surface temperature field distribution images obtained from simulations at 10, 20, and 30 s are consistent with the experimental results. Three points are selected on the rubber wheel for their tested temperature to compare with the

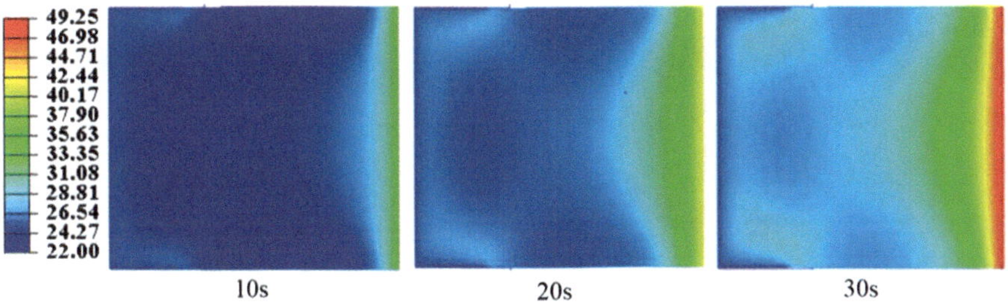

Fig. 19 Internal temperature change process of rubber wheel

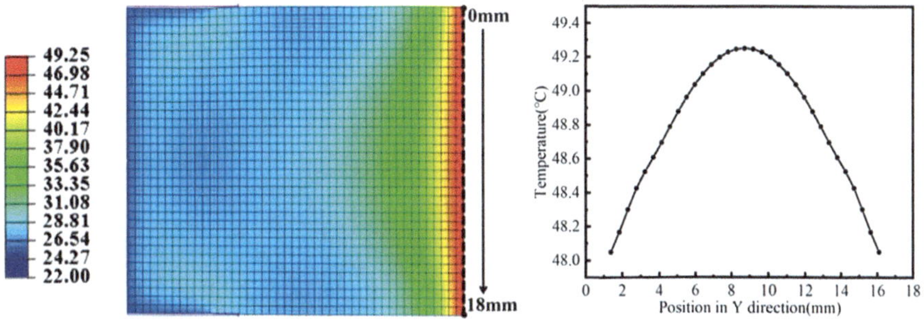

Fig. 20 Temperature variation with position at the straight line on the outside of the rubber wheel

simulated results of the corresponding points. The comparison results show that the temperature data at the three points obtained from the simulation at each time point are in good agreement with the experimental data, and the maximum error is 0.7°C, which does not exceed 1.5%. Therefore, the simulation method can achieve relatively accurate prediction, which can be used for verifying the correctness of the thermo-mechanical coupling process considering both types of heat generation simultaneously.

The internal temperature change process of the rubber wheel is shown in Fig. 19. It can be seen that the highest temperature appears in the outer side area, which gradually decreases from the outer side to the inner cylindrical surface, and increases when it reaches the clamping area of the fixture. This may be because the friction heat source is concentrated on the outer surface of the rubber wheel, and the hysteresis heat source is distributed in the stress concentration area mentioned above. The clamping area of the fixture also generates heat due to the existence of stress. Due to the influence of heat conduction and convection heat transfer, the heat diffuses to the inner cylindrical surface, the upper and lower surfaces of the rubber wheel, and the air. The final formed temperature distribution is shown in Fig. 19. It is difficult to test the distribution of the internal temperature field of the rubber wheel, simulation methods can be used to predict and simulate the internal temperature field.

A straight line is selected on the inner section of the rubber wheel to study the temperature rise characteristics as shown in Fig. 20.

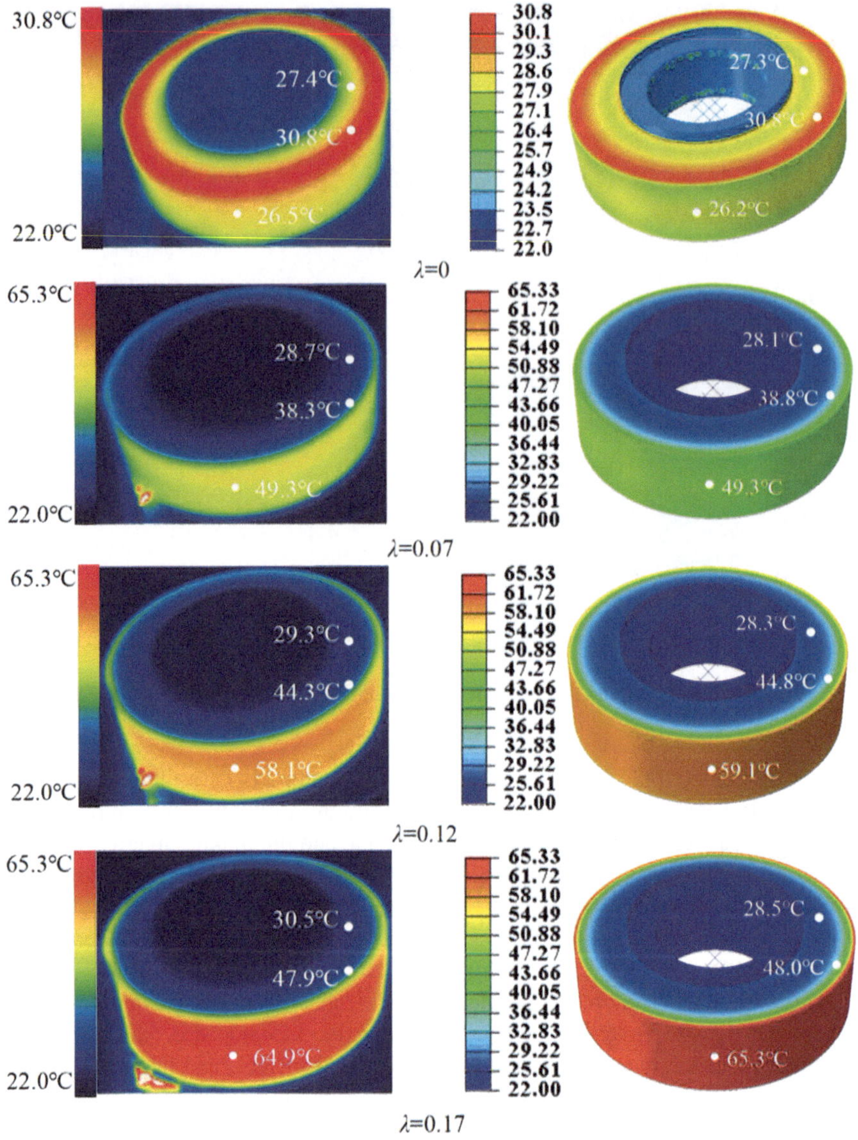

Fig. 21 Comparison of experiments and simulations under different slip rates

The straight line is located at the outermost side of the section, and the ordinate of the top point of the straight line is set to be 0. The ordinate value gradually increases from top to bottom. As the ordinate value increases, the temperature first increases and then decreases, meaning that the temperature reaches its highest point at the center and decreases from the center to both ends. This is because the heat dissipation at the center of the outer side is poor, while the two ends are close to the upper and lower surfaces, resulting in better heat dissipation conditions.

The simulation results after 30 s of motion under four conditions of slip rate 0, 0.07, 0.12, and 0.17 are shown in Fig. 21. The lower limit of temperature display for both the experiment and simulation is 22°C, and the upper limit of display is 65.3°C. Under different slip rates, the simulated and measured temperatures at three points on the same position on the surface of the rubber wheel are well corresponding, with a maximum error of 2°C, which is not more than 7%.

5 Conclusion

The frictional heat generation characteristics of rubber blocks under pure sliding conditions are tested through high-speed friction tests. The highest temperature occurred in the contact area of the outer surface of the rubber block. The temperature decreases with the distance away from the contact area. The highest temperature on the rubber block surface is raised rapidly from 20 to 109°C within 4 s under sliding friction conditions. A test program for different slip conditions of rubber wheels is established to explore the effect of slip rate on the temperature field. The results indicate that heat generation is jointly generated by friction loss on the contact surface and internal hysteresis loss. The highest temperature occurs on the side of the contact between the rubber wheel and the grinding wheel due to the heat generated by the friction loss. The slip rates of 0.07, 0.12, and 0.17 ultimately reach temperatures of 49, 59, and 65°C. Compared with the pure rolling condition of 30°C, the temperature with slip significantly increases. The larger the slip rate, the faster the temperature rises, and the higher the final temperature reached on the surface.

The calculation theories of hysteresis heat generation and friction heat generation of rubber are discussed, and a thermo-mechanical coupling simulation method of rubber material of aircraft tire is established. The temperature field results and temperature rise curves of the rubber block and rubber wheel were obtained through simulation and compared with previous experimental data. The results show that the deviation between simulation and experimental data is acceptable. The distribution patterns of temperature fields obtained by the simulation and experiment at various times have achieved good agreement. The maximum errors are lower than 7%, which supports the correctness of the finite element simulation.

Acknowledgments This research work received financial support from the National Natural Science Foundation of China (52075119).

References

1. Alroqi AA, Wang W (2017) Investigation of the heat and wear of aircraft landing gear Tyres [D]. University of Sussex
2. Makhovskaya Y (2019) Modeling sliding friction of a multiscalewavy surface over a vis-coelastic foundation taking into account adhesion. Lubricants 7
3. Mahjouri S, Shabani R, Skote M (2022) Reducing temperature, drag load and wear during aircraft tyre spin-up. J Aircr Eng Aerosp Technol
4. Bennett M, Christie SM, Graham A et al (2011) Composition of smoke generated by landing aircraft. J Environ Sci Technol 45(8):3533–3538
5. Kahms S, Wangenheim M (2021) Experimental investigation and simulation of aircraft tire wear. J Tire Sci Technol 49(1):55–74
6. Wu J, Chen L, Wang Y et al (2019) Effect of temperature on wear performance of aircraft tire tread rubber. J Polym Test 79:106037
7. Rosu I, Elias-Birembaux HL, Lebon F (2018) Finite element modeling of an aircraft tire rolling on a steel drum: experimental investigations and numerical simulations. J Appl Sci 8(4):593
8. Korunović N, Trajanović M, Stojković M et al (2011) Finite element analysis of a tire steady rolling on the drum and comparison with experiment. J Mech Eng 57(12):888–897
9. Behnke R, Kaliske M (2015) Thermo-mechanically coupled investigation of steady state rolling tires by numerical simulation and experiment. J Int J Nonlin Mech 68:101–131
10. Wallaschek J, Wies B (2013) Tyre tread-block friction: modelling, simulation and experimental validation. J Veh Syst Dyn 51(7):1017–1026
11. Rosu I, Elias-Birembaux HL, Lebon F et al (2016) Experimental and numerical simulation of the dynamic frictional contact between an aircraft tire rubber and a rough surface. J Lubr 4(3):29
12. Tong X, Chen X, Xu J et al (2018) The heat build-up of a polymer matrix composite under cyclic loading: experimental assessment and numerical simulation. J Int J Fatigue 116:323–333
13. Peter O, Stoček R, Kratina O (2022) Experimental and numerical description of the heat build-up in rubber under cyclic loading. M.//degradation of elastomers in practice, experiments and modeling. Springer, Cham, pp 121–141
14. Linke T, Wangenheim M, Lind H et al (2014) Experimental friction and temperature investigation on aircraft tires. J Tire Sci Technol 42(3):116–144
15. Whicker D, Browne AL, Segalman DJ et al (1981) A thermomechanical approach to tire power loss modeling. J Tire Sci Technol 9(1):3–18
16. Alroqi AA, Wang W (2016) The prevention of aircraft tires overheating by pre-rotating the wheels. C.//2016 7th international conference on mechanical and aerospace engineering (ICMAE). IEEE, pp 443–447
17. Li F, Liu F, Liu J et al (2018) Thermo-mechanical coupling analysis of transient temperature and rolling resistance for solid rubber tire: numerical simulation and experimental verification. J Compos Sci Technol 167:404–410
18. Kondé AK, Rosu I, Lebon F et al (2013) Thermomechanical analysis of an aircraft tire in cornering using coupled ale and lagrangian formulations. J Cent Eur J Eng 3(2):191–205
19. Song HS, Jung SP, Park TW (2018) Simulation of temperature rise within a rolling tire by using FE analysis. J Mech Sci Technol 32(7):3419–3425

Adv Polym Sci (2024) 294: 95–124
https://doi.org/10.1007/1345_173

Published online: 28 April 2024

Phase-Field Fracture Modeling of Polymeric Materials Considering Thermo-Viscoelastic Constitutive Behavior at Finite Strains

Bo Yin, Le Zhang, and Michael Kaliske

Contents

Abstract Phase-field modeling has been intensively studied in the recent past and has been recognized as an established method for fracture analysis over the last decade. An increasing number of researchers has scientific interests in this methodology due to the main advantage that it does not depend on any explicit criterion for fracture evolution. In the meantime, the simulation results also show good agreement comparing to experimental evidence. For the majority of polymers, both elastic and

M. Kaliske (✉)
Institute for Structural Analysis, Technische Universität Dresden, Dresden, Germany
e-mail: michael.kaliske@tu-dresden.de

viscous behavior are investigated simultaneously. The latter phenomenon naturally leads to rate-dependent response for both deformation and fracture aspects. Furthermore, polymeric materials are also very sensitive to temperature loading, which presents a volumetric expansion and shrinking effect. In this contribution, a thermo-viscoelastic rheological model based on multiplicative decomposition of the deformation gradient is coupled to phase-field modeling to investigate temperature-dependent and rate-dependent fracture within polymeric materials. Regarding the phase-field driving force, only the elastic strain energy potential is supposed to evolve fracture. It comes from both the equilibrium and non-equilibrium branches. The formulation is consistently derived and implemented into an in-house coding platform. Several representative numerical studies demonstrate the capabilities of the present model. Last, a summary with related findings and potential perspectives close the paper.

Keywords Finite strains, Multiplicative decomposition, Phase-field fracture, Thermo-viscoelasticity

1 Introduction

Polymeric materials are of significant importance in engineering applications. As a special class of polymers, which, for example, has entangled but un-crosslinked macromolecules, is the so-called thermoplastics. In regard to their performance, significant creep phenomena can be observed especially when they are subjected to static mechanical loads. Such a response is largely relying on the special characteristics of the micro-structure within the materials. A large number of experimental investigations indicates that the majority of polymers exhibit both elastic nonlinearity and rate-dependent inelasticity simultaneously, see [1, 2] for a general view. To address this constitutive property, various numerical rheological models are proposed to constitute the viscoelastic relations. An early presentation of linear and nonlinear viscoelasticity is outlined in [3]. Specifically, the linear viscoelastic models, based on the convolution integral algorithm posited in [4–9], prove inadequate for the comprehensive assessment of intricate viscoelastic attributes. Consequently, advancements in the area of nonlinear viscous evolution laws are explored. Bergström & Boyce [10] introduce an evolution law by regulating the rate of inelastic deformation, later scrutinized and incorporated into the Finite Element (FE) framework by [11]. This algorithmic framework is similarly applied in constitutive finite elastoplasticity [12] and viscoplasticity [13]. Reese & Govindjee [14] propose a model for finite viscoelasticity that extends beyond states close to thermodynamic equilibrium.

Simultaneously, thermal effects bring substantial influence on the material's mechanical response. Various published models adeptly tackle these complex multi-physics coupling problems. While some of them rely on established phenomenological or empirical formulations for simplification, others contribute meaningfully by developing thermodynamically consistent formulations, particularly within a finite deformation framework. Generally, these models consider the mature concept of splitting the deformation gradient into volumetric and isochoric components. Additionally, a further split of the isochoric part into elastic and viscous contributions is considered. A multiplicative kinematics is assumed [15], and the temperature evolution law incorporates the specific heat capacity of polymeric materials into the Helmholtz energy derivative. Typically, the heat flux vector follows Fourier's law and is a function of the thermal conductivity coefficient for the relevant thermomechanical coupling problem [16]. Introducing a specific formulation of the Helmholtz energy [17], which consists of volumetric, isochoric, thermal, and latent components, enables the derivation of the temperature governing equation. Furthermore, this energy is employed to delineate the dissipative behavior of the material, considering a transition from mechanical to thermal energetic parts [18].

Turning attention to the fracture or failure of polymeric materials, numerous studies classify them into two main categories: nucleation of cavities (cavitation) and fully evolved crack propagation. For the cavitation investigation, an early finding [19] reveals sudden internal flaws and cavitations within polymers are studied in [20] with specimens which consist of rigid spherical inclusions inside. Then, mathematical analyses of the singular solutions to the equations of a nonlinear elastic solid in tension states are depicted in [21]. More recently, the physical mechanism of cavitation at microscale with respect to experimental investigation is analyzed in [22, 23]. A number of numerical models are developed to capture the failure mechanism of crack propagation in studies. For an overall review, the reader is referred to [24]. Classical Linear Elastic Fracture Mechanics (LEFM) [25–28] is also employed for predicting crack growth direction, yet struggles with the singularity at the crack tip. The J-integral formulation [29], incorporated within finite element analysis for rubber materials [30, 31], is extended to a rate-dependent fracture evolution in viscoelastic solids [32–34]. Alternatively, the material force method is also used [35, 36], and Geißler et al. [37] formulate a viscoelastic cohesive zone model for rubber tearing tests.

Regarding the phase-field modeling for rate-dependent fracture, a common method is to include an artificial fracture resistance in the phase-field governing equation, resulting in a delayed fracture evolution. This approach is not restricted to characterizing solids as viscoelastic, with applications ranging from quasi-static analysis [38] to transient solutions [39]. Recent contributions aim to model rate-dependent phase-field fracture within viscoelastic materials [40]. Shen et al. [41] develop a phase-field model in viscoelastic solids, while Schänzel [42] postulates a phase-field model in elastomers. Loew et al. [43] postulate linear viscoelasticity based on the convolution integration algorithm at finite strain.

This work employs the viscoelastic formulation of Reese & Govindjee [14], coupled with a standard phase-field method. Due to the constitutive independency of the equilibrium and non-equilibrium branches, the isochoric energy density functions for both branches can be flexibly defined. This allows for the analytical computation of elastic energy stored in both branches based on their definitions. Consequently, the phase-field driving force is explicitly defined by the strain energy potential, aligning with the classical Griffith-type fracture definition. In contrast to models in [41, 43], the dissipated energy is not assumed to drive the fracture. Additionally, the temperature evolution is also coupled, enabling the calculation of volumetric expansion and shrinking based on the temperature degree of freedom. Thus, a three-field global finite element framework is consistently derived and the Newton-type solver relies on a staggered scheme to solve the unknown displacement, temperature field, and phase-field from the global equation system. All computational developments and constitutive descriptions are successfully validated through a representative numerical study.

2 Constitutive Law of Finite Thermo-Viscoelasticity

2.1 Preliminaries and Kinematics

This section outlines the fundamental theoretical basis of the constitutive framework for isotropic thermo-viscoelasticity. Let $\mathscr{B}$ be the solid body under consideration and its reference configuration $\mathscr{B}_0$ as a subset of the Euclidean space $\mathscr{B}_0 \subset \mathbb{R}^3$ at $t_0 \in \mathcal{T} \mid \mathcal{T} \subset \mathbb{R}_+$. Every material point P of $\mathscr{B}$ has a position vector at reference configuration $\mathbf{X} \in \mathscr{B}_0$, and a position vector $\mathbf{x} \in \mathscr{B}_t$, where $\mathscr{B}_t$ denotes the current configuration of $\mathscr{B}$ at time $t \in \mathcal{T}$. The mapping $\boldsymbol{\varphi}_t : \mathscr{B}_0 \times \mathcal{T} \to \mathbb{R}^3$ denotes the motion of the solid domain at the time interval $\mathcal{T}$, which is a nonlinear and bijective mapping, reading

$$\boldsymbol{\varphi}_t : \mathbf{X} \mapsto \mathbf{x} = \boldsymbol{\varphi}_t(\mathbf{X}). \tag{1}$$

The deformation gradient $\mathbf{F}$ is now defined as

$$\mathbf{F}(\mathbf{X}) := \mathrm{Grad}(\boldsymbol{\varphi}_t(\mathbf{X})) = \frac{\partial \boldsymbol{\varphi}_t(X)}{\partial \mathbf{X}} = \mathbf{g}_i \otimes \mathbf{G}^i, \tag{2}$$

where $\mathbf{g}_i$ and $\mathbf{G}^i$ are basis vectors defined with respect to the reference and the current configuration, respectively. The determinant of deformation gradient is given as $J(\mathbf{X}) := \det(\mathbf{F})$.

Furthermore, the absolute temperature $\theta \geq 0\,\mathrm{K}$ for a point of $\mathscr{B}_t$ is introduced, while the reference temperature for a material point of $\mathscr{B}_0$ is denoted as θ_0 and the change of temperature $\boldsymbol{\vartheta}$ is defined as

$$\boldsymbol{\vartheta} : = \theta - \theta_0 \, . \tag{3}$$

Considering a finite thermo-viscoelastic formulation, the deformation gradient $\mathbf{F}$ is usually split into the thermal and viscoelastic parts

$$\mathbf{F} = \mathbf{F}_{\vartheta} \cdot \mathbf{F}_0. \tag{4}$$

The thermal part of the deformation is constitutively assumed to be of purely volumetric nature. This implies that an increase in temperature leads to an expansion of the solid body's volume and vice versa. Here, the volumetric thermal expansion is modeled by

$$\mathbf{F}_{\vartheta} = (J_{\vartheta})^{\frac{1}{3}}\mathbf{1} , \qquad \text{where} \qquad J_{\vartheta} = \exp\,(3\;\alpha_t\;\vartheta) , \tag{5}$$

where $\mathbf{1}$ is the identity tensor and α_t depicts the isotropic thermal expansion coefficient. Herein an exponential form of the thermal expansion is introduced based on [18]. The non-thermal part

$$\mathbf{F}_0 = (J_0)^{\frac{1}{3}}\overline{\mathbf{F}}_0 , \qquad \text{where} \qquad J_0 = \det(\mathbf{F}_0) , \tag{6}$$

can also be further multiplicatively split regarding a volume preserving formulation, namely

$$\overline{\mathbf{F}}_0 = \overline{\mathbf{F}}_e \cdot \overline{\mathbf{F}}_i \tag{7}$$

for the elastic and viscous parts; see, e.g., [15, 44–48]. Consequently, the determinant of the deformation gradient J adheres to $J = J_{\vartheta} J_0$.

The viscous component of the deformation, denoted as $\overline{\mathbf{F}}_i$, characterizes the irreversible and inelastic part of the total isochoric deformation. This portion can undergo evolution, such as governed by the creep law, resulting in micro-structural rearrangements of the material and consequently contributing to dissipation and temperature alterations. Once $\overline{\mathbf{F}}_i$ evolves, the incompressibility condition

$$\det\left(\overline{\mathbf{F}}_i\right) = J_i = 1 \;\forall\; \overline{\mathbf{F}}_i \tag{8}$$

has to be ensured for most of the materials, particularly for metals, see [49].

The left Cauchy-Green tensors $\overline{\boldsymbol{b}}_0 = \overline{\boldsymbol{F}}_0 \cdot \overline{\boldsymbol{F}}_0^T$ and $\overline{\boldsymbol{b}}_e = \overline{\boldsymbol{F}}_e \cdot \overline{\boldsymbol{F}}_e^T$ are two strain quantities for the elastic response at the equilibrium and non-equilibrium branches, respectively. Alternatively, $\overline{\boldsymbol{b}}_e = \overline{\boldsymbol{F}} \cdot \overline{\mathbf{C}}_i^{-1} \cdot \overline{\boldsymbol{F}}^T$ can be obtained with the definition of $\overline{\boldsymbol{C}}_i^{-1} = \overline{\boldsymbol{F}}_i^{-1} \cdot \overline{\boldsymbol{F}}_i^{-T}$

The heat flow q_n out of the surface $\partial \mathscr{B}_t$ of the current configuration can be described as $q_n = \;: \mathbf{q} \cdot \mathbf{n}$. Here, $\mathbf{q}$ represents the spatial heat flux through a point $\mathbf{x} \in \partial \mathscr{B}_t$ and $\mathbf{n}$ is the current outward normal at the point under observation. Within the solid body, the spatial heat flux vector $\mathbf{q}$ at point $\mathbf{x} \in \mathscr{B}_t$, describing the heat conduction, is assumed to follow Fourier's law

$$\mathbf{q} = -\frac{k}{J}\,\mathrm{grad}(\boldsymbol{\vartheta})\;, \tag{9}$$

where k is the thermal conductivity of material. Equation (9) effectively describes the phenomena of spatial heat conduction as determined by the current state of the temperature gradient $\mathrm{grad}(\boldsymbol{\vartheta})$. Furthermore, the conductive dissipation $\mathcal{D}_{con} := -\frac{1}{\theta}\mathbf{q}\cdot\mathrm{grad}(\boldsymbol{\vartheta})$ is also fulfilled.

2.2 Helmholtz Free Energy and Kirchhoff Stress

The Helmholtz free energy density function is isothermally defined as

$$\varphi_0^e = U(J_0) + \overline{\varphi}^{eq}\left(\overline{\boldsymbol{b}}_0\right) + \overline{\varphi}^{ne}\left(\overline{\boldsymbol{b}}_e\right), \tag{10}$$

where $U(J_0)$ represents the volumetric energy. A common form of it is given here

$$U(J_0) = \kappa(J_0 - \ln(J_0) - 1). \tag{11}$$

The model parameter κ depicts the bulk modulus. $\overline{\varphi}^{eq}\left(\overline{\boldsymbol{b}}_0\right)$ and $\overline{\varphi}^{ne}\left(\overline{\boldsymbol{b}}_e\right)$ in Eq. (10) are denoted as the isochoric free energy at the equilibrium and non-equilibrium branches, respectively. A simple Neo-Hookean model [50] is chosen in this contribution for both branches, i.e.,

$$\overline{\varphi}^{eq}\left(\overline{\boldsymbol{b}}_0\right) = \frac{\mu^{eq}}{2}(\bar{I}_1 - 3) \qquad \text{and} \qquad \overline{\varphi}^{ne}\left(\overline{\boldsymbol{b}}_e\right) = \frac{\mu^{ne}}{2}\left(\bar{I}_1^e - 3\right). \tag{12}$$

The shear moduli μ^{eq} and μ^{ne} are introduced here as model parameters for the equilibrium and the non-equilibrium branches, respectively. $\bar{I}_1$ and $\bar{I}_1^e$ depict the first invariants of $\overline{\boldsymbol{b}}_0$ and $\overline{\boldsymbol{b}}_e$, respectively. Based on the presented form of energy decomposition in Eq. (10), the Kirchhoff stress yields

$$\boldsymbol{\tau} = \boldsymbol{\tau}_{vol} + \boldsymbol{\tau}_{iso}^{eq} + \boldsymbol{\tau}_{iso}^{ne}, \tag{13}$$

where the volumetric Kirchhoff part denotes $\boldsymbol{\tau}_{vol} = J_0\, p\, \mathbf{1}$ with $p = \partial_{J_0} U(J_0)$. The isochoric Kirchhoff stress tensors $\boldsymbol{\tau}_{iso}^{eq}$ and $\boldsymbol{\tau}_{iso}^{ne}$ are given here as

$$\boldsymbol{\tau}_{iso}^{eq} = \mathbb{P} : \overline{\boldsymbol{\tau}}^{eq} \qquad \text{with} \qquad \overline{\boldsymbol{\tau}}^{eq} = 2\,\frac{\partial\,\overline{\varphi}^{eq}\left(\overline{\boldsymbol{b}}_0\right)}{\partial\,\overline{\boldsymbol{b}}_0}\cdot\overline{\boldsymbol{b}}_0 \tag{14}$$

and

$$\boldsymbol{\tau}_{iso}^{ne} = \mathbb{P} : \overline{\boldsymbol{\tau}}^{ne} \qquad \text{with} \qquad \overline{\boldsymbol{\tau}}^{ne} = 2\,\frac{\partial\,\overline{\varphi}^{ne}(\overline{\boldsymbol{b}}_e)}{\partial\,\overline{\boldsymbol{b}}_e}\cdot\overline{\boldsymbol{b}}_e, \tag{15}$$

respectively, where $\mathbb{P}$ denotes the fourth order projection tensor.

2.3 Inelastic Dissipation

The dissipation inequality of a closed system is described by the second axiom of thermodynamics as

$$\mathcal{D} = \boldsymbol{\tau} : \boldsymbol{d} - \dot{\varphi}_0^e \geq 0, \tag{16}$$

where $\boldsymbol{d}$ is the symmetric Eulerian rate of the deformation tensor, which can be further split into spherical and unimodular parts as $\boldsymbol{d} = \boldsymbol{d}_{vol} + \boldsymbol{d}_{iso}$, see [10, 11, 14]. As indicated in the research of [51], the dissipation inequality is split into volumetric and isochoric components as

$$\mathcal{D} = \mathcal{D}_{vol} + \mathcal{D}_{iso}^{eq} + \mathcal{D}_{iso}^{ne} \geq 0. \tag{17}$$

Given that the dissipation rates $\mathcal{D}_{vol}$ and $\mathcal{D}_{iso}^{eq}$ are derived from the purely elastic response, it follows $\mathcal{D}_{vol} = \mathcal{D}_{iso}^{eq} = 0$ under isothermal condition. Thus, Eq. (17) is further simplified as

$$\mathcal{D} = \mathcal{D}_{iso}^{ne} = \boldsymbol{\tau}_{iso}^{ne} : \boldsymbol{d}_{iso} - \dot{\overline{\varphi}}^{ne} \geq 0. \tag{18}$$

The material time derivative of $\overline{\varphi}^{ne}$ in Eq. (18) is $\partial_{\overline{\boldsymbol{b}}_e}\,\overline{\varphi}^{ne} : \dot{\overline{\boldsymbol{b}}}_e$, where the rate of $\overline{\boldsymbol{b}}_e$ is according to [14] defined as

$$\dot{\overline{\boldsymbol{b}}}_e = \boldsymbol{l}_{iso}\cdot\overline{\boldsymbol{b}}_e + \overline{\boldsymbol{b}}_e\cdot\boldsymbol{l}_{iso}^T + \overline{\mathcal{L}}_v(\overline{\boldsymbol{b}}_e). \tag{19}$$

In Eq. (19), $\boldsymbol{l}_{iso}$ is the isochoric velocity gradient with respect to the current configuration and the term $\overline{\mathcal{L}}_v(\overline{\boldsymbol{b}}_e)$ denotes the spatial time derivative of the inelastic response. According to [11], it is defined as $\overline{\mathcal{L}}_v(\overline{\boldsymbol{b}}_e) = \overline{\boldsymbol{F}}\cdot\dot{\overline{\boldsymbol{C}}}_i^{-1}\cdot\overline{\boldsymbol{F}}^T$. With further simplification, the total dissipation inequality yields

$$\mathcal{D} = \boldsymbol{\tau}_{iso}^{ne} : \overline{\boldsymbol{d}}_i, \tag{20}$$

where the derivation in detail is referred to [10, 14]. $\overline{\boldsymbol{d}}_i$ is defined as $\overline{\boldsymbol{d}}_i = -\frac{1}{2}\overline{\mathcal{L}}_v(\overline{\boldsymbol{b}}_e)\overline{\boldsymbol{b}}_e^{-1}$ at the isochoric level.

2.4 Viscoelastic Evolution Law

In this study, disregarding the viscous evolution of the volumetric response, the rate of the inelastic deformation tensor in the current configuration according to [14] is explicitly defined as

$$\overline{\boldsymbol{d}}_i = \eta_0 \ \boldsymbol{\tau}_{iso}^{ne}, \tag{21}$$

where η_0 is a model parameter to describe the isochoric viscous response. To effectively evaluate the time integration, several numerical assumptions are proposed. One key assumption during the elastic trial phase is that the inelastic deformation remains static or frozen, leading to $\dot{\overline{\boldsymbol{C}}}_i^{-1} = \boldsymbol{0}$. Therefore, the inelastic right Cauchy-Green tensor is supposed to remain unchanged from the previous computational step, i.e., $\overline{\boldsymbol{C}}_i^{-1,tr} = \overline{\boldsymbol{C}}_i^{-1,t_n}$. Consequently, the elastic trial for $\overline{\boldsymbol{b}}_e^{tr}$ at time t_{n+1} is

$$\overline{\boldsymbol{b}}_e^{tr} = \overline{\boldsymbol{F}}_0^{t_{n+1}} \cdot \overline{\boldsymbol{C}}_i^{-1,tr} \cdot \overline{\boldsymbol{F}}_0^{T,t_{n+1}} = \overline{\boldsymbol{F}}_0^{t_{n+1}} \cdot \overline{\boldsymbol{C}}_i^{-1,t_n} \cdot \overline{\boldsymbol{F}}_0^{T,t_{n+1}}. \tag{22}$$

$\overline{\boldsymbol{F}}_0^{t_{n+1}}$ is assumed to be fixed at the current step t_{n+1}. This assumption leads to $\boldsymbol{l}_{iso} = \boldsymbol{0}$. Following this, the elastic predictor part $\left(\boldsymbol{l}_{iso} \cdot \overline{\boldsymbol{b}}_e + \overline{\boldsymbol{b}}_e \cdot \boldsymbol{l}_{iso}^T\right)$ in Eq. (19) eventually vanishes, yielding

$$\dot{\overline{\boldsymbol{b}}}_e = \overline{\mathcal{L}}_v\left(\overline{\boldsymbol{b}}_e\right) = -2\,\overline{\boldsymbol{d}}_i \cdot \overline{\boldsymbol{b}}_e^{tr}. \tag{23}$$

As proposed in [13, 52], an exponential mapping for $\overline{\boldsymbol{b}}_e$ is introduced and further approximated as

$$\overline{\boldsymbol{b}}_e = \exp\left(-2\int_{t_n}^{t_{n+1}} \eta_0 \ \boldsymbol{\tau}_{iso}^{ne} \ dt\right)\overline{\boldsymbol{b}}_e^{tr} \approx \exp\left(-2\ \eta_0 \ \boldsymbol{\tau}_{iso}^{ne} \ \Delta t\right)\overline{\boldsymbol{b}}_e^{tr}, \tag{24}$$

while the alternative expressions of $\overline{\boldsymbol{b}}_e$ and $\overline{\boldsymbol{b}}_e^{tr}$ read

$$\overline{\boldsymbol{b}}_e = \sum\nolimits_{a=1}^{3} \overline{\lambda}_a^{e^2} \boldsymbol{n}_a \otimes \boldsymbol{n}_a \qquad \text{and} \qquad \overline{\boldsymbol{b}}_e^{tr} = \sum\nolimits_{a=1}^{3} \overline{\lambda}_a^{e,tr^2} \boldsymbol{n}_a \otimes \boldsymbol{n}_a, \tag{25}$$

respectively. The eigenvalues of $\overline{\boldsymbol{b}}_e$ and $\overline{\boldsymbol{b}}_e^{tr}$ are expressed by $\overline{\lambda}_a^{e^2}$ and $\overline{\lambda}_a^{e,tr^2}$, meanwhile, they have the same eigenvectors $\boldsymbol{n}_a$ with $a = 1, 2, 3$. Using the eigenvalues and the logarithmic operation of Eq. (24), a simplified equation

$$\varepsilon_a^e \approx -\eta_0 \ \mathrm{dev}\left(\overline{\tau}_a^{ne}\right)\Delta t + \varepsilon_a^{e,tr} \tag{26}$$

is obtained, where the principal stretch notations $\varepsilon_a^e = \ln\overline{\lambda}_a^e$ and $\varepsilon_a^{e,tr} = \ln\overline{\lambda}_a^{e,tr}$ are given here and the operator $\mathrm{dev}(*)$ is defined as

Table 1 Local Newton-Raphson iteration steps

Initiation	$k=0,\ \varepsilon_a^{e,k}=\varepsilon_a^{e,tr}$
Do while	
Residual	$R_a=\varepsilon_a^e+\eta_0\ \mathrm{dev}\left(\overline{\tau}_a^{ne}\right)\Delta t-\varepsilon_a^{e,tr}$
Linearization	$\mathrm{Lin}\ R_a=R_a\big\vert_{\varepsilon_a^{e,k}}+\partial_{\varepsilon_b^e}\ R_a\big\vert_{\varepsilon_a^{e,k}}\Delta\varepsilon_b^{e,k}$
Tangent	$K_{ab}=-\ \partial_{\varepsilon_b^e}\ R_a\big\vert_{\varepsilon_a^{e,k}}$
Solving	$\Delta\varepsilon_b^{e,k}=K_{ab}^{-1}\ R_a$
Update	$\varepsilon_b^{e,k+1}=\varepsilon_b^{e,k}+\Delta\varepsilon_b^{e,k},\ k=k+1$
While	$\mathrm{Tol}\leq\Vert\ R_a\ \Vert$

$$\mathrm{dev}\left(\overline{\tau}_a^{ne}\right)=\overline{\tau}_a^{ne}-\frac{1}{3}\left(\overline{\tau}_1^{ne}+\overline{\tau}_2^{ne}+\overline{\tau}_3^{ne}\right), \tag{27}$$

where the term $\overline{\tau}_a^{ne}$ is obtained by

$$\overline{\tau}_a^{ne}=2\ \overline{\lambda}_a^{e^2}\ \frac{\partial\ \overline{\varphi}^{ne}}{\partial\ \overline{I}_1^e}. \tag{28}$$

Consequently, Eq. (26) is characterized as nonlinear owing to the dependence of $\overline{\tau}_a^{ne}$ on $\overline{\lambda}_a^e$. Therefore, the residual form

$$R_a=\varepsilon_a^e+\eta_0\ \mathrm{dev}\left(\overline{\tau}_a^{ne}\right)\Delta t-\varepsilon_a^{e,tr}=0 \tag{29}$$

is proposed to find the solution of ε_a^e using an internal Newton-Raphson iteration process. With a consistent linearization of Eq. (29), the local tangent K_{ab} is derived as

$$K_{ab}=-\frac{\partial\ R_a}{\partial\ \varepsilon_b^e}=-\left(\delta_{ab}+\eta_0\ \frac{\partial\ \mathrm{dev}\left(\overline{\tau}_a^{ne}\right)}{\partial\ \overline{\tau}_c^{ne}}\frac{\partial\ \overline{\tau}_c^{ne}}{\partial\ \varepsilon_b^e}\ \Delta t\right). \tag{30}$$

In order to achieve further simplifications, with the definition of T_{ab} as

$$T_{ab}=\frac{\partial\ \overline{\tau}_a^{ne}}{\partial\ \varepsilon_b^e}=4\ \overline{\lambda}_a^e\ \overline{\lambda}_b^e\ \delta_{ab}\ \frac{\partial\ \overline{\varphi}^{ne}}{\partial\ \overline{I}_1^e}+4\ \overline{\lambda}_a^{e^2}\overline{\lambda}_b^{e^2}\ \frac{\partial^2\ \overline{\varphi}^{ne}}{\partial\ \overline{I}_1^e\ \partial\ \overline{I}_1^e} \tag{31}$$

and the definition of the operator $\overline{T}_{ab}=T_{ab}-\frac{1}{3}\sum_{c=1}^{3}T_{cb}$, the local stiffness K_{ab} in Eq. (30) consequently yields

$$K_{ab}=-\ \left(\delta_{ab}+\eta_0\ \Delta t\ \overline{T}_{ab}\right). \tag{32}$$

Thus, the straightforward internal iteration in Table 1 can be conducted. As a result, this algorithmic manipulation yields the numerical solution for the left Cauchy-

Green tensor at the non-equilibrium branch $\overline{\boldsymbol{b}}_e$. In the sequel, the Kirchhoff stress at the non-equilibrium branch $\boldsymbol{\tau}^{ne}$ is obtained as well.

2.5 Consistent Material Tangent

Having all the Kirchhoff stresses at hand, the consistent Eulerian tangent moduli can be straightforwardly derived. The volumetric tangent is obtained as

$$\mathbb{C}_{vol} = \Big(J_0\, \partial_{J_0}\, U(J_0) + (J_0)^2\, \partial^2_{J_0}\, U(J_0)\Big)\mathbf{1} \otimes \mathbf{1} - 2\, J_0\, \partial_{J_0}\, U(J_0)\, \mathbb{I}, \tag{33}$$

and the isochoric tangents for both the equilibrium and non-equilibrium branches are

$$\mathbb{C}^{eq}_{iso} = \mathbb{P} : \overline{\mathbb{C}}^{eq}_{iso} : \mathbb{P} + \frac{2}{3}\Big((\overline{\boldsymbol{\tau}}^{eq} : \mathbf{1})\mathbb{P} - \boldsymbol{\tau}^{eq}_{iso} \otimes \mathbf{1} - \mathbf{1} \otimes \boldsymbol{\tau}^{eq}_{iso}\Big) \tag{34}$$

and

$$\mathbb{C}^{ne}_{iso} = \mathbb{P} : \overline{\mathbb{C}}^{ne}_{iso} : \mathbb{P} + \frac{2}{3}\Big((\overline{\boldsymbol{\tau}}^{ne} : \mathbf{1})\mathbb{P} - \boldsymbol{\tau}^{ne}_{iso} \otimes \mathbf{1} - \mathbf{1} \otimes \boldsymbol{\tau}^{ne}_{iso}\Big), \tag{35}$$

respectively. The terms $\boldsymbol{\tau}^{eq}_{iso}$, $\overline{\boldsymbol{\tau}}^{eq}$, $\boldsymbol{\tau}^{ne}_{iso}$, and $\overline{\boldsymbol{\tau}}^{ne}$ are already introduced in Eqs. (14) and (15). The isochoric tangent for the equilibrium branch in Eq. (34) is given as

$$\overline{\mathbb{C}}^{eq}_{iso} = 4\, \overline{\boldsymbol{b}}_0 \cdot \frac{\partial^2\, \overline{\varphi}^{eq}_{iso}}{\partial\, \overline{\boldsymbol{b}}_0\, \partial\, \overline{\boldsymbol{b}}_0} \cdot \overline{\boldsymbol{b}}_0 \tag{36}$$

according to [50]. However, deriving a closed form of $\overline{\mathbb{C}}^{ne}_{iso}$ in Eq. (35) is not straightforward, see [11, 14]. Therefore, with respect to the inelastic response of the non-equilibrium branch, a fictitious intermediate configuration is assumed between the initial and current configuration, where the eigenvectors are depicted by $\tilde{\boldsymbol{N}}^a$ with $a = 1, 2, 3$. The trial deformation tensor $\overline{\boldsymbol{F}}^{tr}_e$ represents the kinetic relations from the fictitious intermediate configuration to the current configuration, which is expressed as

$$\overline{\boldsymbol{F}}^{tr}_e = \sum\nolimits_{a=1}^{3} \overline{\lambda}^{e,tr}_a\, \boldsymbol{n}^a \otimes \tilde{\boldsymbol{N}}^a. \tag{37}$$

In the sequel, using a pullback operation, the second Piola-Kirchhoff stress tensor $\overline{\boldsymbol{S}}^{ne}_{iso}$ is obtained as

$$\overline{\boldsymbol{S}}_{iso}^{ne} = \overline{\boldsymbol{F}}_{e}^{tr^{-1}} \cdot \overline{\boldsymbol{\tau}}^{ne} \cdot \overline{\boldsymbol{F}}_{e}^{tr^{-T}} = \sum\nolimits_{a=1}^{3} \frac{\overline{\tau}_{a}^{ne}}{\overline{\lambda}_{a}^{e,tr^2}} \tilde{\boldsymbol{N}}_a \otimes \tilde{\boldsymbol{N}}_a. \tag{38}$$

The consistent tangent modulus $\widetilde{\mathbb{C}}_{iso}^{ne}$ at the intermediate configuration reads

$$\begin{aligned}\widetilde{\mathbb{C}}_{iso}^{ne} &= 2 \frac{\partial \overline{\boldsymbol{S}}_{iso}^{ne}}{\partial \overline{\boldsymbol{C}}_{e}^{tr}} \\ &= \sum\nolimits_{a=1}^{3} \sum\nolimits_{b=1}^{3} \frac{\zeta_{ab} - 2\, \overline{\tau}_{a}^{ne}\, \delta_{ab}}{\overline{\lambda}_{a}^{e,tr^2}\, \overline{\lambda}_{b}^{e,tr^2}} \widetilde{\mathbb{M}}^{ab} + \sum\nolimits_{a=1}^{3} \sum\nolimits_{b \neq a}^{3} \frac{\overline{\tau}_{a}^{ne}/\overline{\lambda}_{a}^{e,tr^2} - \overline{\tau}_{b}^{ne}/\overline{\lambda}_{b}^{e,tr^2}}{\overline{\lambda}_{a}^{e,tr^2} - \overline{\lambda}_{b}^{e,tr^2}} \widetilde{\mathbb{G}}^{ab},\end{aligned} \tag{39}$$

with the definitions

$$\begin{aligned}\widetilde{\mathbb{M}}^{ab} &= \tilde{\boldsymbol{N}}^{a} \otimes \tilde{\boldsymbol{N}}^{a} \otimes \tilde{\boldsymbol{N}}^{b} \otimes \tilde{\boldsymbol{N}}^{b} \quad \text{and} \\ \widetilde{\mathbb{G}}^{ab} &= \tilde{\boldsymbol{N}}^{a} \otimes \tilde{\boldsymbol{N}}^{b} \otimes \left(\tilde{\boldsymbol{N}}^{a} \otimes \tilde{\boldsymbol{N}}^{b} + \tilde{\boldsymbol{N}}^{b} \otimes \tilde{\boldsymbol{N}}^{a} \right).\end{aligned} \tag{40}$$

The quantity $\overline{\tau}_{a}^{ne}$ in Eq. (39) is illustrated in Eq. (28) and ζ_{ab} is defined as

$$\zeta_{ab} = \frac{\partial\, \overline{\tau}_{a}^{ne}}{\partial\, \varepsilon_{b}^{e,tr}} = \frac{\partial\, \overline{\tau}_{a}^{ne}}{\partial\, \varepsilon_{c}^{e}} \frac{\partial\, \varepsilon_{c}^{e}}{\partial\, \varepsilon_{b}^{e,tr}} = T_{ac} \frac{\partial\, \varepsilon_{c}^{e}}{\partial\, \varepsilon_{b}^{e,tr}}, \tag{41}$$

where the term $\frac{\partial\, \varepsilon_{c}^{e}}{\partial\, \varepsilon_{b}^{e,tr}}$ is not directly determined. Owing to the condition $\frac{\partial\, r_a}{\partial\, \varepsilon_{b}^{e,tr}} = 0$, it holds

$$-\delta_{ab} - K_{ac} \frac{\partial\, \varepsilon_{c}^{e}}{\partial\, \varepsilon_{b}^{e,tr}} = 0, \tag{42}$$

which leads to $\frac{\partial\, \varepsilon_{c}^{e}}{\partial\, \varepsilon_{b}^{e,tr}} = -K_{cb}^{-1}$. Thus, according to Eq. (41), the coefficient $\zeta_{ab} = -T_{ac}\, K_{cb}^{-1}$ is eventually obtained.

Considering the singularity of Eq. (39), e.g., equal eigenvalues $\overline{\lambda}_a = \overline{\lambda}_b$, L' Hospital's rule is applied to address this issue, which yields

$$\lim_{\overline{\lambda}_b \to \overline{\lambda}_a} \frac{\overline{\tau}_{a}^{ne}/\overline{\lambda}_{a}^{e,tr^2} - \overline{\tau}_{b}^{ne}/\overline{\lambda}_{b}^{e,tr^2}}{\overline{\lambda}_{a}^{e,tr^2} - \overline{\lambda}_{b}^{e,tr^2}} = \frac{1}{2} \frac{\zeta_{aa} - 2\, \overline{\tau}_{a}^{ne}}{\overline{\lambda}_{a}^{e,tr^4}}. \tag{43}$$

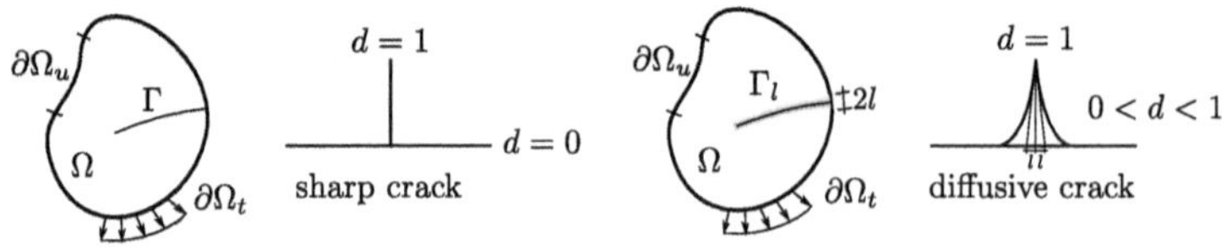

Fig. 1 Crack topologies

As a result, the consistent tangent in the current configuration is obtained by using push-forward operation of $\widetilde{\mathbb{C}}^{ne}_{iso}$, which reads

$$\overline{\mathbb{C}}^{ne}_{iso_{ijkl}} = \overline{\boldsymbol{F}}^{tr}_{e_{iI}} \overline{\boldsymbol{F}}^{tr}_{e_{jJ}} \overline{\boldsymbol{F}}^{tr}_{e_{kK}} \overline{\boldsymbol{F}}^{tr}_{e_{lL}} \widetilde{\mathbb{C}}^{ne}_{iso_{IJKL}}. \tag{44}$$

3 Thermo-Viscoelastic Phase-Field Fracture Model

3.1 *Phase-Field Topology*

In a continuous solid domain, a sharp crack can be approximated numerically using a diffusive phase-field distribution, as illustrated in Fig. 1. The introduced phase-field variable, denoted as $d(\boldsymbol{x}, t)$, serves to identify the material state. In this representation, the sound material corresponds to $d = 0$, while the fully cracked state is indicated by $d = 1$. This mathematical formulation is inspired by a one-dimensional bar with infinite length $L \in [-\infty, +\infty]$, assumed to be cracked at position $x = 0$. The closed-form solution for a continuous phase-field, as outlined in [38], is approximated by an exponential function

$$d(x) = \exp\left(-\frac{|x|}{l}\right). \tag{45}$$

Nevertheless, it is crucial to note that this function is not the exclusive approximation for a one-dimensional representation of the phase-field profile. This solution provided in Eq. (45) is appropriately and naturally bounded by $(0 \le d \le 1)$. Another pivotal parameter, the length scale l, is utilized to control the width of the transition zone between fractured and sound states of the material. Expanding to a two- or three–dimensional framework, the second-order functional of the crack surface density is defined as

$$\gamma_l = \frac{1}{2l}\left(d^2 + l^2|\nabla_{\boldsymbol{x}} d|^2\right), \tag{46}$$

where $\nabla_{\boldsymbol{x}}(*)$ represents a spatial gradient operator. According to [53], the crack domain Γ can be approximately calculated by the phase-field crack domain Γ_l within the entire material space $\mathscr{B}$ as long as $l \rightarrow 0$, i.e.,

$$\Gamma \approx \lim_{l \rightarrow 0} \Gamma_l = \lim_{l \rightarrow 0} \int_{\mathscr{B}} \gamma_l(d, \nabla_{\boldsymbol{x}} d)\, dV, \tag{47}$$

which is known as the Γ-convergence condition for fracture.

The crack surface density function defined in Eq. (46) is commonly known as the AT2 model, resulting in an exponentially shaped crack profile as described by Eq. (45). The application of this function facilitates straightforward and efficient numerical derivation and implementation, given the differentiable and continuous properties of the phase-field solution in Eq. (45). However, it is important to note that this model lacks an elastic limit, and the simulations are significantly influenced by the length scale parameter l. As demonstrated by [54] in the context of FEM applications, achieving realistic results necessitates a minimum element size $h_e \leq l/2$ in the potential crack region. Recent studies, such as [55, 56], have also delved into the precision and dependability of phase-field modeling concerning length scale and mesh sensitivity. In addition to the AT2 model, another commonly employed alternative for approximating the crack surface density function is the classical AT1 model. A key distinction between these models lies in the fact that the AT1 model comprises only the linear term d rather than the quadratic term d^2 in Eq. (46), as discussed in [57]. The AT1 model ensures an ideally linear elastic response up to the elastic stress limit. However, due to the absence of differential continuity, additional numerical treatments become crucial to prevent negative phase-field evolution, which could lead to a nonlinear phase-field evolution. For the purpose of simplification in this contribution, we consider a standard AT2 model based on Eq. (46).

3.2 Thermo-Viscoelastic Phase-Field Evolution

A natural choice of the viscoelastic phase-field formulation at hand is the classical volumetric-isochoric split due to the isochoric inelastic response, hence, the Helmholtz strain energy density function coupled with the phase-field degradation function reads

$$\varphi(\boldsymbol{F}) = g(d)\underbrace{\left([J_0 - 1]_+ U(J_0) + \bar{\varphi}^{eq}_{iso}(\bar{\boldsymbol{b}}_0) + \bar{\varphi}^{ne}_{iso}(\bar{\boldsymbol{b}}_e)\right)}_{\varphi_e^+} + \underbrace{[J_0 - 1]_- U(J_0)}_{\varphi_e^-} + \varphi_\vartheta(\vartheta) + \mathcal{G}_c \gamma_l, \tag{48}$$

where the operator $[*]_\pm$ is defined as

$$[J_0-1]_+=\begin{cases}1, J_0\geq 1\\ 0, J_0<1\end{cases} \quad \text{and} \quad [J_0-1]_-=\begin{cases}0, J_0\geq 1\\ 1, J_0<1\end{cases}. \tag{49}$$

Therefore, the elastic strain energy stored in both the equilibrium and the non-equilibrium branches is characterized to evolve the phase-field, which consequently results in a rate-dependent driving force for the phase-field fracture evolution. As a result, the overall Kirchhoff stress and the consistent tangent are given as

$$\boldsymbol{\tau}=g(d)\underbrace{\left([J_0-1]_+\boldsymbol{\tau}_{vol}+\boldsymbol{\tau}^{eq}_{iso}+\boldsymbol{\tau}^{ne}_{iso}\right)}_{\boldsymbol{\tau}^+}+\underbrace{[J_0-1]_-\boldsymbol{\tau}_{vol}}_{\boldsymbol{\tau}^-} \tag{50}$$

and

$$\mathbb{C}=g(d)\underbrace{\left([J_0-1]_+\mathbb{C}_{vol}+\mathbb{C}^{eq}_{iso}+\mathbb{C}^{ne}_{iso}\right)}_{\mathbb{C}^+}+\underbrace{[J_0-1]_-\mathbb{C}_{vol}}_{\mathbb{C}^-}, \tag{51}$$

respectively.

Moreover, a purely thermal contribution φ_{ϑ} captures the variation of free energy resulting from any change of temperature. The thermal part $\rho_0\varphi_{\vartheta}=\rho_0\varphi_{\vartheta}(\vartheta)$ is defined as

$$\rho_0\varphi_{\vartheta}(\vartheta)=-\rho_0 c\left(\theta\ln\left(\frac{\theta}{\theta_0}\right)-\vartheta\right)+C(\vartheta). \tag{52}$$

The corrector function $C(\vartheta)$ is necessary to guarantee the assumed constant heat capacity at constant deformation. The specific heat capacity c is defined as

$$\rho_0 c:=-\theta\frac{\partial^2\rho_0\varphi(\boldsymbol{F})}{\partial\vartheta\partial\vartheta}=-\theta g(d)\frac{\partial^2\rho_0\varphi^+_e(J,J_{\vartheta})}{\partial\vartheta\partial\vartheta}-\theta\frac{\partial^2\rho_0\varphi^-_e(J,J_{\vartheta})}{\partial\vartheta\partial\vartheta}-\theta\frac{\partial^2\rho_0\varphi_{\vartheta}(\vartheta)}{\partial\vartheta\partial\vartheta} \tag{53}$$

and the temperature-dependent part of the Helmholtz energy is consequently determined by

$$\frac{\partial^2\rho_0\varphi_{\vartheta}(\vartheta)}{\partial\vartheta\partial\vartheta}=-g(d)\frac{\partial^2\rho_0\varphi^+_e(J,J_{\vartheta})}{\partial\vartheta\partial\vartheta}-\frac{\partial^2\rho_0\varphi^-_e(J,J_{\vartheta})}{\partial\vartheta\partial\vartheta}-\frac{\rho_0 c}{\theta}. \tag{54}$$

An integration over the temperature domain using the constraints $\partial_{\theta}\varphi_{\vartheta}(\vartheta=0):=0$ and $\partial^2_{\theta\theta}\varphi_{\vartheta}(\vartheta=0):=0$ as well as considering $\theta=\theta_0\ \forall\ \vartheta=0$ would yield the specific formulation of φ_{ϑ} and $C(\vartheta)$, respectively. This work considers the heat capacity as a material constant, which differs according to different material intrinsic characteristics. The external power

$$\rho\, w_{\text{ext}} = \rho\theta\left(2\frac{\partial^2\varphi(\boldsymbol{F})}{\partial\mathbf{g}\partial\boldsymbol{\vartheta}}\right) \tag{55}$$

signifies the quantity of energy that alters the temperature at $\mathbf{x}$, attributed to an arbitrarily applied deformation rate. This part of the entropy change is a function of the volumetric energy, since temperature changes primarily impact the changes in volumetric deformation of most materials. The internal power is expressed as

$$\rho\, w_{\text{int}} = \rho\left(\frac{\partial\varphi(\boldsymbol{F})}{\partial\overline{\mathbf{F}}_i} - \theta\frac{\partial^2\varphi(\boldsymbol{F})}{\partial\overline{\mathbf{F}}_i\partial\boldsymbol{\vartheta}}\right) : \dot{\overline{\mathbf{F}}}_i\,, \tag{56}$$

which captures any variation of temperature at $\mathbf{x} \in \mathscr{B}_t$ due to the evolution of internal variables in an irreversible manner, whenever $\dot{\overline{\mathbf{F}}}_i \neq \mathbf{0}$. Equation (55) captures the reversible change of temperature at any deformation rate, whereas Eq. (56) can be construed as the fundamental response to changes in the ground state temperature. If $\dot{\overline{\mathbf{F}}}_i = \mathbf{0}$ and $\dot{\mathbf{F}} \neq \mathbf{0}$, an entropic cooling at tension and heating at compression is present at $\mathbf{x} \in \mathscr{B}_t$ and, if $\dot{\overline{\mathbf{F}}}_i \neq \mathbf{0}$ and $\dot{\mathbf{F}} \neq \mathbf{0}$, the temperature is increased as time elapses.

For simplicity without showing the derivation process, the strong forms of temperature, the elastic, and inelastic response with respect to the current configuration are eventually obtained as

$$\rho c\dot{\theta} = -\nabla_{\boldsymbol{x}} \cdot \mathbf{q} + \rho w_{\text{ext}} - \rho w_{\text{int}}, \tag{57}$$

$$\rho\, \ddot{\boldsymbol{u}} - \nabla_{\boldsymbol{x}} \cdot \left(g(d)\boldsymbol{\tau}^+ + \boldsymbol{\tau}^-\right) - \mathfrak{b} = \mathbf{0}, \tag{58}$$

$$\partial_d\, g(d)\mathcal{H} + \frac{\mathcal{G}_c}{l}d - \mathcal{G}_c\, l\, \Delta_{\boldsymbol{x}} d + \chi\, \dot{d} = 0. \tag{59}$$

It is notable that the phase-field strong form incorporates a significantly small artificial mobility number χ to enhance the numerical stability. For the time discretization of $\dot{d}$, it is referred to a simple implicit method. The consequent phase-field strong form is similar to the aforementioned one, where the history variable is redefined as

$$\mathcal{H} = \underset{\tau \leq t_{n+1}}{\text{Max}}\left([J_0 - 1]_+ U(J_0, \tau) + \overline{\varphi}^{eq}_{iso}\left(\overline{\boldsymbol{b}}_0, \tau\right) + \overline{\varphi}^{ne}_{iso}\left(\overline{\boldsymbol{b}}_e, \tau\right)\right). \tag{60}$$

4 Numerical Simulation

This section presents three meaningful numerical studies to demonstrate the capability of the presented model. Firstly, a three-dimensional double-edge-notched tensile (DENT) test is taken into account. In this example, the temperature is considered to be frozen. Since the comparing reference in [43] does not provide sufficient experimental data, the first example, for sake of simplicity, excludes temperature evolution by fixing the temperature degree of freedom for all the nodes in the FEM model. The second and third examples consider a thermo-mechanically coupled simulation. Therefore, not only displacement governed but also temperature loading controlled fracture mechanisms are studied based on the two fictitious numerical examples.

4.1 Fracture of a Viscoelastic Rubbery Specimen

The problem at hand involves a three-dimensional boundary value study, depicting the plane stress characteristics of ethylene propylene diene monomer (EPDM) rubber. This particular rubber setup has been investigated both numerically and experimentally by Loew et al. in [43]. They propose a phase-field model for rubber damage that is dependent on the rate, and the corresponding material parameters are determined through experimental studies. In contrast to the viscoelastic formulation outlined in this study, [43] establishes viscoelasticity using the classical convolution law, depicting a linear viscous evolution under finite deformation. Their phase-field model for fracture evolution incorporates a driving force defined by both the total strain potential and dissipated energy. However, it is important to note that the current model does not incorporate dissipated energy for driving the phase-field fracture. Nevertheless, the present model at hand does not include the dissipated energy to drive the phase-field fracture. This example aims to contrast the rate-dependent fracture evolution proposed in the present study with the numerical and experimental findings outlined in [43].

As illustrated in Fig. 2a, tensile loading is applied to the specimen by clamping the top and bottom edges. The simulation focuses on a quarter of the initial specimen owing to its symmetric properties, and this portion is discretized by 9856 8-node linear elements. In the vicinity of the potential crack area, the element size is set to $h_e = 0.25$ mm. It is noteworthy, as specified in [43], that the impact of temperature on material properties is not considered in this analysis. All experiments are carried out at a consistent temperature 20 °C and the material parameters are detailed in Table 2.

The investigation comprises three principal comparisons: varying loading rates, different geometries, and the creep test. Firstly, the evolution of fracture and the peak values of the reaction force at various loading rates are compared with the experimental results, as depicted in Figs. 2b and 3, respectively. Here, the initial crack

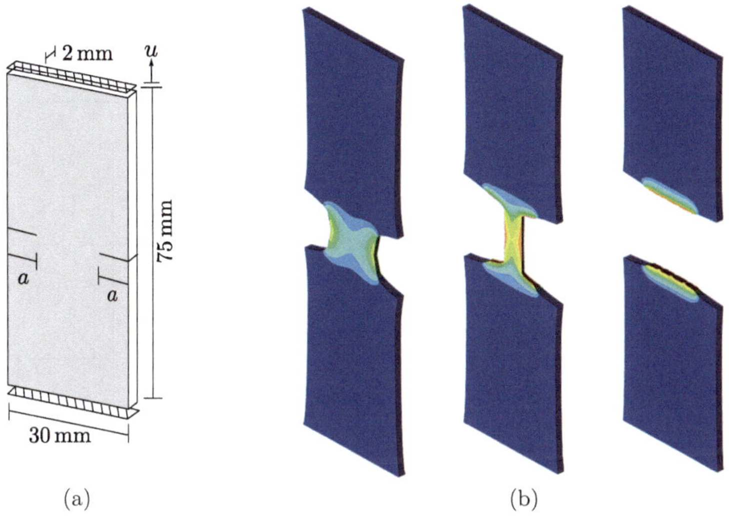

Fig. 2 Failure simulation with plane stress setup for a double-notch specimen

Table 2 Material parameters

κ	μ_{eq}	μ_{ne}	θ_0	η_0	$\mathcal{G}_c$	l
18.96 MPa	0.936 MPa	0.624 MPa	20° C	0.088 $(\text{s}\cdot\text{MPa})^{-1}$	12.22 J/mm^2	0.5 mm

Fig. 3 Peak loads at different clamp velocities of the specimen

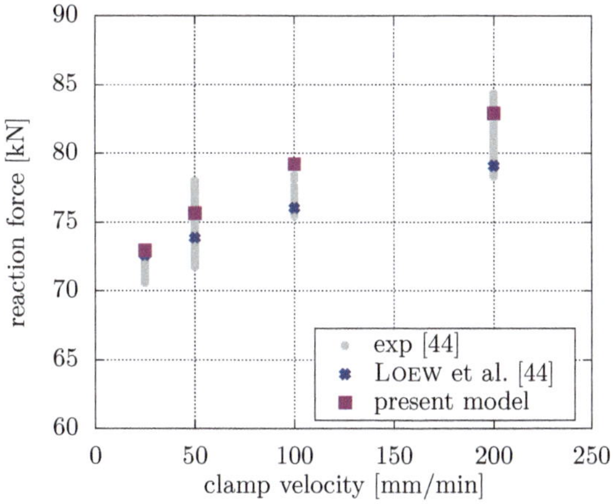

length is fixed at $a = 7$ mm and the loading rates are set at $\dot{u} = [25, 50, 100, 200]$ mm/min. The observation that a higher loading rate results in a larger reaction force, as seen in the experiments, is properly replicated by the

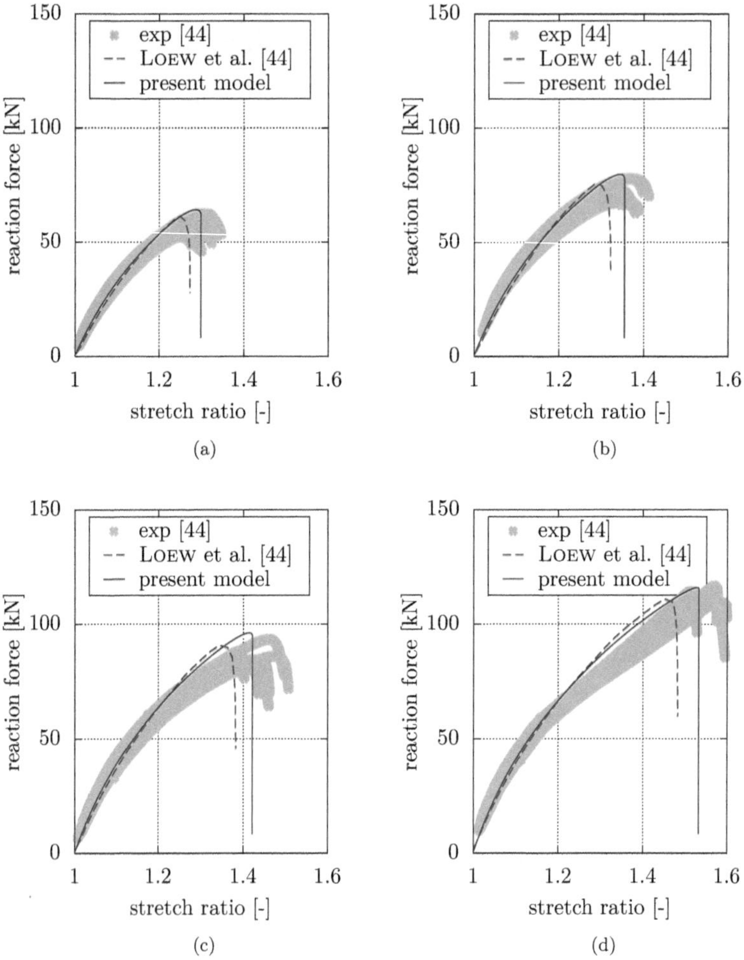

Fig. 4 Relation between the reaction forces and the stretch ratios with different prescribed notches

current phase-field model in this study. Secondly, the relationship between the reaction force and the stretch ratio, considering different initial crack lengths, is compared in Fig. 4. In these testes, all samples are subjected to a clamp velocity $\dot{u} = 75$ mm/min with initial crack length set at $a = [3, 5, 7, 9]$ mm. Good agreement is observed between the experimental results in [43] and the numerical results obtained from the present fracture model. It includes proper estimations of both the final stretch at fully cracked states and the peak load values. Additionally, the model successfully simulates creep, a significant characteristic of rubber materials. Following the setup in [43], the initial crack length is set to $a = 7$ mm and a resultant force of 65 kN is applied to the top edge, with a clamp velocity of 100 mm/min. The loading function is illustrated in Fig. 5a. The initial displacement at $t = 0$ s is noted

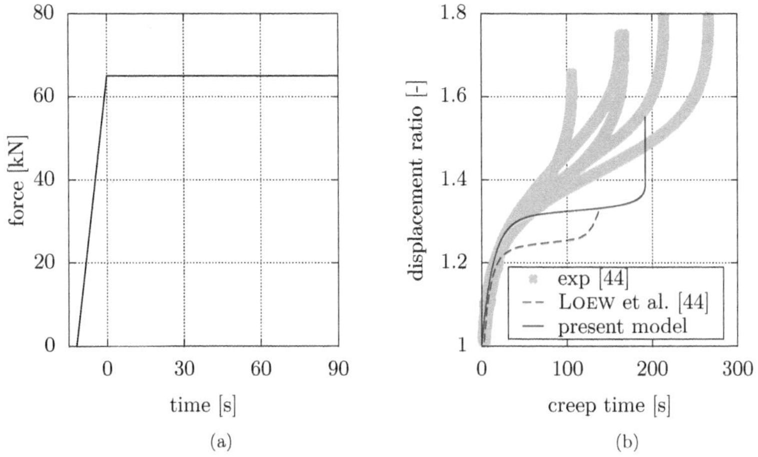

Fig. 5 Investigation of the fracture evolution by creep test

as $u_{t=0}$. Figure 5b depicts the ratio of the subsequent displacement u to $u_{t=0}$ over time. The comparison of numerical and experimental results reveals that the failure behavior and the maximum time to failure are properly predicted. However, both the present model and the numerical model described in [43] exhibit a consistent pattern of unstable crack propagation. This instability arises because once the specimen is fully cracked, it loses all resistance. Consequently, the application of a constant force on the top edge may cause unlimited displacement, leading to a range of numerical problems.

4.2 *Fracture of a DENT Test with Non-Aligned Notches*

This example takes a two-dimensional boundary value problem into consideration. The numerical specimen has two non-aligned straight notches and it is subjected to a displacement-controlled tensile loading at the upper and bottom boundaries, see the representation in Fig. 6 for the geometric setup and loading conditions.

In order to validate the proposed constitutive thermo-viscoelastic model, a set of numerical simulations is performed, which are based on two categories of material parameters (set A and set B) for both the equilibrium and non-equilibrium responses. For each set, e.g., A and B, it contains three different shear moduli for the non-equilibrium response to mimic different material types that have specific ratios between μ_{eq} and μ_{ne}. They are given in Table 3. Besides the deformation parameters, the critical energy release rate $\mathcal{G}_c$ is also an important parameter for fracture study. This specific material parameter is not always constant and might be influenced by various factors particularly for polymeric materials. In general, several inelastic

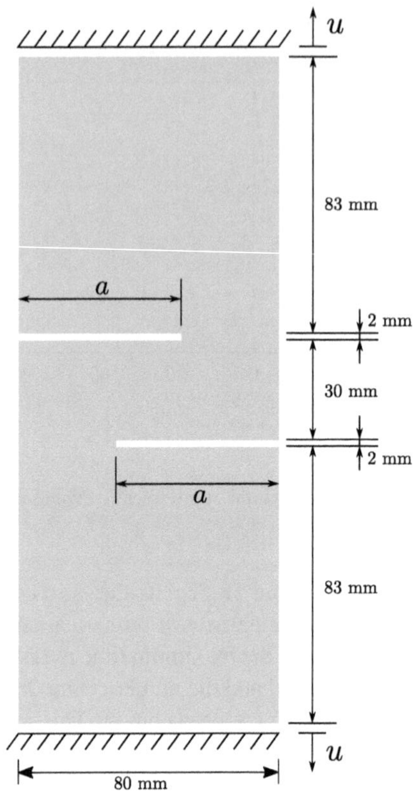

Fig. 6 Geometry and boundary setup

characteristics, e.g., plasticity, viscosity, or viscoplasticity, naturally lead to a portion of energy dissipation. Therewith, the internal dissipation rate is derived as a local heat resource that drives heat transfer. A large dissipation rate certainly yields higher local temperature concentration. From this sense, the experimental evaluation always shows different critical energy release rates $\mathcal{G}_c$ for cooled and warmed specimens. Nevertheless, in this work, $\mathcal{G}_c$ is assumed to be constant and independent of other variables for the sake of simplifications. Herein, two different magnitudes of fracture toughness are also considered for each of the material set A and B, namely, $\mathcal{G}_c = 2.67e^{-1}$ J/mm^2 and $\mathcal{G}_c = 2.67e^{-2}$ J/mm^2. The purpose is to investigate the fracture occurring at different loading states, i.e., larger value of $\mathcal{G}_c$ allows the eventual material failure at a relatively larger deformation. Taking the material set A3 and set B3 with the notch length $a = 50$ mm as examples, the phase-field evolution for both fracture toughness values are shown in Figs. 7 and 8, respectively. Furthermore, the relationship between the displacement and reaction force is also depicted in Fig. 9. It is evident that, with a larger value of $\mathcal{G}_c$, the model needs more elastic energy to evolve the crack, yielding a larger deformation of the whole specimen until its failure. The peak load and the maximum displacement at fracture

Table 3 Material parameter sets for simulation

$\alpha_t = 2.0e^{-3}$, $k_t = 0.3$ N/s · K, $\rho_0 c = 1.8$ N/(mm^2 · K)		
$\eta_0 = 5$ (s · MPa)$^{-1}$, $l = 4$ mm, $\kappa = 1.960$ MPa		
Material A1	$\mu_{eq} = 0.203$ MPa	$\mu_{ne} = 0.0203$ MPa
Material A2		$\mu_{ne} = 0.1015$ MPa
Material A3		$\mu_{ne} = 0.1827$ MPa
Material B1	$\mu_{eq} = 0.406$ MPa	$\mu_{ne} = 0.0406$ MPa
Material B2		$\mu_{ne} = 0.2030$ MPa
Material B3		$\mu_{ne} = 0.3654$ MPa

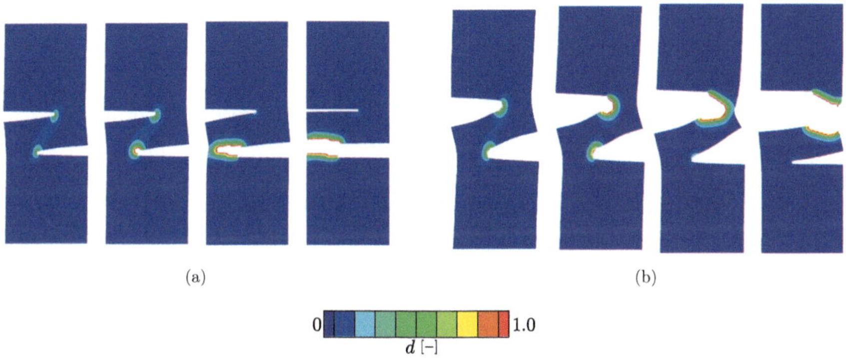

Fig. 7 Phase-field evolution using material A3 with (**a**) $\mathcal{G}_c = 2.67e^{-2}$ J/mm^2, (**b**) $\mathcal{G}_c = 2.67e^{-1}$ J/mm^2

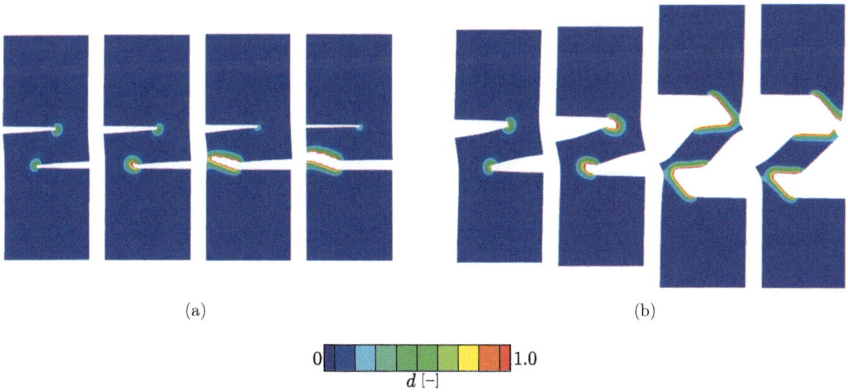

Fig. 8 Phase-field evolution using material B3 with (**a**) $\mathcal{G}_c = 2.67e^{-2}$ J/mm^2, (**b**) $\mathcal{G}_c = 2.67e^{-1}$ J/mm^2

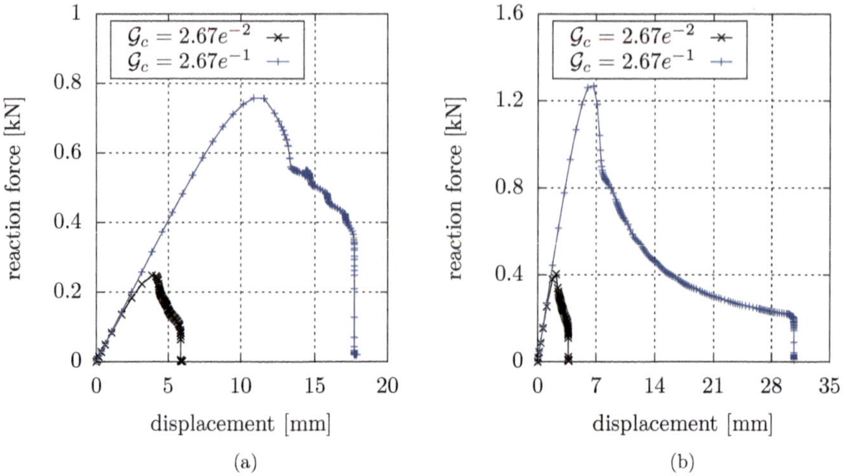

Fig. 9 Comparison of different $\mathcal{G}_c$-parameters in J/mm^2 for material A3 and B3

both increase with a larger fracture toughness value as well. Since the notches are non-aligned, the area between the two side notches undergoes a considerable rotation during the phase-field crack development. From the loading curve investigation of the case $\mathcal{G}_c = 2.67e^{-1}$ J/mm^2 in Fig. 9b, a phenomenological softening is observed after the load reaches the peak value. Nevertheless, as long as the rotation is stabilized, brittle fracture is obtained again.

In addition, the influence of rotation is also studied by setting three different notch lengths, namely, $a = [10, 30, 50]$ mm. The fracture toughness $\mathcal{G}_c$ with a larger value is chosen for material A3 and B3, respectively. The results are depicted in Figs. 10, 11. It clearly shows that the rotation and corresponding softening phenomenon reduces along with the notch length decreases. For the shorter initial notch, e.g., $a = 10, 30$ mm, the eventual crack merges from the prescribed two notches. The fracture behavior represents brittle failure from the loading-displacement curve evaluation.

Moreover, the shear modulus ratio between the equilibrium and non-equilibrium branches is also comprehensively studied. Using all material sets A1-3 and B1-3 with a larger value of $\mathcal{G}_c = 2.67e^{-1}$ J/mm^2, the simulation results based on the notch length $a = 50$ mm are shown in Figs. 12, 13. In general, an increase of μ_{ne} also increases the isochoric free energy at the non-equilibrium branch. Meanwhile, the peak load also increases but the maximum displacement at full fracture decreases. An interesting finding for material parameter set B3 is that the peak load is significantly larger than others and the maximum displacement for full fracture also increases. This is different from another investigation, since the phase-field crack evolves at both two notches until the upper one leads to final rupture. Hence, the central area of the middle of specimen largely rotates, which naturally delays the full degradation in stiffness of the entire structure.

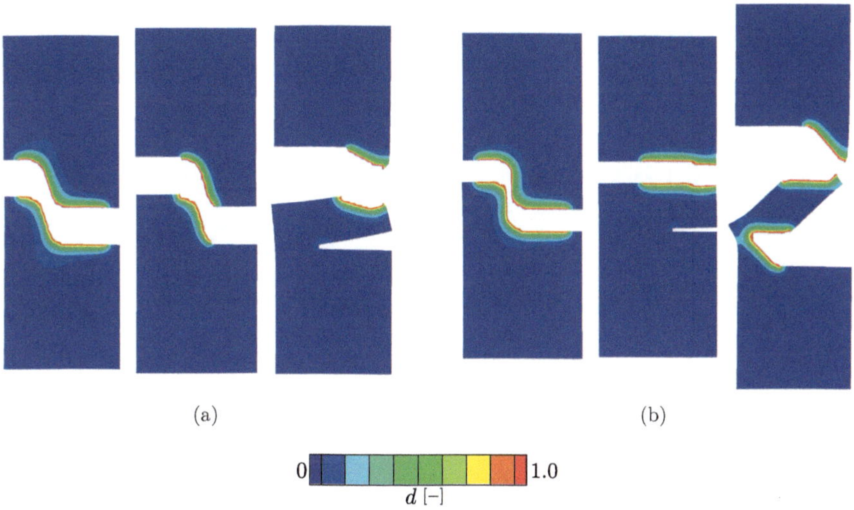

Fig. 10 Phase-field evolution for $a = [10, 30, 50]$ mm using materials A3 and B3 with $\mathcal{G}_c = 2.67e^{-1}$ J/mm^2

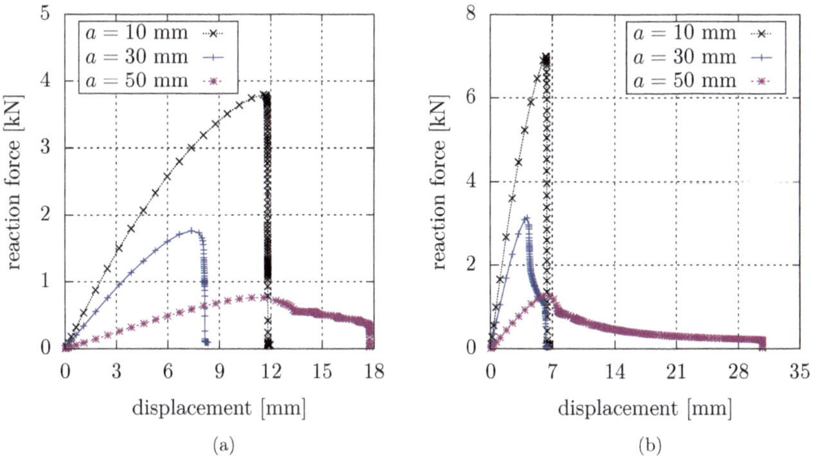

Fig. 11 Analysis of notch-test $a = [10, 30, 50]$ mm using materials A3 and B3 with $\mathcal{G}_c = 2.67e^{-1}$ J/mm^2

The last investigation of this example is the temperature evolution as the presented model is constituted as a thermo-mechanically coupled formulation. The temperature naturally increases based on the dissipation rate, which comes from the material's inelastic response. The simulation results according to the material sets A3

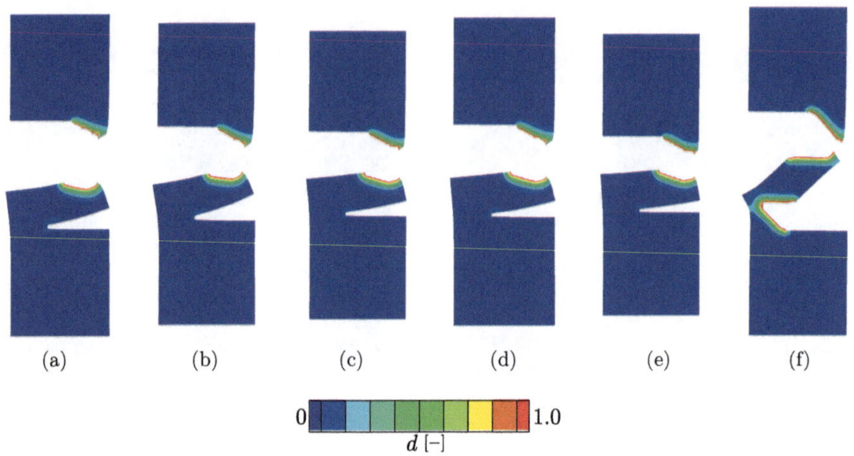

Fig. 12 Phase-field evolution for material sets A and B with $\mathcal{G}_c = 2.67e^{-1}$ J/mm^2

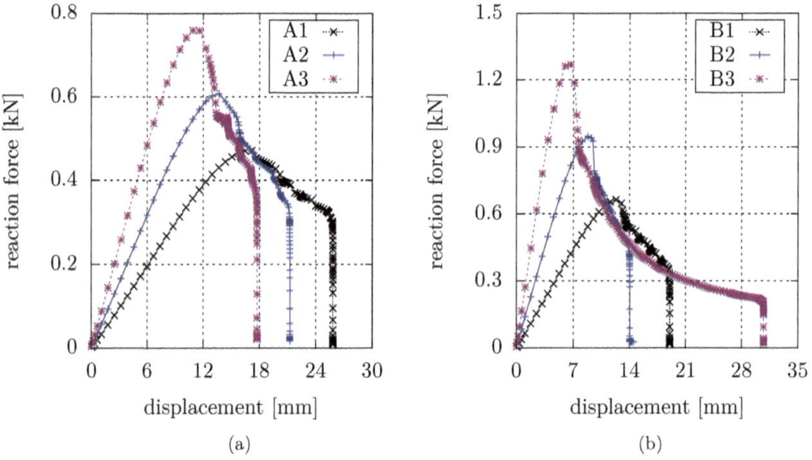

Fig. 13 Analysis of μ_{ne} for material sets A and B with $\mathcal{G}_c = 2.67e^{-1}$ J/mm^2

and B3 are illustrated in Fig. 14. This temperature increment is relatively small with a maximum value of 0.5 °C for material A3 and 0.95 °C for material B3. The local maximum deformation generally occurs at the vicinity of crack tip region, where the inelastic dissipation rate concentrates at. Therewith, the contour plot of temperature increment distribution also shows the peak values at the crack tips, similar to the results shown by Alfat in [58] but with materials at small strains.

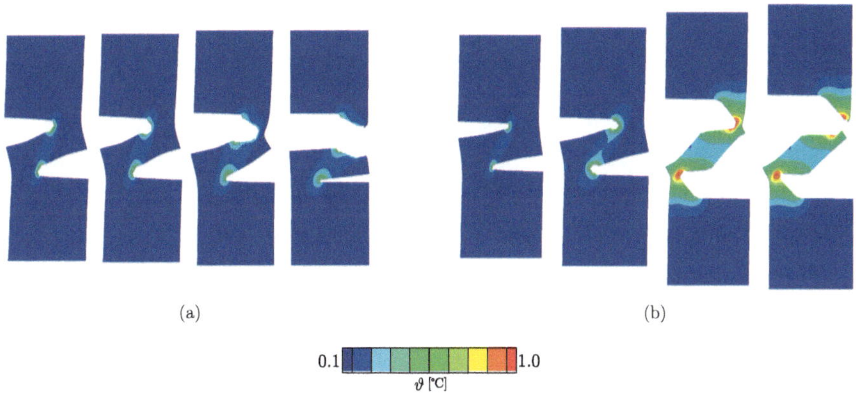

Fig. 14 Temperature evolution with $a = 50$ mm and $\mathcal{G}_c = 2.67e^{-1}$ J/mm^2 for (**a**) material A3 and (**b**) material B3

4.3 *Structural Failure with Temperature Loading*

This numerical study focuses on the material failure by both excessive temperature loading as well as temperature/displacement loading. The former one considers a cooling process that leads to a shrinkage deformation, which eventually yields the phase-field crack initiation and propagation till complete rupture. The latter one takes an isothermal temperature as a pre-loading condition into account. Then, a displacement-controlled loading is subjected to the specimen, resulting in a tensile fracture. These two fictitious tests are performed aiming at mimicking several potential engineering phenomena in a general sense. The geometrical setup is still based on the example in Sect. 4.2, with initial crack length $a = 50$ mm.

Considering that a nozzle continuously spraying liquid nitrogen on the central region of specimen, the temperature in that part certainly decreases. Therefore, in the finite element modeling, the cooling down process in the central region of the structure can be simulated by applying Dirichlet boundary conditions to the temperature degrees of freedom. The upper and lower edges are fully clamped. Using the parameter $\mathcal{G}_c = 2.67e^{-2}$ J/mm^2, $\mu_{eq}/\mu_{ne} = 0.203/0.1827$ MPa, $\alpha_t = 6.0e^{-4}$ (other material parameters are from Table 3), the results of phase-field evolution and temperature distribution are shown in Fig. 15, respectively. Since the temperature cooling process is applied relative fast, the temperature gradient still can be observed in Fig. 15b. Nevertheless, if the temperature loading is slow or the conductivity coefficient is relatively large, the temperature gradient cannot be observed in a straightforward manner. When the temperature reaches approximately -66 °C in the central region, the structure reaches a fully ruptured state.

The second test considers a mixed temperature and displacement loading pattern. First, the whole specimen is cooled down till -30 °C. The shrink deformation can drive the phase-field evolution at the notch tip, nevertheless, the material status

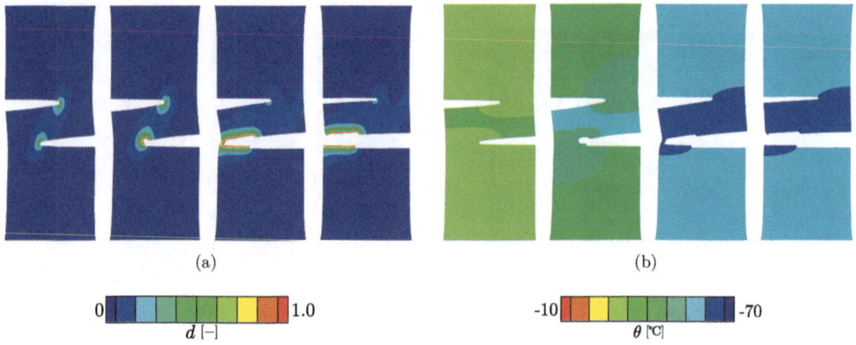

Fig. 15 Fracture only induced by excessive temperature cooling: Phase-field evolution and temperature distribution

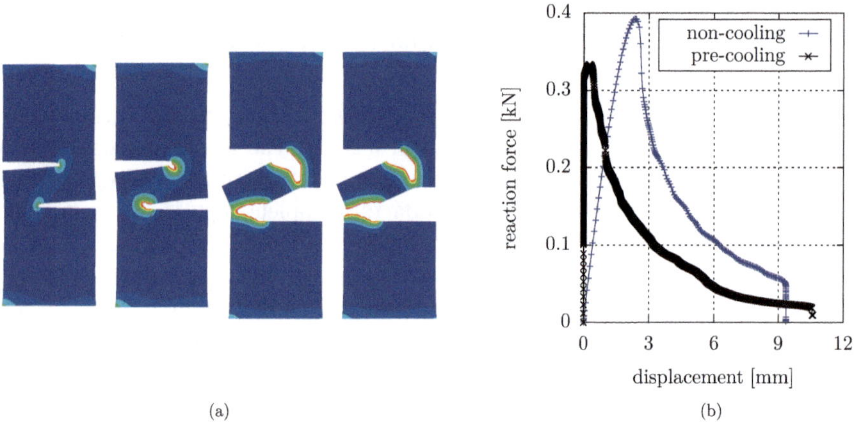

Fig. 16 Fracture induced by pre-cooling and displacement loading: phase-field evolution and comparison of loading states

therein is far away from fully damaged state, i.e., $d \ll 1$. In the sequel, tensile displacement load is applied, which leads to further material deformation till the phase-field crack initiates and propagates. Using the parameters $\mathcal{G}_c = 2.67e^{-2}$ J/mm^2, $\mu_{eq}/\mu_{ne} = 0.406/0.3654$ MPa, $\alpha_t = 6.0e^{-4}$ (other material parameters are from Table 3), the phase-field evolution is depicted in Fig. 16. As a reference, pure tensile loading failure is also considered, i.e., there is no cooling process before applying the displacement-controlled load till tensile rupture. The results of reaction force versus displacement are compared in Fig. 16. For the pre-cooling case, the temperature loading leads to thermal shrinkage. Therefore, the reaction force increases while the displacement remains the same. Furthermore, the peak value of the reaction force for the pre-cooling case is smaller than that non-cooling case.

5 Summary

This work at hand formulates a thermo-viscoelastic phase-field model for fracture simulation of polymeric materials. The finite viscoelastic constitutive model of the material is based on multiplicative deformation gradient kinematics [14, 47]. The driving force for fracture evolution in the phase-field model includes the pure elastic strain energy derived from both the equilibrium and the non-equilibrium branches of the rheological model. Unlike [41, 43], the dissipated energy is not taken into account to evolve fracture. The temperature evolution law is also presented, which eventually formulates a three-field finite element model. The governing equations for the mechanical deformation, temperature response, and phase-field evolution are consistently derived by means of thermodynamic analysis. It is implemented within the FE framework. The capability of the presented model is demonstrated for both displacement-controlled rate-dependent fracture and temperature-loading (cooling) induced fracture failure. Furthermore, for future research interests, comprehensive parametric studies and validations need be conducted comparing to related experimental evidences. Other inelastic properties, like plasticity or viscoplasticity, are also necessarily to be taken into account as well.

References

1. Miehe C, Keck J (2000) Superimposed finite elastic–viscoelastic–plastoelastic stress response with damage in filled rubbery polymers. experiments, modelling and algorithmic implementation. J Mech Phys Solids 48:323–365
2. Sullivan JL (1986) The relaxation and deformational properties of a carbon-black filled elastomer in biaxial tension. J Polym Sci B Polym Phys 24:161–173
3. Christensen RM (1982) Theory of viscoelasticity: An introduction 2nd edn. Academic, New York
4. Govindjee S, Simo JC (1992) Mullins' effect and the strain amplitude dependence of the storage modulus. Int J Solids Struct 29:1737–1751
5. Govindjee S, Simo JC (1993) Coupled stress-diffusion: Case ii. J Mech Phys Solids 41:863–887
6. Holzapfel GA (1996) On large strain viscoelasticity: continuum formulation and finite element applications to elastomeric structures. Int J Numer Methods Eng 39:3903–3926
7. Kaliske M, Rothert H (1997) Formulation and implementation of three-dimensional viscoelasticity at small and finite strains. Comput Mech 19:228–239
8. Simo JC (1987) On a fully three-dimensional finite-strain viscoelastic damage model: formulation and computational aspects. Comput Methods Appl Mech Eng 60:153–173
9. Simo JC, Hughes TJR (1998) Computational inelasticity. Springer, New York
10. Bergström JS, Boyce MC (1998) Constitutive modeling of the large strain time-dependent behavior of elastomers. J Mech Phys Solids 46:931–954

11. Dal H, Kaliske M (2009) Bergström–boyce model for nonlinear finite rubber viscoelasticity: theoretical aspects and algorithmic treatment for the FE method. Comput Mech 44:809–823
12. Simo JC, Miehe C (1992) Associative coupled thermoplasticity at finite strains: Formulation, numerical analysis and implementation. Comput Methods Appl Mech Eng 98:41–104
13. Simo JC (1992) Algorithms for static and dynamic multiplicative plasticity that preserve the classical return mapping schemes of the infinitesimal theory. Comput Methods Appl Mech Eng 99:61–112
14. Reese S, Govindjee S (1998) A theory of finite viscoelasticity and numerical aspects. Int J Solids Struct 35:3455–3482
15. Miehe C (1992) Kanonische modelle multiplikativer elasto-plastizität. Thermodynamische Formulierung und numerische Implementation. Habilitation Thesis. Universität Hannover
16. Fleischhauer R, Platen J, Kato J, Terada K, Kaliske M (2022) A finite anisotropic thermo-elasto-plastic modeling approach to additive manufactured specimens. Eng Comput (under review)
17. Zerbe P, Schneider B, Moosbrugger E, Kaliske M (2017) A viscoelastic-viscoplastic-damage model for creep and recovery of a semicrystalline thermoplastic. Int J Solids Struct 110:340–350
18. Miehe C (1988) Zur numerischen behandlung thermomechanischer Prozesse. PhD Thesis, Universität Hannover
19. Gent AN, Lindley PB (1959) Internal rupture of bonded rubber cylinders in tension. Proc Roy Soc Lond. A Math Phys Sci 249:195–205
20. Gent AN, Park B (1984) Failure processes in elastomers at or near a rigid spherical inclusion. J Mater Sci 19:1947–1956
21. Ball JM (1982) Discontinuous equilibrium solutions and cavitation in nonlinear elasticity. Philos Trans Roy Soc Lond. A Math Phys Sci 306:557–611
22. Euchler E, Bernhardt R, Wilde F, Schneider K, Heinrich G, Tada T, Wießner S, Stommel M (2021a) First-time investigations on cavitation in rubber parts subjected to constrained tension using in situ synchrotron x-ray microtomography (srμct). Adv Eng Mater 23:2001347
23. Euchler E, Bernhardt R, Schneider K, Heinrich G, Tada T, Wießner S, Stommel M (2021b) Cavitation in rubber vulcanizates subjected to constrained tensile deformation. Fatigue Crack Growth Rubber Mater Exp Modell, 203–224
24. Knauss WG (2015) A review of fracture in viscoelastic materials. Int J Fract 196:99–146
25. Cooke ML, Pollard DD (1996) Fracture propagation paths under mixed mode loading within rectangular blocks of polymethyl methacrylate. J Geophys Res Solid Earth 101:3387–3400
26. Gol'dstein RV, Salganik RL (1974) Brittle fracture of solids with arbitrary cracks. Int J Fract 10: 507–523
27. Lazarus V, Buchholz FG, Fulland M, Wiebesiek J (2008) Comparison of predictions by mode ii or mode iii criteria on crack front twisting in three or four point bending experiments. Int J Fract 153:141–151
28. Pons AJ, Karma A (2010) Helical crack-front instability in mixed-mode fracture. Nature 464: 85–89
29. Rice JR (1968) A path independent integral and the approximate analysis of strain concentration by notches and cracks. J Appl Mech 35:379–386
30. Hocine NA, Abdelaziz MN, Mesmacque G (1998) Experimental and numerical investigation on single specimen methods of determination of J in rubber materials. Int J Fract 94:321–338
31. Hocine NA, Abdelaziz MN, Imad A (2002) Fracture problems of rubbers: J-integral estimation based upon η factors and an investigation on the strain energy density distribution as a local criterion. Int J Fract 117:1–23
32. Kroon M (2011) Steady-state crack growth in rubber-like solids. Int J Fract 169:49–60
33. Kroon M (2014) Energy release rates in rubber during dynamic crack propagation. Int J Solids Struct 51:4419–4426
34. Schapery RA (1984) Correspondence principles and a generalized j integral for large deformation and fracture analysis of viscoelastic media. Int J Fract 25:195–223

35. Özenç K, Kaliske M (2014) An implicit adaptive node-splitting algorithm to assess the failure mechanism of inelastic elastomeric continua. Int J Numer Methods Eng 100:669–688
36. Özenç K (2016) Approaches to model failure of materials by configurational mechanics: theory and numerics. PhD Thesis, Technische Universität Dresden
37. Geißler G, Kaliske M, Nase M, Grellmann W (2007) Peel process simulation of sealed polymeric film computational modelling of experimental results. Eng Comput 24:586–607
38. Miehe C, Welschinger F, Hofacker M (2010) Thermodynamically consistent phase-field models of fracture: Variational principles and multi-field fe implementations. Int J Numer Methods Eng 83:1273–1311
39. Kuhn C, Müller R (2010) A continuum phase field model for fracture. Eng Fract Mech 77: 3625–3634
40. Yin B, Steinke C, Kaliske M (2020) Formulation and implementation of strain rate-dependent fracture toughness in context of the phase-field method. Int J Numer Methods Eng 121:233–255
41. Shen R, Waisman H, Guo L (2019) Fracture of viscoelastic solids modeled with a modified phase field method. Comput Methods Appl Mech Eng 346:862–890
42. Schänzel LM (2015) Phase field modeling of fracture in rubbery and glassy polymers at finite thermo-viscoelastic deformations. PhD Thesis, Universität Stuttgart
43. Loew PJ, Peters B, Beex LAA (2019) Rate-dependent phase-field damage modeling of rubber and its experimental parameter identification. J Mech Phys Solids 127:266–294
44. Lee EH (1969) Elastic-plastic deformation at finite strains. J Appl Mech 36:1–6
45. Moran B, Ortiz M, Shih CF (1990) Formulation of implicit finite element methods for multiplicative finite deformation plasticity. Int J Numer Methods Eng 29:483–514
46. Lubarda VA (2004) Constitutive theories based on the multiplicative decomposition of deformation gradient: Thermoelasticity, elastoplasticity, and biomechanics. Appl Mech Rev 57:95–108
47. Yin B, Kaliske M (2020) Fracture simulation of viscoelastic polymers by the phase-field method. Comput Mech 65:293–309
48. Yin B (2022) Phase-field fracture description on elastic and inelastic materials at finite strains. PhD Thesis, Technische Universität Dresden
49. Simo JC (1998) Numerical analysis and simulation of plasticity. Handbook Numer Anal 6:183–499
50. Holzapfel GA (2000) Nonlinear solid mechanics: a continuum approach for engineering. Wiley, Chichester
51. Miehe C (1998) A constitutive frame of elastoplasticity at large strains based on the notion of a plastic metric. Int J Solids Struct 35:3859–3897
52. Weber G, Anand L (1990) Finite deformation constitutive equations and a time integration procedure for isotropic, hyperelastic-viscoplastic solids. Comput Methods Appl Mech Eng 79: 173–202
53. Borden MJ, Hughes TJR, Landis CM, Verhoosel CV (2014) A higher-order phase-field model for brittle fracture: Formulation and analysis within the isogeometric analysis framework. Comput Methods Appl Mech Eng 273:100–118
54. Hofacker M (2013) A thermodynamically consistent phase field approach to fracture. PhD Thesis, Universität Stuttgart
55. Zhang X, Vignes C, Sloan SW, Sheng D (2017) Numerical evaluation of the phase-field model for brittle fracture with emphasis on the length scale. Comput Mech 59:737–752
56. Mandal TK, Nguyen VP, Wu JY (2019) Length scale and mesh bias sensitivity of phase-field models for brittle and cohesive fracture. Eng Fract Mech 217:106532
57. Pham K, Amor H, Marigo JJ, Maurini C (2011) Gradient damage models and their use to approximate brittle fracture. Int J Damage Mech 20:618–652
58. Alfat S (2023) On phase field approach for thermal fracturing in thermoviscoelasticity solids and its application for studying thermal response due to crack growth: Part I. Kelvin-Voigt Type.

Adv Polym Sci (2024) 294: 125–144
https://doi.org/10.1007/12_2024_171

Published online: 31 March 2024

Effect of Carbon Black on Heat Build-up and Energy Dissipation in Rubber Materials

W. A. Kyei-Manu, L. B. Tunnicliffe, C. R. Herd, K. Akutagawa, O. Kratina, R. Stoček, and J. J. C. Busfield

Contents

Abstract The heat build-up and energy dissipation of carbon black reinforced natural rubber compounds are studied using a rebound resilience tester, Goodrich

---

Supplementary Information The online version contains supplementary material available at https://doi.org/10.1007/12_2024_171.

W. A. Kyei-Manu, K. Akutagawa, and J. J. C. Busfield (✉)
Queen Mary University of London, School of Engineering and Materials Science, London, UK
e-mail: j.busfield@qmul.ac.uk

L. B. Tunnicliffe and C. R. Herd
Birla Carbon, Marietta, Georgia, USA

O. Kratina and R. Stoček
Centre for Polymer Systems, Tomas Bata University in Zlín, Zlín, Czech Republic

flexometer and a Coesfeld heat build-up analyser. Natural rubber compounds containing eight different carbon blacks at 50 parts per hundred (phr) were studied. The carbon blacks varied widely in their structure and surface area properties allowing quantitative correlations to the heat build-up and rebound resilience measurements to be extracted from their colloidal properties. The results show the nature of deformation dictates different levels of influence of the carbon black properties. Carbon black surface area dominates in influencing the heat build-up and energy dissipation in the Goodrich flexometer, Coesfeld heat build-up analyser and rebound resilience measurements. The Goodrich flexometer heat build-up measurement, however, is the only experimental measurement where carbon black structure plays a significant role in determining the heat build-up. The differences in the levels of influence of carbon black colloidal properties can be understood when the deformation index concept is applied to the experimental results. Goodrich flexometer measurements are predominantly strain-controlled measurements while rebound resilience and Coesfeld heat build-up are a complex combination of strain and energy-controlled deformations. Understanding the type of deformation that dominates in various heat build-up and energy dissipation tests and how carbon black properties affect the results of the tests are of practical importance. It enables the heat build-up and energy dissipation tests that closely represent the type of deformation that will be predominant in field applications to be selected during compound development and characterization. It also gives an understanding of how carbon black properties can be leveraged to optimize desired heat build-up and energy dissipation properties depending on the type of deformation during compound development.

Keywords Carbon black · Coesfeld heat build-up analyser · Deformation index · Energy dissipation · Goodrich flexometer · Heat build-up · Hysteresis · Rebound resilience · Temperature rise

1 Introduction

In many engineering applications, rubber is subjected to dynamic loading conditions which lead to mechanical hysteresis [1] because of reversible and irreversible process [2]. Mechanical hysteresis most often causes heat generation [3] and the poor thermal conductivity of rubber leads to progressive heat build-up. Heat build-up has been shown to have detrimental effects on mechanical properties such as tensile strength, abrasion resistance, fatigue and fracture properties [4–6]. Heat build-up also leads to chemical degradation such as effects on crosslink system and oxidation which may cause scission or formation of additional crosslinks depending on the type of polymer [1]. In thick rubber components, heat build-up can generate volatile compounds resulting in an increase in the internal pressure and eventual blowout of the component [7].

Despite the detrimental effects of heat build-up, it is unavoidable in many engineering applications. For example, the operational temperatures of dump truck tires increase to between 80 and 110 °C [8, 9]. Various numerical and experimental studies have recently been performed to elucidate our understanding of heat build-up in operation. Understanding heat build-up in rubber can be used to explain observations during tire deformations such as abrasion [10] and cut and chip [11]. Heat build-up in solid tires has also been studied using numerical [12] and experimental [12, 13] techniques. Liu et al. [14] provide an in-depth review of recent heat generation papers and attribute heat generation to viscoelasticity and friction from the polymer chains. Luo et al. [15] developed an analytical method to calculate heat build-up that considers the effect of rising temperature on the loss modulus of rubber. They validated this method using sinusoidal compression measurements of carbon black (CB) reinforced cylindrical samples on a dynamic mechanical analyser (DMA) at different frequencies and low strains. Esmaeli et al. [16] used a finite element analysis model to study the heat build-up in rubber during DMA testing and determined the best shape of specimen to minimize temperature rise during testing. Schieppati et al. [17] similarly studied the temperature build-up for two specimen geometries; a 3D dumbbell specimen used for crack nucleation and a planar strain specimen used for fatigue crack growth tests. Their experiments contributed to the differentiation between failures related to fatigue and the ones related to local temperature increase. Dedova et al. [18] studied energy dissipation under different states of deformation (equibiaxial, asymmetrical biaxial, pure shear and uniaxial) and discovered a linear dependence of log crack rate with the dissipated energy. He et al. [19] predicted the steady-state temperature distribution and rolling resistance using three different finite element models and ranked them according to their accuracy and average predicted errors.

Although the influence of CB on heat build-up is fairly well known, there has at yet not been a quantitative study mapping carbon black's morphological properties to heat build-up. This work bridges this gap with experimental studies of different types of CB in natural rubber. Conventional experimental techniques are performed including dynamic strain amplitude sweeps, rebound resilience and Goodrich flexometer measurements which measure energy dissipation and heat build-up in rubber composites. The results are quantitatively correlated to determine the influence of CB structure and surface area. Heat build-up measurements are also performed using a novel heat build-up machine that closely replicates in-service conditions [20, 21].

The paper also determines the predominant type of deformation (strain, energy or stress) control in the different experimental techniques using the deformation index concept. The deformation index concept was developed over 30 years ago by Futamura [22] for determining the relevant dynamic viscoelastic property for the hysteretic energy response of deformed elastomers. Since then, the method has been further refined and applied to rolling resistance of tires [23], steady state and transient thermal analysis of pneumatic [23] and non-pneumatic [24] tires, impact [25] and fatigue [26]. The deformation index is used here to compare the different heat build-up and energy dissipation techniques and determine which type of deformation is dominant for each experimental technique. Different sections of a tire experience different types of deformation. For example, while the tread section of a radial medium traction (RMT) tire generally shows equal stress deformation, the side wall shows equal strain deformation and other parts show an equal energy deformation (mixed equal stress and equal strain) [27]. Understanding the predominant deformation type in commonly used heat build-up and energy dissipation tests is therefore important to determine which type of test is most appropriate for translating heat build-up measurements in the lab to heat build-up of engineering applications in the field based on the type of deformation they are likely to face.

2 Materials and Methods

2.1 Materials

Eight Standard Malaysian Rubber (SMR) CV 60 natural rubber (NR) compounds each reinforced with a different grade of CB at 50 parts per hundred (phr) loading were studied. An unreinforced equivalent was included for reference. The CBs were selected to cover a wide colloidal space of varying structure and surface area relevant for rubber applications, as shown by the coloured symbols in Fig. 1. The most basic building block of CB is the primary particle with diameters ranging from ~200 nm to ~5 nm. Primary particles assemble and fuse together during production to form the CB aggregate. The complexity (i.e., the number and spatial arrangement) of the aggregates describes the structure of the CB. Particle size which generally has an

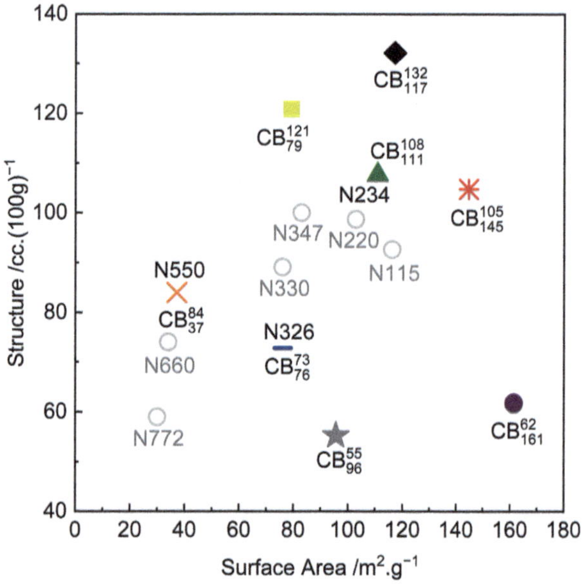

Fig. 1 Colloidal plot of the carbon black grades that were incorporated into the tested NR compounds

Table 1 Formulation of compounds

Component	Loading/Parts per hundred rubber (phr)	Manufacturer of component
NR-SMR CV-60	100	Herman Weber & Co.
Carbon black	50	Birla Carbon
Zinc oxide	5	Akrochem
Stearic acid	3	PMC Biogenix
Anti-ozonant/ Antioxidant	3	Americas International
Micro-wax	2	Strahl & Pitsch
Sulphur	2.5	R.E. Carroll
TBBS[a]-75	0.8	Akrochem

[a] N-Tertiarybutyl-2-benzothiazolesulfenamide, 75% assay

Table 2 CB naming convention adopted for the paper, CB structure and surface area

Grade Name per ASTM D1765	Carbon Black/ Compound Code	Structure (COAN)/cc. (100 g)$^{-1}$	Surface area (STSA)/m^2 g^{-1}
NA	Unfilled NR	NA	NA
NA	CB_{117}^{132}	132	117
NA	CB_{145}^{105}	105	145
NA	CB_{79}^{121}	121	79
N234	CB_{111}^{108}	108	111
N326	CB_{76}^{73}	73	76
NA	CB_{96}^{55}	55	96
NA	CB_{161}^{62}	62	161
N550	CB_{37}^{84}	84	37

inverse relationship with surface area per unit mass of CB can be deduced from measurements of surface area using gas adsorption techniques [28]. Structure can be measured from oil absorption tests [29, 30]. Figure 1 also includes some commonly used CB grades indicated in the grey symbols for reference. Table 1 shows the compound formulation used in this study.

Table 2 shows the structure and surface area of the different CB grades. For this work, a naming convention that refers to the structure (measured by compressed oil absorption number, COAN) and surface area (measured by statistical thickness surface area, STSA) of the CB is adopted. For example, N234 which has a structure of 108 cc. (100 g)$^{-1}$ and a surface area of 111 m^2 g^{-1} is referred to as CB_{111}^{108}. The compounds show very good dispersion of CB in all the rubber compounds based on previously published interferometric microscope (IFM) dispersion index (DI) values [31].

The compounds were prepared by Birla Carbon in Marietta, GA, USA using a 1.6 L capacity Banbury mixer. The T_{90} at 150 °C was determined using a moving die rheometer (MDR) from Alpha Technologies. Vulcanized sheets measuring

110 mm × 110 mm × ~2 mm were prepared at 150 °C for a time of T_{90} + 5 min for uniaxial tensile to break and dynamic strain sweep testing.

2.2 Tensile to Break Tests

Five dumbbell specimens for each compound were stretched until failure using a five station United Tensile Tester equipped with a 1 kN load cell at an extension rate of 500 mm/min (~strain rate of 0.19/s). ASTM D412 die C dumbbells were used with approximate gauge length, measurement width and thickness dimensions of 25 mm, 6 mm and 2 mm, respectively. The dumbbells were cut from the vulcanized sheets so that the longitudinal axis was in the rolling direction. The strain was measured accurately using travelling non-contact extensometers and all tests were performed at 21 ± 1 °C, 55% relative humidity and atmospheric pressure conditions.

2.3 Dynamic Strain Amplitude Sweep Tests

Dynamic strain sweeps between 0.1 and 62.5% single strain amplitude were performed using an ARES G2 torsional rheometer at 10 Hz with zero mean strain at 60 °C. The specimens used were cylinders measuring ~8 mm in diameter and ~2 mm in thickness. They were bonded to the rheometer parallel plates using Loctite 480 adhesive. The tests were performed in torsional shear with a compressive force of 100 g applied to the cylinders. The strain sweep tests were performed by pre-conditioning the specimen six times at the specific dynamic strain amplitude before collecting torque-time data from low to high strains.

2.4 Rebound Resilience Tests

Rebound resilience was measured at room temperature (22 °C) and 60 °C using a Zwick Roell 5109 rebound resilience tester according to ASTM D 1054-Standard Test Method for Rubber Property-Resilience Using a Rebound Pendulum. The specimens were buttons measuring ~25.4 mm (1″) in diameter and ~12.7 mm (0.5″) in thickness which were cured at 150 °C for a time of T_{90} + 10 min. The buttons were held in position at the bottom of the tester against a stainless-steel base. A free-falling hemispherical pendulum hammer with diameter ~15 mm was dropped from an angle of 90° and the extent to which the pendulum rebounds was used to calculate the percentage rebound resilience. For the tests done at 60 °C, the specimens were first pre-conditioned in an oven for 1 h at 60 °C, removed from the oven and immediately tested at room temperature.

2.5 *Heat Build-up Measurements*

2.5.1 Goodrich Flexometer Heat Build-up

Heat build-up was measured at room temperature using a BF Goodrich Flexometer at a constant dynamic peak-to-peak displacement amplitude of 4.45 mm (0.175″) for a period of 30 min according to ASTM D623-14 Standard Test Methods for Rubber Property—Heat Generation and Flexing Fatigue in Compression. A static compressive pre-load of ~200 N was applied prior to applying the cyclic displacement. The frequency of the cyclic displacement was 30 ± 0.2 Hz. The equilibrium temperature rise was measured using a radial thermocouple on the surface of the specimens. The specimens were cylinders each having a diameter of 17.8 ± 0.1 mm (0.700″ ± 0.005″) and a height of 25 ± 0.15 mm (1.000″ ± 0.010″) which were cured at 150 °C for a time of T_{90} + 10 min. Figure 2 shows a schematic of the loading conditions on the rubber specimen of the Goodrich Flexometer.

2.5.2 Coesfeld Heat Build-up

Heat build-up was measured using a Coesfeld Heat Build-up analyser. This is a novel machine manufactured by Coesfeld GmbH & Co. KG, Dortmund, Germany which characterizes the heat build-up inside and on the surface of rubber sample in-situ with the specimen subjected to multi-axial loading conditions [20, 21]. Figure 3 shows a schematic of the loading conditions of the specimen. The specimens were cylinders of diameter 20 mm and length 100 mm. During the moulding process, a cylindrical hole with diameter 5 mm and length 50 mm from the end to the centre of the specimens was made on one end using an insert. The specimens were cured at 150 °C for a time of T_{90} + 20 min.

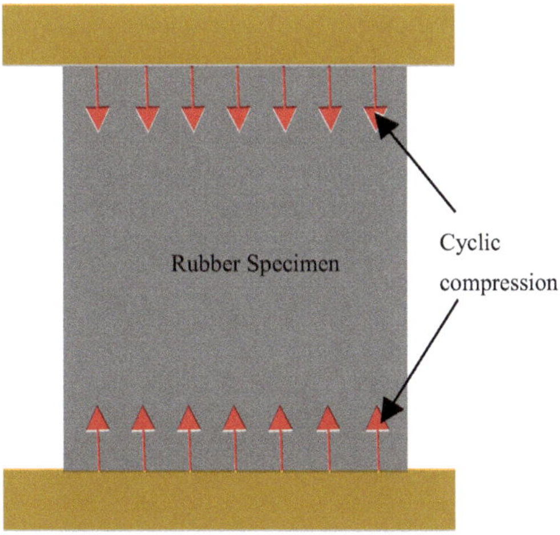

Fig. 2 Schematic of loading conditions of Goodrich Flexometer specimen

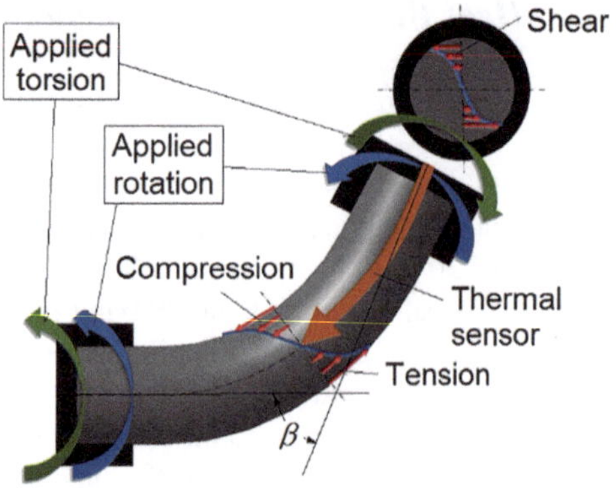

Fig. 3 Schematic of loading conditions of Coesfeld heat build-up specimen (Reproduced from Stoček et al. [20] and Kratina et al. [21])

The specimens were rotated at 1200 rpm for 30,000 cycles. The specimens were bent at an angle, β, of 50°. The internal temperature of the specimens was measured using a flexible thermocouple, which was inserted into the specimens through the cylindrical hole at the one end of the specimen. The external temperature on the surface of the specimens was measured using a Pyrometer Serie Metis MY84 IR sensor manufactured by Sensortherm GmbH, Germany.

2.6 *Deformation Index*

The deformation index concept was developed by Futamura to correlate relevant dynamic viscoelastic properties with hysteretic energy-loss processes in elastomers [22]. The concept is applied in this work to determine which dynamic viscoelastic property best predicts the measured heat build-up and rebound resilience of the compounds.

Elastomer properties such as strength, tear resistance, wear and rolling resistance have been shown to be closely related to and enhanced by internal energy dissipation and hysteretic energy loss in rubber materials. Correlating relevant dynamic visco-elastic properties with heat build-up and energy dissipation experiments is therefore of practical importance. In service, elastomer components are deformed in strain, stress or energy control, or more commonly in a complex combination of all three. The relevant dynamic property that best predicts the energy dissipation mechanism is determined by the type of deformation the elastomer component undergoes during operation.

To determine the deformation index and hence the type of deformation in the heat build-up and rebound resilience measurements, the relevant parameter is plotted as a function of the ratio of the loss modulus and complex modulus.

$$H = k_1 \frac{G''}{G^{*n}} + k_2 \quad (1)$$

where H = Heat build-up or rebound resilience percentage

k_1 and k_2 = constants

G'' = Dynamic loss modulus

G^* = Complex modulus which for elastomers well above their glass transition temperature is nearly equivalent to the storage modulus

n = Deformation index.

The value of the deformation index, n, which yields the best correlation of the measured, H, characterizes the nature of the deformation. The relevant parameter is plotted as a function of the ratio of $\frac{G''}{G^{*n}}$ using values of $0 \leq n \leq 2$ in increments of 0.1 to determine the value of n that provides the best goodness of fit (best R^2-value). Values of n = 0, 1 and 2 are special deformation cases where deformation and energy losses are under strain control, energy control and stress control, respectively. Hysteresis scales with loss modulus (n = 0), loss tangent (n = 1) and loss compliance (n = 2) for the cases of equal strain, equal energy and equal stress control, respectively. Non-integer values are also possible for the more common combination of deformation processes.

3 Results and Discussion

3.1 Tensile Stress-Strain Measurements

Table 3 summarizes some of the key tensile mechanical properties of the curves. The compounds show significant differences in the static mechanical properties. For example, there is a factor of 3 difference between the stress at 300% for CB^{62}_{161}

Table 3 Tensile to failure properties of tested compounds (Data reproduced from Kyei-Manu et al. [31])

Compound code	Tensile strength/MPa	Percent elongation at break	Stress at 100% strain/MPa	Stress at 300% strain/MPa
Unfilled NR	14.6	598	0.91	2.23
CB^{132}_{117}	26.6	402	5.41	19.89
CB^{121}_{79}	27.1	389	6.35	21.76
CB^{108}_{111}	27.3	491	4.04	16.69
CB^{105}_{145}	27.7	530	3.39	14.58
CB^{84}_{37}	23.1	461	4.10	15.38
CB^{73}_{76}	27.2	536	2.90	12.77
CB^{62}_{161}	30.8	679	1.90	7.82
CB^{55}_{96}	28.0	620	2.09	8.33

(which has the lowest stress at 300% of the CB reinforced compounds) and the stress at 300% for CB_{79}^{121} (which has the highest stress at 300% of the CB reinforced compounds). Table 3 shows that the modulus (stress at large strains) of the compounds is influenced by the structure of the CB. At equivalent volume fraction of CB, higher CB structure causes higher incremental modulus of the compounds due to strain amplification/overstraining of the rubber matrix through occlusion of a certain volume of the rubber by the CB aggregate structure [31].

3.2 Dynamic Strain Amplitude Sweeps

Figure 4a shows the storage modulus of the tested compounds as a function of strain amplitude. The storage modulus is generally representative of the level of reinforcement in the rubber material. The influence of CB properties on the storage modulus is dependent on the strain level [31, 32]. At low strains, CB surface area plays the most significant role in determining the level of reinforcement most likely due to particle networking. At higher strains though, the main reinforcing mechanism is that of strain amplification which is influenced by CB structure level. This is consistent with the earlier observation where the tensile stress-strain moduli were seen to be influenced by CB structure, which is measured at high strain.

Loss modulus, tan delta and loss compliance can be used to predict compounds' viscoelastic energy dissipation under strain, energy and stress-controlled deformations, respectively [22]. There is a complex relationship between CB structure and surface area on each of these predictive parameters dependent on the strain level [31].

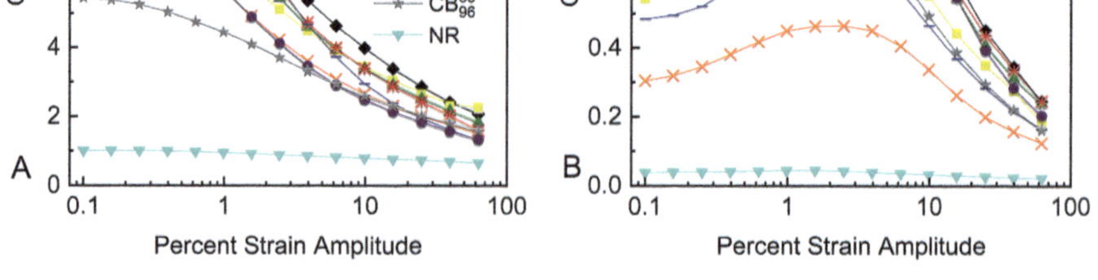

Fig. 4 Storage modulus and loss modulus as a function of percent strain amplitude (Reproduced from Kyei-Manu et al. [31])

3.3 *Rebound Resilience and Heat Build-up Measurements*

3.3.1 Coesfeld Heat Build-up Measurements

Figure 5a, b shows the temperature rise measured in the middle of the specimens using an inserted thermocouple and externally on the surface of the specimens using a Pyrometer Serie Metis MY84 IR sensor, respectively. The curves are an average of three repeats for each sample. The curves show a temperature rise with the number of cycles that reaches equilibrium after about 10,000 cycles for the internal, and after about 7500 cycles for the external temperature measurements. The rate of temperature rise is faster at the surface due to a higher deformation level at the surface compared to the internal section of the sample [21].

The equilibrium temperature rise measured by the internal thermocouple is generally higher compared to the external temperature rise. This can be attributed to heat accumulation inside the sample due to low thermal conductivity of rubber and hence higher temperature attained compared to the external surface where heat is dissipated by convection [21].

There is a difference of about a factor of two between the CB with lowest temperature rise $\left(CB_{37}^{84}\right)$ and the CB with the highest temperature rise $\left(CB_{145}^{105}\right)$.

To determine the equilibrium temperature rise of the samples, the curves were fitted in Origin 2022 using an asymptotic function of the form

$$y = a - bc^x \tag{2}$$

where y is the temperature rise at a specific cycle number, x, a is the equilibrium temperature rise and b and c are constants. The results of the fit are shown in the Supplementary Tables S1 and S2 with all the samples showing very good fits. The lowest R-squared values are 0.993 and 0.984 for the internal and external

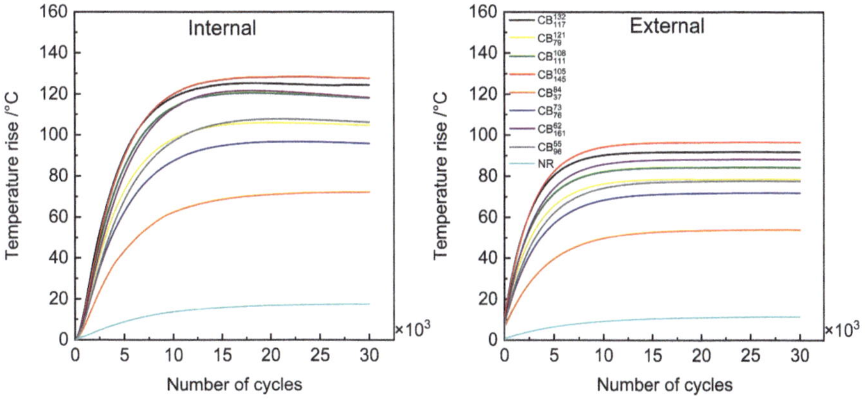

Fig. 5 Coesfeld heat build-up measured internally inside the specimen using a thermocouple inserted into the specimen (left) and externally on the surface of the specimen using a Pyrometer Serie Metis MY84 IR sensor (right)

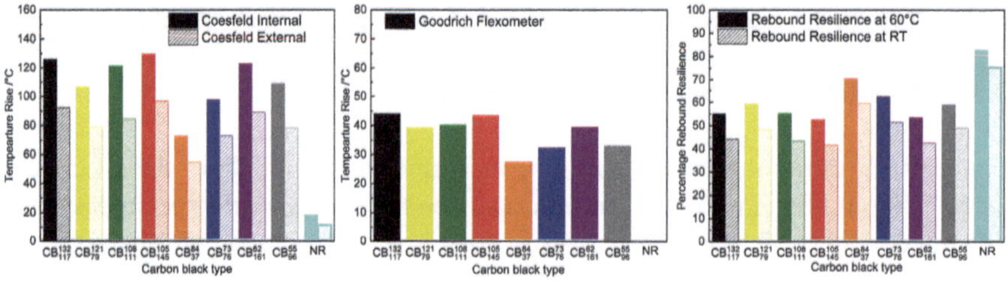

Fig. 6 Summary of Coesfeld heat build-up (left), Goodrich flexometer heat build-up (middle) and rebound resilience measurements (right). The y-axis on the Goodrich flexometer scales to 80 °C while the y-axis on the Coesfeld heat build-up scales to160 °C

temperature rise fits, respectively. The equilibrium internal and external temperature rise derived from the fitting are plotted in Fig. 6 with the results of the Goodrich Flexometer heat build-up measurements and the rebound resilience measurements.

3.3.2 Carbon Black Effect on Heat Build-up and Rebound Resilience

Figure 6 and Table S3 in the supplementary information summarize the rebound resilience and heat build-up measurements of the tested CBs. The rebound resilience percent is inversely proportional to the hysteretic energy loss, that is the higher the rebound percentage, the lower the hysteretic energy loss/dissipation. From the measurements, the samples tested at elevated temperature generally dissipate less energy as the polymer matrix becomes more elastic as it moves further from the glass transition temperature. This reduction in energy dissipation may also be due to decreased polymer–filler and polymer–polymer interaction due to more free volume in the polymer. At elevated temperatures, the filler networking is also reduced leading to lower Payne effect.

The heat build-up values are directly proportional to the energy dissipation. The Coesfeld heat build-up measurements show larger differentiation between the compound types compared to the Goodrich flexometer measurements. Even though the actual values differ, the general trends in the heat build-up behaviour of the compounds appear to be similar.

To analyse the effect CB properties play on heat build-up and rebound resilience, multiple regression analyses were performed for each measurement compared to structure and surface area. The results of the regression analysis including *p*-values and coefficients are shown in supplementary Table S4. Table 4 summarizes the effect of increasing CB structure and surface area on the heat build-up and rebound measurements based on the regression analysis.

From the regression analysis on the experimental results, CB surface area leads to higher energy dissipation and heat build-up. CB surface area is negatively correlated with rebound percentage, that is, the higher the surface area, the lower the rebound percentage (higher energy dissipation). CB surface area is positively correlated with

Table 4 Effect of CB properties on heat build-up and rebound resilience measurements

Parameter	Experimental technique	Increasing CB structure	Increasing CB surface area
Percentage rebound resilience	Room temperature (22 °C)	No effect	Decreases
	Elevated temperature (60 °C)	No effect	Decreases
Heat build-up	Goodrich flexometer temperature rise	Increases (Strong)	Increases (Strong)
	Coesfeld internal temperature rise	No effect	Increases
	Coesfeld external temperature rise	Increases (Weak)	Increases

the heat build-up, that is, the higher the surface area, the higher the heat build-up in the rubber.

CB structure plays a significant influence in the Goodrich flexometer temperature rise and has minimal influence on the external temperature measurements of the Coesfeld heat build-up measurements. This gives some insight into the type of deformation predominant in the specimen in these tests. Previous results by Mars [26] and Futamura [27] showed that the level of deformation affects the level of heat build-up in rubber materials. High structure CB enhances the stress level of the compound at a given deformation and therefore increases the absolute amount of energy dissipated. The strong influence of CB structure on the Goodrich flexometer measurements suggests deforming the samples to equal strain levels in the Goodrich flexometer test impacts the test results. The deformation index concept which was introduced in Sect. 2.6 is applied to the heat build-up and rebound resilience measurements in the next section to determine if the measurements are strain, energy or stress controlled or a complex combination of all three.

3.4 *Deformation Index Applied to Heat Build-up and Rebound*

To determine the type of deformation predominant in the heat build-up and rebound resilience measurements, the deformation index concept is applied. The dynamic mechanical parameters of the strain sweep test are used to compute the deformation index values. To apply the deformation index concept, the loss modulus and loss compliance of the compounds should be independent of each other to make it mathematically and experimentally feasible [22] and allow discrimination between strain and stress-controlled deformations. Supplementary table S5 shows the R-squared values of the correlation between loss modulus and loss compliance at each specific strain level for the tested compounds. There is minimal correlation between the loss compliance and loss modulus at strain levels, $\varepsilon \geq 2.5\%$ single strain amplitude meaning that dynamic mechanical values beyond this strain level can be

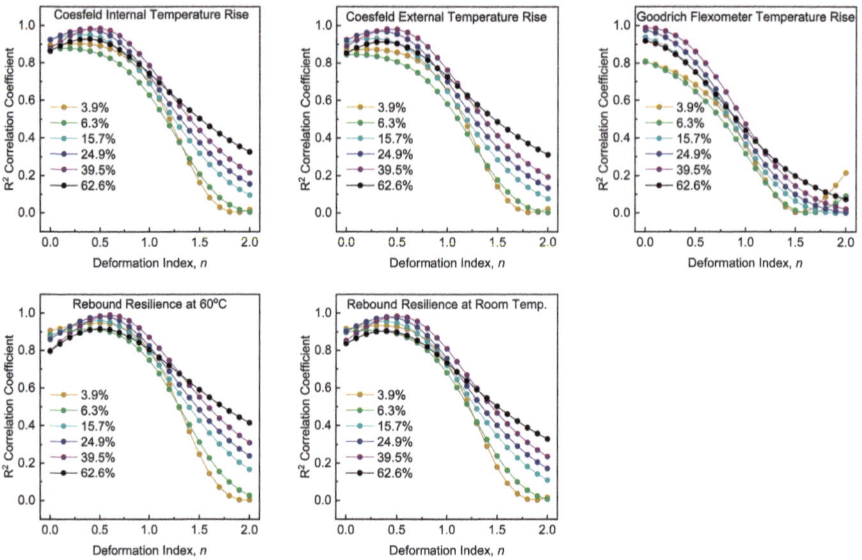

Fig. 7 Correlation coefficient as a function of the deformation index values, n, for each of the heat build-up and rebound resilience measurements at various strain levels

used. The appropriate strain level also needs to be used to determine the deformation index value. For this work, the dynamic mechanical properties at strain levels, $\varepsilon \geq 2.5\%$ are plotted for each experimental measurement and the strain level which provides the highest correlation is chosen as the appropriate strain level. Figure 7 shows the correlation coefficient between the heat build-up and rebound resilience as a function of deformation index at various dynamic strain levels. The dynamic strain levels which generally give the highest correlation coefficient ($R^2 \geq 0.98$) are at 24.5% and 39.5% single strain amplitude. This would suggest the specimens in these measurements deform to similar strain levels in the heat build-up and rebound resilience measurements.

Table 5 summarizes the deformation index values that give the best fit at these strain levels for the heat build-up and rebound resilience tests.

The maximum correlation coefficient value of the Goodrich flexometer temperature rise measurements occurs at a deformation index value of $n = 0$ which corresponds to an equal strain/strain-controlled deformation. The deformation index values are 0.5 and 0.6 for rebound resilience measurements at room and elevated temperatures, respectively, while the deformation index values of the Coesfeld internal temperature rise and external temperature rise are 0.5 and 0.4, respectively. These deformation index values suggest the rebound resilience and Coesfeld heat build-up measurements are a complex combination of strain and energy control deformations. The heat build-up and rebound resilience values are plotted against the maximum loss modulus and maximum tan delta values to determine which is the more predictive of the hysteresis and energy dissipation.

Table 5 Deformation index values, *n*, of rebound resilience and heat build-up measurements

Experimental measurement		Value of *n* that gives best fit to experimental data using viscoelastic parameters measured at 24.5% and 39.5% single strain amplitude (Deformation Index)
Rebound resilience	Room temperature (22 °C)	0.5
	Elevated temperature (60 °C)	0.6
Heat Build-up	Goodrich flexometer temperature rise	0
	Coesfeld internal temperature rise	0.5
	Coesfeld external temperature rise	0.4

Table 6 Correlation of maximum loss modulus and maximum tan delta values with rebound resilience and heat build-up measurements

Experimental measurement		Correlation coefficient	
		Maximum loss modulus	Maximum tan delta
Rebound resilience	Room temperature (22 °C)	0.95	0.69
	Elevated temperature (60 °C)	0.94	0.75
Heat build-up	Goodrich flexometer temperature rise	0.86	0.32
	Coesfeld internal temperature rise	0.98	0.63
	Coesfeld external temperature rise	0.81	0.63

This data is shown in Supplementary Figs. 1, 2, 3, 4 and 5 in the supplementary information. Table 6 summarizes the correlation factor (R^2 value) between the heat build-up/rebound resilience measurements and the maximum loss modulus and maximum tan delta values.

From the table, the maximum loss modulus is the best predictor of the heat build-up and rebound resilience measurements. Tan delta provides some correlation with the rebound resilience and Coesfeld heat build-up measurements but not the Goodrich flexometer heat build-up measurements. This suggests the Goodrich flexometer measurements is purely a strain-controlled deformation. The rebound resilience and Coesfeld measurements are, however, a complex combination of strain and energy-controlled deformation, with the strain-controlled deformation having a greater influence.

The above observations can be rationalized by considering the conditions under which each of the tests is conducted. Goodrich flexometer experiments are conducted by applying a compressive oscillating deflection force to the specimen thus making it a predominantly strain-controlled experiment. The novel Coesfeld heat build-up analyser puts the specimen under a multi-axial loading of tension, compression and torsion. This puts the sample under an equal energy deformation (mix of equal stress and equal strain deformations). However, the level of deformation in the specimen is influenced by the compression and tension in the specimen and hence the domination of strain control. The deformation index value from the external temperature measurements ($n = 0.4$) is closer to 0 (equal strain deformation) compared to the deformation index value obtained from the internal temperature measurements ($n = 0.5$). This is most likely because the specimen is deformed more on the surface and hence the level of deformation/strain plays a more significant role.

Rebound resilience measurements are generally considered to be equal energy deformation measurements, and thus a deformation index value of 1 is anticipated. However, Futamura [22] and Ransom et al. [25] determined deformation index values of 0.8 and 0.7, respectively, from their rebound resilience measurements. In Ransom et al.'s experiments, a steel projectile was dropped from a fixed height onto a rubber disc. FEA analysis of Ransom et al.'s results showed that at the vicinity of impact, the deformation index value of the rubber specimen varied strongly with position. FEA analysis showed if the steel projectile's cross-sectional area was significantly increased to make it flat, then the compressed region would approach the shape of a rubber disk. The rubber disk would become compressed more uniformly and give a net deformation index value closer to 1 (equal energy deformation). A similar explanation can rationalize why the rebound resilience measurements are not purely equal energy deformation but instead a complex mix of equal strain and equal energy. In the rebound resilience measurements, a free-falling pendulum steel hammer was dropped from an angle of 90° to hit the rubber specimen. Also, it is worth highlighting the rubber specimen was constrained with a metallic substrate along the back face. At the point of contact, the energy from the pendulum is split. Some of the energy goes into deforming the rubber specimen and some deforms the much more rigid substrate. The rigid substrate deforms elastically albeit at much smaller strains and some of the energy from the pendulum is lost to the rigid substrate. The test is therefore limited by the thickness of the specimen as well, with a thicker specimen more likely to show the as expected pure energy-controlled type of deformation.

It is worth highlighting the slight difference in the deformation index values of the room temperature ($n = 0.5$) and elevated temperature ($n = 0.6$) rebound resilience measurements. The elevated temperature measurements are more shifted towards an equal energy deformation most likely because more energy is transferred and then returned from the heated rubber specimen compared to the unheated specimen. The minimal difference ($\Delta n = 0.1$) between the deformation index values of the heated and unheated rubber specimens suggests the influence of CB colloidal properties in rubber is not significantly affected by change in temperature as much as it is affected by change in polymer.

Understanding the type of deformation that dominates in a particular heat build-up or energy dissipation condition has practical implications for material design. By combining the knowledge of the type of deformation and how CB properties influence heat build-up and energy dissipation, compounds can be designed to maximize or minimize heat build-up and energy dissipation. For example, if looking to minimize rolling resistance (an example of an energy dissipative mechanism) for a rubber component that will be used at a medium-high strain level (~25–40%) under a mix equal strain and equal energy deformation mode (deformation index value $n = 0.6$, similar to that obtained from the elevated temperature rebound resilience measurements), CB surface area should be minimized. For a component that will be used under the special case of an equal strain condition (deformation index, $n = 0$ similar to that obtained from the Goodrich flexometer measurements), at a similar strain level, both CB surface area and structure should be reduced to minimize rolling resistance of the component.

4 Conclusion

There were two main objectives in this paper:

1. To study the effect of CB structure and surface area on heat build-up and energy dissipation mechanisms in rubber. To achieve this objective, eight natural rubber compounds with CBs that varied widely in structure and surface area were tested using a rebound resilience tester, Goodrich flexometer tester and a novel Coesfeld heat build-up analyser. The CBs had similar loading of 50 phr.
2. To determine the predominant deformation mode in each of these heat build-up and energy dissipation tests by applying the deformation index concept.

CB surface area plays a dominant role in increasing the heat build-up and energy dissipation in rubber materials. Increasing CB surface area increases heat build-up and energy dissipation most likely due to the increased energy dissipative mechanisms from the increased polymer–filler and filler–filler interaction. CB structure also has some influence on the heat build-up depending on the type of heat build-up measurements. In Goodrich flexometer heat build-up measurements where the samples are deformed to medium to high strain levels, CB structure increases stress level in the compound at a given deflection and therefore increases the absolute amount of energy dissipated.

These observations can be rationalized by looking at the types of deformation (equal strain, equal energy or equal stress) dominant in each of the heat build-up and energy dissipation tests. It was determined that the Goodrich flexometer heat build-up measurements are predominantly strain-controlled based on deformation index concept. Conversely, rebound resilience and Coesfeld heat build-up measurements are a combination of strain and energy-controlled deformations. It is clear that to predict the level of heat build-up encountered in practice in a given application that the deformation index has to be mapped between the test used to characterize the rubber and the service conditions that are encountered in practice.

Acknowledgements One of the authors (W.A.K-M) would like to thank Birla Carbon for the financial support throughout his PhD studies and providing the materials studied in this article. The authors would like to thank Mahatab Chowdhury and Quillan McGlynn (both Birla Carbon) for support in conducting some of the experiments. This work was also supported by the Ministry of Education, Youth and Sports of the Czech Republic—DKRVO (RP/CPS/2022/006).

References

1. Medalia AI (1991) Heat generation in elastomer compounds: causes and effects. Rubber Chem Technol 64(3):481–492. https://doi.org/10.5254/1.3538565
2. Akutagawa K, Hamatani S, Nashi T (2015) The new interpretation for the heat build-up phenomena of rubbery materials during deformation. Polymer 66(1):201–209. https://doi.org/10.1016/j.polymer.2015.04.040
3. Kyei-Manu WA, Tunnicliffe LB, Plagge J, Herd CR, Akutagawa K, Pugno NM, Busfield JJ (2021) Thermomechanical characterization of carbon black reinforced rubbers during rapid adiabatic straining. Front Mater 421. https://doi.org/10.3389/fmats.2021.743146
4. Lake G, Lindley P (1964) Cut growth and fatigue of rubbers. II. Experiments on a non-crystallizing rubber. J Appl Polym Sci 8:707–721
5. Young D (1986) Dynamic property and fatigue crack propagation research on tire sidewall and model compounds. Rubber Chem Technol 59:785–805
6. Schrittesser B, Pinter G, Major Z (2012) Temperature dependent fracture behavior of Rubbers used in the oil and gas industry. 19th European Conference on Fracture: Fracture Mechanics for Durability, Reliability and Safety, ECF 2012
7. Gent A, Hindi M (1988) Heat build-up and blowout of rubber blocks. Rubber Chem Technol 61(5):892–905
8. Goryunov S, Khoreshok A, Grigoryeva N, Preis E, Alitkina O (2019) The research of operational temperatures of dump trucks tires. E3S Web Conf 134 (2019) 01014, 134. https://doi.org/10.1051/e3sconf/201913401014
9. Conant F (1971) Tire temperatures. Rubber Chem Technol 44(2):397–439. https://doi.org/10.5254/1.3547374
10. Chen D, Wu J, Su B, Cui B, Teng F, An S, Wang Y et al (2023) Thermo-mechanical-abrasive coupling analysis of solid rubber tire under high-speed rolling. Wear 512–513(204546). https://doi.org/10.1016/j.wear.2022.204546
11. Stoček R, Heinrich G, Kipscholl R, Kratina O (2021) Cut & chip wear of rubbers in a range from low up to high severity conditions. Appl Surface Sci Adv 6(100152). https://doi.org/10.1016/j.apsadv.2021.100152
12. Behnke R, Kaliske M (2015) Thermo-mechanically coupled investigation of steady state rolling tires by numerical simulation and experiment. Int J Non-Linear Mech 68:101–131. https://doi.org/10.1016/j.ijnonlinmec.2014.06.014
13. Rukkur S, Dechwayukul C, Thongruang W, Patarapaiboolchai O (2013) Heat build-up of industrial solid tires in Thailand. Adv Mater Res 844:445–449. https://doi.org/10.4028/www.scientific.net/amr.844.445
14. Liu Y, Chen W, Jiang D (2023) Review on heat generation of rubber composites. Polymers 15(2). https://doi.org/10.3390/polym15010002
15. Luo W, Yin B, Hu X, Zhou Z, Deng Y, Song K (2018) Modeling of the heat build-up of carbon black filled rubber. Polymer Testing 69:116–124. https://doi.org/10.1016/j.polymertesting.2018.05.017
16. Esmaeeli R, Nazari A, Aliniagerdroudbari H, Hashemi S, Alhadri M, Zakri W, Farhad S (2018) Heat build-up during dynamic mechanical analysis (DMA) testing of rubber specimens. In: Proceedings of the ASME 2018 International Mechanical Engineering Congress and

Exposition, vol 9: Mechanics of Solids, Structures, and Fluids, 9. https://doi.org/10.1115/IMECE2018-88627
17. Schieppati J, Schrittesser B, Wondracek A, Robin S, Holzner A, Pinter G (2019) Heat build-up of rubbers during cyclic loading. Conference: 2019 European Conference on Constitutive Models for Rubber - ECCMR 2019. Nantes
18. Dedova S, Schneider K, Stommel M, Heinrich G (2020) Dissipative heating, fatigue and fracture behaviour of rubber under multiaxial loading. In: Heinrich IG, Kipscholl R, Stoček R (eds) Fatigue crack growth in rubber materials. Advances in polymer science, vol 286. Springer, Cham, pp 421–443
19. He H, Liu J, Zhang Y, Han X, Mars W, Zhang L, Li F (2022) Heat build-up and rolling resistance analysis of a solid tire: experimental observation and numerical simulation with thermo-mechanical coupling method. Polymers 14(11). https://doi.org/10.3390/polym14112210
20. Stoček R, Stěnička M, Kipscholl R (2019) Heat build-up characterization under realistic load. Constitutive Models for Rubber XI - Proceedings of the 11th European Conference on Constitutive Models for Rubber, 2019. CRC Press/Balkema, Nantes, pp 157–162. https://doi.org/10.1201/9780429324710-28
21. Kratina O, Voldánová J, Stoček R (2022) How the rubber compounds of different tire's components heat up? Constitutive Models for Rubber XII Proceedings of the 12th European Conference on Constitutive Models for Rubber (ECCMR 2022). Milano
22. Futamura S (1991) Deformation index-concept of hysteretic energy-loss process. Rubber Chem Technol 64(1):57–64. https://doi.org/10.5254/1.3538540
23. Futamura S, Goldstein A (2004) A simple method of handling thermomechanical coupling for temperature computation in a rolling tire. Tire Sci Technol 32(2):56–68
24. Gibert J, Ananthasayanam B, Joseph P, Rhyne T, Cron S (2013) Deformation index-based modelling of transient, thermo-mechanical rolling resistance for a nonpneumatic tire. Tire Sci Technol 41(2):82–108
25. Ransom T, Gamache R, Mason B, Ladouceur H, Roland C (2020) Deformation index applied to impact. Rubber Chem Technol 93(2):261–273
26. Mars W (2011) Analysis of stiffness variations in context of strain-, stress-, and energy-controlled processes. Rubber Chem Technol 84(2):178–186
27. Futamura S, Goldstein A (2016) Prediction and simulation of tire performance characteristics based on deformation index concept. Rubber Chem Technol 89(1):1–21
28. ASTM International (2019) ASTM Standard D6556 Standard test method for carbon black-Total and external surface area by nitrogen adsorption. Standard Test Method for Carbon Black—Total and External Surface Area by Nitrogen. West Conshohocken, Pennsylvania: Annu. Book ASTM Stand. 09.01
29. ASTM International (2019) ASTM Standard D2414 Standard test method for carbon black oil absorption number (OAN). Standard Test Method for Carbon Black–Oil Absorption Number (OAN). West Conshohocken, Pennsylvania: Annu. Book ASTM Stand. 09.01
30. ASTM International (2019) ASTM Standard D3493 Standard test method for carbon black - Oil Absorption Number of Compressed Sample (COAN). Standard Test Method for Carbon Black–Oil Absorption Number of Compressed Sample (COAN). West Conshohocken, Pennsylvania: Annu. Book ASTM Stand. 09.01
31. Kyei-Manu W, Herd C, Chowdhury M, Busfield J, Tunnicliffe L (2022) The influence of colloidal properties of carbon black on static and dynamic mechanical properties of natural rubber. Polymers 14(1194)
32. Rutherford K, Akutagawa K, Ramier J, Tunnicliffe L, Busfield J (2023) The influence of carbon black colloidal properties on the parameters of the Kraus model. Polymers 15(1675)

Adv Polym Sci (2024) 294: 145–170
https://doi.org/10.1007/12_2023_164

Published online: 7 October 2023

Characterisation of Crack Propagation in Natural Rubber Under Complex Loading Conditions Using Thermographic Methods

S. Dedova and K. Schneider

Contents

Abstract The present work describes the formation and propagation of double cracks in elastomer materials under complex multiaxial loading. Mechanical measurements are complemented by optical measurements to determine local strain

S. Dedova and K. Schneider (✉)
Leibniz-Institut für Polymerforschung Dresden, Dresden, Germany
e-mail: schneider@ipfdd.de

fields and thermal measurements to detect local temperature changes and energy dissipation in the material, respectively. The results are applied to analyse the crack growth behaviour in the case of double cracks and compared with the classical method of Rivlin and Thomas (J Polym Sci 3:291–318, 1952), with which they show good agreement. As material a natural rubber, loaded with 20 phr carbon black, is used. The experiments were performed on a biaxial testing machine using an optical digital image correlation system to measure the local strain. Simultaneously, the thermal behaviour was determined by infrared thermography.

Keyword Crack deviation · Dissipative heating · Double crack formation · Fatigue · Fracture · Multiaxial testing of rubber · Tearing energy

1 Introduction

Elastomers are a widely used group of materials that can be used in a variety of ways in technical components due to their specific properties. This specific property profile includes, above all, the hyperelastic and viscoelastic properties that enable elastomers to be used, for example, for tires, vibration dampers, sealing parts, V-belts, diaphragms, etc. These parts are often subjected to multiaxial cyclic loads during operation. These specific loads pose a challenge for the description of elastic behaviour and, in particular, fatigue and fracture properties under a complex load. A special feature here is the crack splitting in natural rubber under certain loading conditions.

Two approaches have proven useful for describing fracture properties.

One is a global approach according to Rivlin and Thomas [1], which is nowadays referred to as "classical TFA" (Tear Fatigue Analysis). The method is based on the Griffith approach [2] and determines the tearing energy during crack growth from the perspective of the global deformation energy. This method is widely used for describing the fatigue crack and fracture properties of elastomers and is applied for different areas of analysis, such as determination of mechanical properties [3, 4], dissipative heating [5–8], crystallisation effects [9], filler influences [10], the type of crack development [11, 12] by different research groups [13–15]. A further development of the method is presented in detail in [16, 17]. In [18–20], the influences of different loading characteristics such as strain, temperature, R-ratio, and waveform and their effects on crack propagation in rubber lends were studied using TFA.

The other approach is a local approach [21, 22], also known as J-integral. This method has the advantages in that it can be applied independently of specimen geometry and crack deflection. It has been applied by different groups to analyse the fracture behaviour of elastomers, e.g., [23–27].

Difficulties in describing the deformation behaviour of elastomers and, in particular, in describing fracture behaviour arise from multiaxial loading in realistically loaded components. Recently, the behaviour of elastomers under biaxial loading has been described [28, 29] and fracture under biaxial loading has been investigated using energy analysis [30].

The objective of the present work is to develop an analytical method to describe the crack growth behaviour of elastomers under a complex cyclic loading for both simple crack propagation and splitting of an initial crack. It has been practically proven and is also applied in industries (https://www.coesfeld.com).

2 Theoretical Background

2.1 *Entropy- and Energy-Elastic Components of Internal Energy and Dissipative Heating*

Kinetic theory in its simplest form describes the elasticity of rubber due to changes in the conformation of the system of long-chain molecules during the transition from the unstressed to the stressed state, associated with changes in the configurational entropy of the system. For such a mechanism (i.e. involving only changes in entropy), the stretching force at a given stretching state is proportional to the absolute temperature. The early experiments of Meyer and Ferri [28, 31] showed that this conclusion was essentially satisfied over a wide range of temperatures, provided the strain was sufficiently large.

For a more detailed investigation of these thermoelastic phenomena, it is necessary to develop relationships between force, length, and temperature, on the one hand, and thermodynamic quantities, namely internal energy and entropy [32]. In the following we shortly summarise the procedure, which we described recently in some more detail [33–35].

According to the first law of thermodynamics, the mechanical work dw, done on a volume element of the system by external forces, and the heat δq, absorbed by the system, change the internal energy du in any process.

$$\Delta w + \Delta q = \Delta u_{\text{therm}} + \Delta u_{\text{pot}} + \Delta w_{\text{kin}} + \Delta w_{\text{irrev}}. \tag{1}$$

Here, Δq the thermal energy (heat) emitted per volume element, $\Delta u_{\text{therm}} = \rho\, c_p \mathrm{d}T$ is the change in the thermal part of the internal energy per volume element with ρ, c_p and $\mathrm{d}T$ as density, specific heat (estimated separately by differential scanning calorimetry measurement) and change in temperature. Δu_{pot} is the change in the elastically stored potential energy per volume element, Δw_{kin} the change of the kinetic energy per volume element, and Δw_{irrev} the irreversible part of energy loss per volume element due to the creation of a new crack surface. In most cases, the kinetic energy can be neglected.

For a closed loop without crack propagation, we have therefore

$$\Delta w + \Delta q = \Delta u_{\text{therm}} = \oint \sigma_{ij} d\varepsilon_{ij} + \Delta q = \int \dot{T} \rho \, c_p \, dt. \tag{2}$$

Assuming constant temperature in the sample, the heat release to the environment of a volume element with two surfaces dA can be approximated by a simple linear expression as

$$\Delta q = -\frac{2\alpha}{dV} \int dA(T(t) - T_0) \, dt = \int \dot{T} \rho \, c_p \, dt - \oint \sigma_{ij} d\varepsilon_{ij} \tag{3}$$

with the heat transfer coefficient

$$\alpha = \frac{d_0 \left(\oint \sigma_{ij} d\varepsilon_{ij} - \rho \, c_p \Delta T \right)}{2 \int \lambda_x \lambda_y (T - T_0) \, dt}. \tag{4}$$

(A previously applied approach to determine the heat transfer coefficient α via the time constant of the free cooling of a dissipative heated sample [33, 34] has proven to be unsuitable, since the convective heat transfer of an unmoved sample differs greatly from the situation of a dynamically deformed sample.) The knowledge of α now allows the determination of all energy components of a volume element, i.e. the energy densities, over the entire loading and unloading cycle, as long as no crack propagation is present.

In particular, the change in the potential fraction of the internal energy can now be determined as

$$\Delta u_{\text{pot}} = \Delta w + \Delta q - \Delta u_{\text{therm}}$$
$$= \int_{\varepsilon_0}^{\varepsilon} \sigma \, d\varepsilon - 2 \int_{t_0}^{t} \alpha \, \lambda_x \lambda_y (T - T_0) \, dt - \int_{T_0}^{T} \rho \, c_p \, dT. \tag{5}$$

If the dissipated energy within a closed loading-unloading cycle is identical to the thermal energy (heat), transmitted from the system to the environment, for the reversible part of the internal energy u_{rev} can be written as

$$du_{\text{rev}} = TdS/V + \sigma_{ij} d\varepsilon_{ij} = c_p \rho dT + d\varepsilon_{ij} \tag{6}$$

with the density ρ, the specific heat capacity c_p, the volume V, the temperature T, the stress tensor σ_{ij}, and the strain tensor ε_{ij}.

We consider here the contribution of internal energy associated with temperature increase as the entropic part of internal energy, while the (the non-volume changing) part associated with stress and strain is considered as the potential part of internal energy.

2.2 Temporary Energy Contributions Within a Loading Cycle

Due to the inhomogeneous deformation near a crack tip, it is helpful to first characterise the viscoelastic behaviour of the material under homogeneous loading conditions and use this to describe the inhomogeneous deformation situation.

For the damaged case (inhomogeneous strain situation), the energy sources are not only located in the crack tip, resulting in a local temperature peak, but they are also distributed over the entire deformed viscoelastic material.

In order to analyse these effects under multiaxial loading independently of individual loading axes, the von-Mises-strain was chosen as the reference value. It was calculated assuming incompressibility and plane strain states in the x-y plane according to

$$\varepsilon_M = \exp\left(\sqrt{\frac{4}{3}\left(\left(\ln\lambda_{\text{major}}\right)^2 + \ln\lambda_{\text{major}}\ln\lambda_{\text{minor1}} + \left(\ln\lambda_{\text{minor1}}\right)^2\right)}\right) - 1 \tag{7}$$

with the principal stretches λ_{major}, λ_{minor1}.

The method of the energy determination by correlation with von-Mises-strain or the local temperature changes is presented in [35] and applied for the analysis of single crack growth in a complex multiaxial loading.

In the present work, this method is compared with the (classical) global method for the determination of the tearing energy and used for analysing of a splitting crack. While the classical approach is relevant only for a "pure shear" geometry with a strait propagating crack, the methods used in this work are applied to multiaxial loaded specimens.

3 Experimental Procedure

3.1 Material and Samples

The investigations were carried out with a natural rubber with 20 phr of carbon black (NR20). The composition of the used material is summarised in Table 1. The samples were vulcanised for 11 min at 150°C.

For the tests square specimens with a framing bulge for a form fit in the clamping system were used. To increase the homogeneity of the strain fields, the bulge between the individual clamps was interrupted. The data of the samples are summarised in Table 2.

The surface of the samples was structured by a random pattern by air brush using a flexible paint containing 50% latex.

Table 1 The composition of investigated material

Ingredient	NR with 20 phr CB
SMR 10[a]	100
CB Corax N330[b]	20
Zinc oxide (Red seal)	3
Stearic acid	2
Vulkanox 4020 (6PPD)	1
Sulphur	1.5
Vulkacit CZ (CBS)	1.5

[a] SMR 10: Standard Malaysian rubber from "Weber & Schaer"
[b] Corax N330: type of carbon black

Table 2 Dimensions of the investigated samples and properties of the material

Initial thickness	mm	1.85
Initial length	mm	77
Initial width	mm	77
Diameter of the bulge	mm	5.5
Density	g/cm^3	1.019
Heat capacity, estimated by DSC	J/g K	1.756

3.2 Equipment and Experimental Set-Up

The experiments were done using a Biaxial Tester (Coesfeld GmbH & Co. KG; Dortmund, Germany) at Leibniz-Institut für Polymerforschung Dresden (IPF) [36] (https://www.coesfeld.com). Images of the flat samples were captured with a frame rate of 1 Hz with an optical camera (JAI BM-500 GE) simultaneous to signals of the motors, the force- and the displacement-sensors by the software of the biax-tester.

During the slow steady loading, the strain fields in the specimen were determined in this way by digital image correlation (DIC) using facets with a size of 19 × 19 pixel and a distance of 15 pixel. During the cyclic loading at 1 Hz, the images were taken in each case at the point of maximum loading and used to follow the crack propagation.

Parallel to the mechanical measurement and the optical recording of the strain fields or the crack propagation, a 2D IR-imaging was carried out by an IR camera VarioCAM from InfraTec, Germany. The temperature measurement was done over a total area of 210.2 × 262.7 mm^2 with a frequency of 20 Hz and a resolution of 0.37 mm/pixel. The IRBIS software was used for the evaluation of measured results.

The complete set-up is schematically shown in Fig. 1. The arrangement is described in more detail in [35, 36].

The optical camera was positioned perpendicular to the sample surface, the axis of the thermal characterisation camera was slightly tilted with respect to the vertical axis.

To avoid metal reflections that could influence the observed temperature data, the clamps were covered with white paper. In addition, it was checked and verified that both the illumination of the sample and the angular arrangement of the IR camera

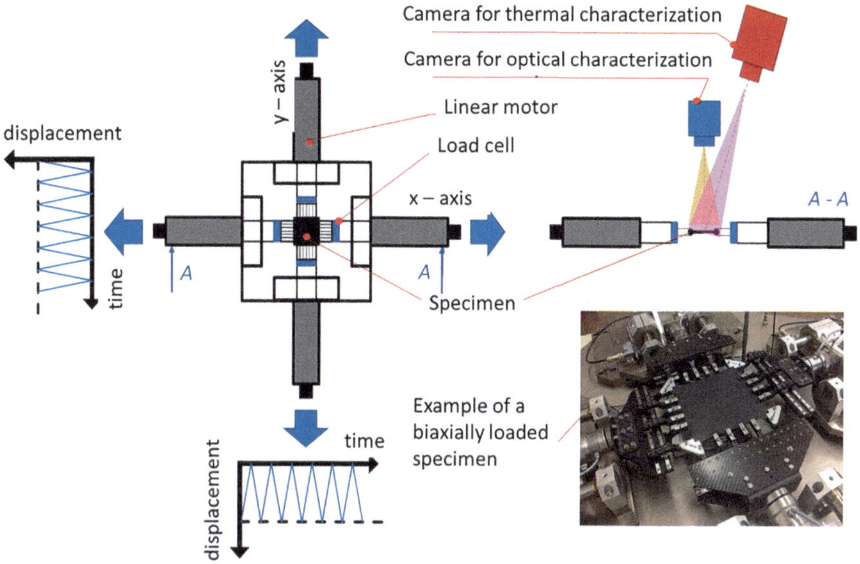

Fig. 1 Scheme of experimental set-up. Left: Principle scheme of biaxial loading (top view). Right: Arrangement of IR- and optical camera (top) and image of a sample in the biaxial tester (bottom)

had no relevant influence on the measured temperature field. The calibration of the camera for thermal characterisation was checked for both the black and the patterned sample with a thermocouple. There was no influence of the pattern on the measured temperature and the temperature values of the IR system.

3.3 Experimental Procedure

The *quasi-static stress-strain behaviour* was estimated by relaxation tests. For this purpose, the specimens were loaded in steps of 10% up to ε_x=120% within 1 s and then relieved in the same steps. The holding time after each loading or unloading step was 1 min.

The *dynamic tests* were performed on unnotched specimens (homogeneous loading) and on notched specimens (inhomogeneous loading).

First, all specimens were preloaded equiaxial up to 120% strain with a 1 Hz triangle load for 30 cycles for removing of Mullins effect ("demullinisation"). The sample state after this conditioning was defined as the "reference state". For the characterisation of the notched specimens, the notch was applied after the conditioning.

At the beginning of dynamic loading, the specimens were loaded once quasi-statically at a rate of 0.13%/s up to the maximum strain for the optical recording of the strain fields up to the target strain.

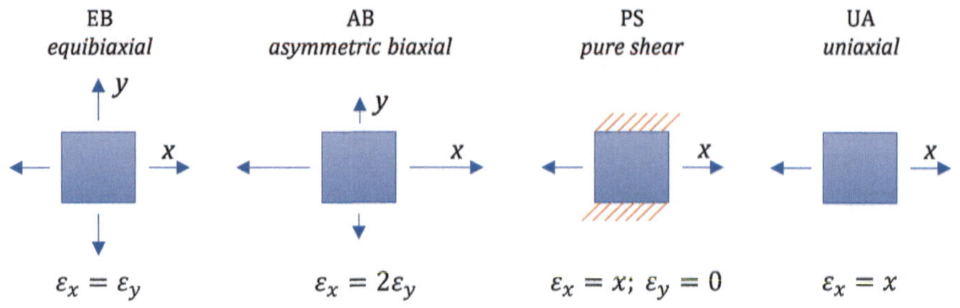

Fig. 2 Schematic representation of experimental regimes

Table 3 Details of the different loading regimes

Triangular load	Unnotched samples (homogeneous stretch)		Notched samples (inhomogeneous stretch)	
1 Hz	Cycles	Main strain = strain amplitude	Cycles	Main strain = strain amplitude
EB ($\varepsilon_x = \varepsilon_y$)	300	20, 30, 40, 60, 80, 90%	1,000	15, 20, 25, 30, 35%
AB ($\varepsilon_x = 2\varepsilon_y$)				
PS ($\varepsilon_x = x;\ \varepsilon_y = 1$)				
UA ($\varepsilon_x = x$)				

Afterwards, cyclic loading at 1 Hz was performed with 4 regimes: equibiaxial (EB), asymmetric biaxial (AB), pure shear (PS), and uniaxial (UA) strain (Fig. 2). Triangular loading was used to allow comparison of the experiments with constant strain rate tests.

Details of the loading regimes are shown in Table 3. In this work, the cyclic tests are always abbreviated by the load type (EB, AB, PS, or UA) followed by the maximum strain amplitude in %, i.e. EB30 describes an equibiaxial test with maximum strain of 30%.

The number of 300 cycles (corresponding to 300 s) for each loading level was selected to achieve thermal stability (temperature plateau).

The upper reversal point was set in relation to the initial sample size as a target strain, the lower reversal point was estimated so that the load should be zero. Due to the "permanent set" of the specimens, the strain at the lower reversal point increased slightly with increasing strain.

The 1,000 cycles for the notched samples were chosen, on the one hand, to reach a temperature plateau and to characterise the type and development of the cracks. On the other hand, this relatively low number of cycles limited crack propagation, especially at low strains, and made it difficult to determine the crack propagation velocity here.

4 Results and Discussion

4.1 Stress Equilibrium at Multiaxial Loading

The performance of the relaxation tests was discussed, for example, in [37, 38]. In this work the equilibrium stress-strain curves were determined by finding, at individual strains, the intersection of the stresses extrapolated over the logarithm of time, see Fig. 3.

So for the equilibrium stress follows:

$$\sigma_K = \sigma_\infty = \sigma_A + (\sigma_B - \sigma_A)\frac{\sigma_D - \sigma_A}{\sigma_B - \sigma_A - \sigma_C + \sigma_D} \tag{8}$$

with the equilibrium stress $\sigma_\infty = \sigma_K$ at point K.

As example the estimation of equilibrium stress for EB loading of NR20 is shown in Fig. 4.

4.2 Dynamic-Mechanical Characterisation and Dissipated Energy

The hysteresis tests were performed for the conditions described in Table 3. NR20 shows a relatively small hysteresis area. The stress-strain behaviour for NR20 for EB loading conditions and amplitudes investigated for the last cycles is shown in Fig. 5.

The investigation of uniaxial deformation (UA) was only possible to a limited extent. For the investigation, 10 mm wide strips were punched out of the square specimens to be clamped in the biaxial tester. The force values showed a high scatter for the specimens due to the not appropriated load cell for this specimen geometry.

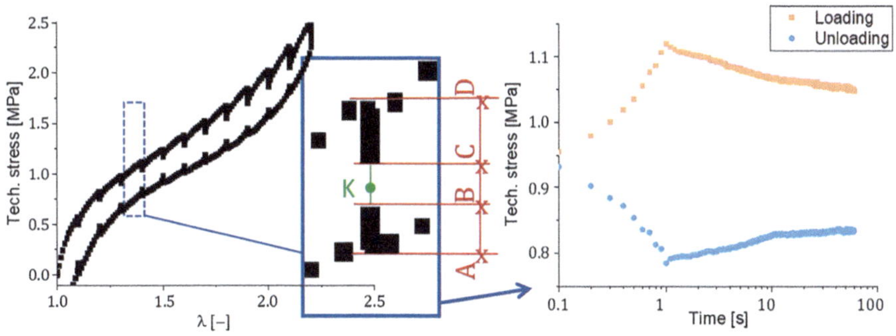

Fig. 3 The principle of estimating the equilibrium stress from the relaxation curves. Left: change of technical stress at different strains. Right: technical stress during the loading and unloading cycle vs. log time for 40% strain

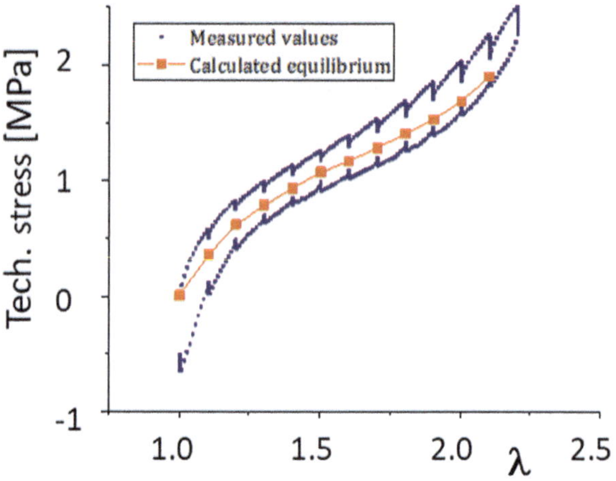

Fig. 4 Estimated equilibrium technical stress by the relaxation experiment of NR20 for EB loading conditions

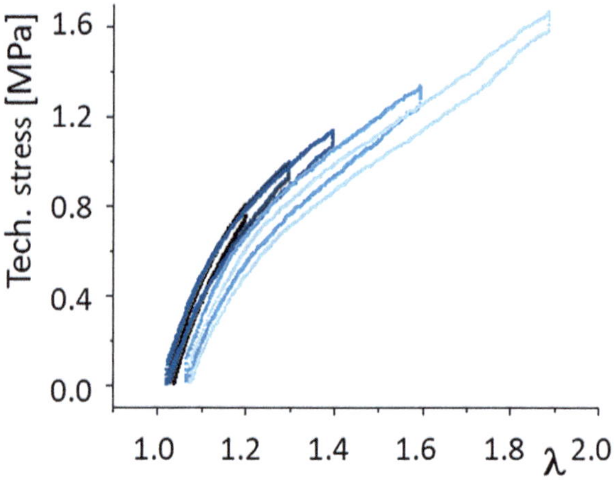

Fig. 5 Technical-stress-stretch curve for the last cycles of NR20 with different maximum strain under EB loading conditions

Unfortunately, due to the concept of the biaxial tester for square specimens, there was no possibility to change the load cell.

For each experiment, the dissipated energy density was calculated by integration according to Eq. 2. The differences in the values of the energy density for the cycles 50, 100, 200, and 300 were marginal. Therefore, the mean value of the four cycles considered. The dependence of the dissipated energy density on the individual load and the von-Mises-strain is shown in Fig. 6. In the diagrams on the left side, the density of dissipated energy is shown as a function of the individual strains, and on the right side as a function of the von-Mises-strains.

It was found that the dissipated energy per volume and cycle correlates well with the von-Mises-strain. The investigation of this correlation was already discussed in [35] for solution styrene-butadiene rubber (SBR) with 50 phr carbon black. This strain is homogeneous over almost the entire equibiaxial stretched specimen (except

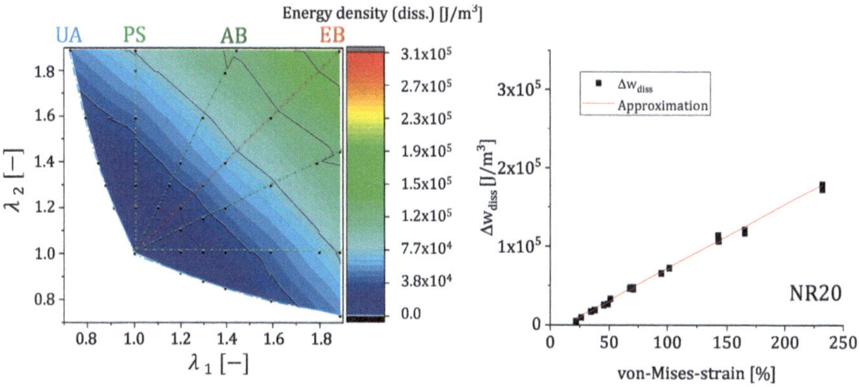

Fig. 6 Density of dissipated energy for NR20 for cycles 50, 100, 200, and 300. Left: as a function of the strains at the two axes; right: as a function of the von-Mises-strain

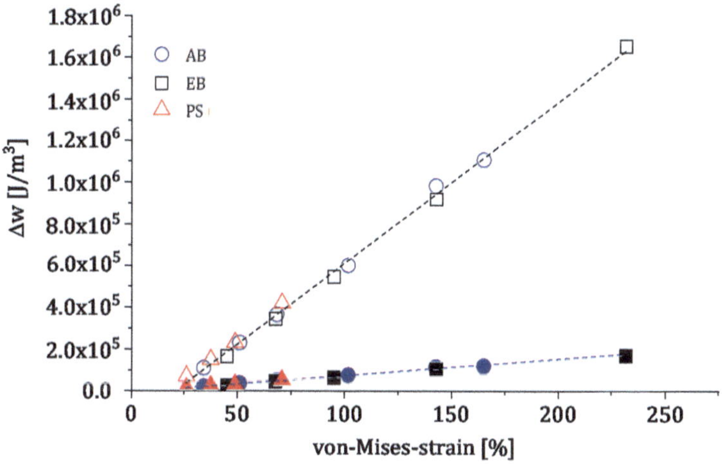

Fig. 7 Total mechanical energy density w_{total} for NR20 during the last cycle (open symbols) and fraction of dissipated energy density w_{diss} (full symbols) as a function of von-Mises-strain

for the edge region of the clamps), and so the temperature distribution is also homogeneous.

Similarly, the elastically stored fraction of the internal energy as well as the dissipated energy per cycle for different loading conditions correlates with the von-Mises-strain, see Fig. 7. Both energy contributions do not depend on the deformation mode as long as the amount of the von-Mises-strain is equal.

4.3 Heat Build-Up and Energy Contributions

The cyclic loading of the unnotched specimens causes homogeneous dissipative heating of the material. The temperature development was recorded parallel to the measurement of the force-displacement data in the middle area of the specimen (approx. 20×20 mm^2, see Fig. 8). The temperature increase can be approximated with an exponential function, to which a smaller oscillation is superimposed (Fig. 8, left).

The temperature profile results from the hysteresis during the saw tooth-like deformation, the non-linear stress-strain behaviour of the specimens, and the heat transfer to the environment.

For each individual cycle it is possible to determine the heat transfer coefficient α according to Eq. (4). Figure 9 shows the heat transfer coefficients for the different loading conditions over the time or the number of the cycles, respectively.

It was found that the heat transfer coefficient is strongly dependent on the total deformation of the sample and the number of cycles. Obviously, the movement of

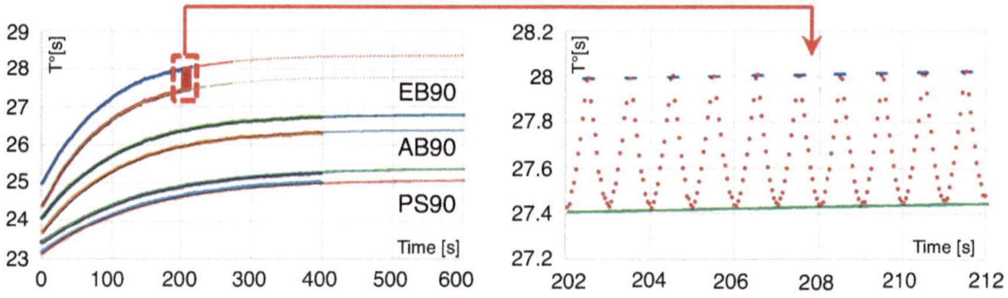

Fig. 8 Dissipative heating for different loading conditions (examples of 90% elongation) for NR20. Left: maximum and minimum temperature of the sample over the time; right: last 10 cycles of EB90. Curves slightly shifted vertically to each other to increase clarity

Fig. 9 Heat transfer coefficients determined for different loading conditions and amplitudes over the cycle number (saw tooth with 1 Hz) for NR20

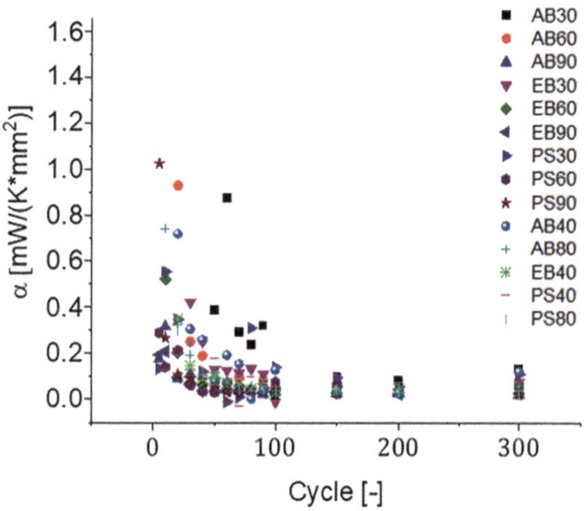

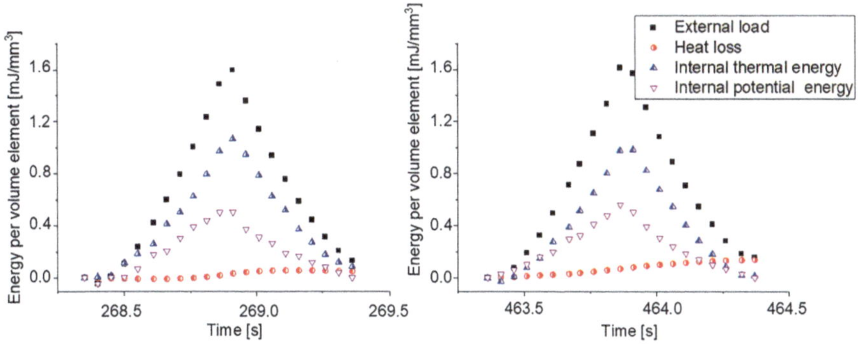

Fig. 10 Contributions of supplied energy and internal energy during a 5th and 200th cycle for EB90 loading

the sample changes the conditions of heat transfer by local convection by the moving air at the surface. After about 100 cycles, the heat transfer coefficient stabilises and remains further in the range of 0.02...0.15 mW/K mm^2 for NR20.

Using the heat transfer coefficients determined in each case, the supplied work was divided into the change in internal energy and the heat dissipation to the environment. The separation of the energy contributions during the individual cycles according to Eq. (5) is shown in Fig. 10 for the 5th and the 200th cycle for EB90.

Figure 10 shows that only a certain part of the supplied mechanical energy can be recovered, another part is used to increase the internal energy in the form of heat or is released directly as heat to the environment. During a deformation cycle, only a small part of the supplied energy is stored elastically as potential energy, another part is stored as reversible thermal energy.

Figure 11 shows the relationship between the potential part u_{pot} of the internal energy and the total internal energy u_{int} for cycles 100, 150, and 200 over the von-Mises-strain for NR20. The mean value of the ratio $\Delta u_{\text{pot}}/\Delta \text{w}_{\text{int}}$ is approx. 0.41 for NR20.

The results can be interpreted as the ratio between the energy-elastic and the total elastic energy fraction described in the literature [39–45]. For the different authors, the ratio between the energy-elastic part of energy and the total elastic fraction was found by means of thermal and mechanical investigations for different materials which was in the range of $\approx$0.2.

For the investigated NR20 in this work, the ratio is slightly higher than most of the results from the literature, but remains in the comparable order for the different multiaxial loading cases.

As long as the bulk of a sample heats up dissipatively, the internal energy is partially stored and leads to a further increase in temperature. More energy is stored at the beginning of the loading process than at the end, which can be observed in the proportion of heat loss.

The curves of the 5th cycle (Fig. 10) show a relatively low heat loss, which is related to the comparable low temperature difference of the sample to ambient

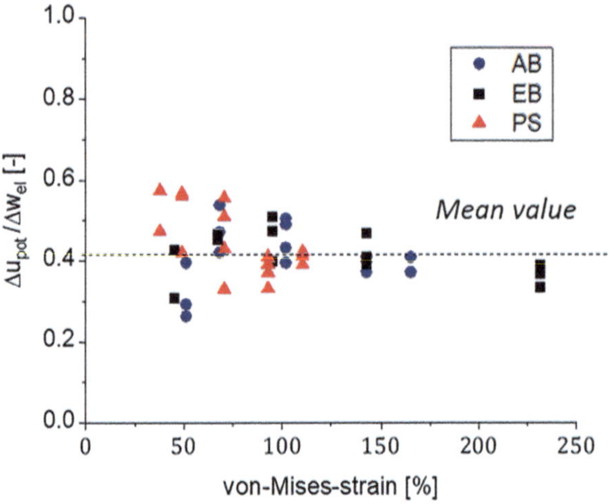

Fig. 11 Relation between the potential part of the internal energy and total elastic energy per cycle for NR20

temperature at the beginning of the loading process $\Delta T = T - T_{\text{ref}}$. As the number of cycles increases, the proportion of dissipated energy released to the environment as heat increases until an equilibrium between dissipation and heat release is reached with a temperature T_{fin} at the bottom inversion point of the oscillating temperature. Reaching T_{fin} with time can be approximated by

$$T = T_{\text{fin}} + (T_{\text{ref}} - T_{\text{fin}}) \exp\left(-\frac{t}{\tau}\right) \tag{9}$$

with a time constant τ. The values for $\Delta T_{\text{fin}} = T_{\text{ref}} - T_{\text{fin}}$ and τ determined are shown in Table 4.

The relationship of the final temperature difference ΔT_{fin} as function von-Mises-strain is shown in Fig. 12, the relationship is almost linear regardless of the loading condition (PS, AB, or EB), as expected.

Without prejudging further investigations, this correlation is a valuable approach to understand the dissipative material behaviour in complex loading situations. The local dissipation behaviour of the material can thus be directly estimated by means of a von-Mises-strain determined, for example, by image correlation analysis. The increase in dissipation with von-Mises-strain depends on the investigated material.

4.4 *Inhomogeneous Deformation and Double Crack Formation*

Double crack formation occurred sometimes at the higher strains. This behaviour can have a material-specific explanation and leads back to the crystallisation behaviour

Table 4 Final temperature differences and time coefficients for cyclic loaded NR20

Loading conditions	PS		AB		EB	
ε_x	ΔT_{fin} [°C]	τ [s]	ΔT_{fin} [°C]	τ [s]	ΔT_{fin} [°C]	τ [s]
20%	–	–	–	–	0.33	166
30%	0.36	163	0.47	187	0.77	148
40%	0.58	192	0.83	157	1.44	159
60%	1.15	180	1.01	118	1.98	112
80%	1.56	143	1.99	118	–	–
90%	1.95	140	2.74	110	3.44	91

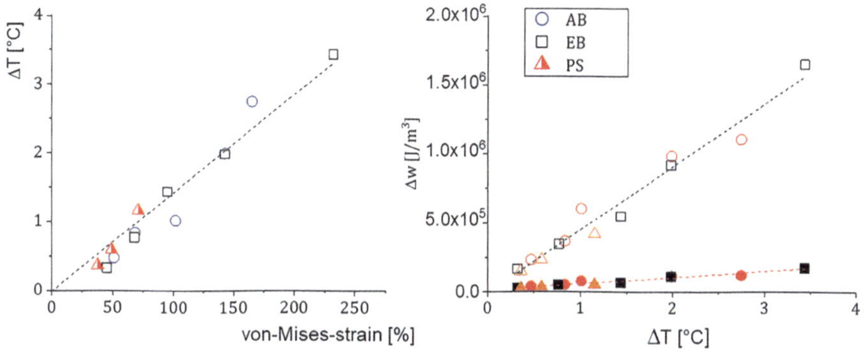

Fig. 12 Left: Relationship between equilibrium temperature differencesΔT_{fin} and von-Mises-strain for different loading conditions. Right: Relationship between dissipative and total parts of energy and equilibrium temperature differencesΔT_{fin}

at the extremely high strain at the crack tip in natural rubber [9, 11, 46]. The influence of the loading type could also contribute to this [11].

In the studies performed on NR20, double crack formation was observed in all equiaxial loaded specimens (EB), as well as under AB35 loading condition.

The formation of a double crack is observed only as continuation of initial single crack. In the process of cyclic loading, either two nearly identical cracks or a smaller branch from the main crack has formed.

So observed double cracks formed in the biaxial loading conditions can be classified into two types:

- Type I: the growth of almost two equal cracks, where the $\varphi \approx 45°$;
- Type II: a main crack located almost in the main growth direction (y), and a smaller branch at an angle φ of approximately 45°.

In Fig. 13 the observed crack types are shown schematically.

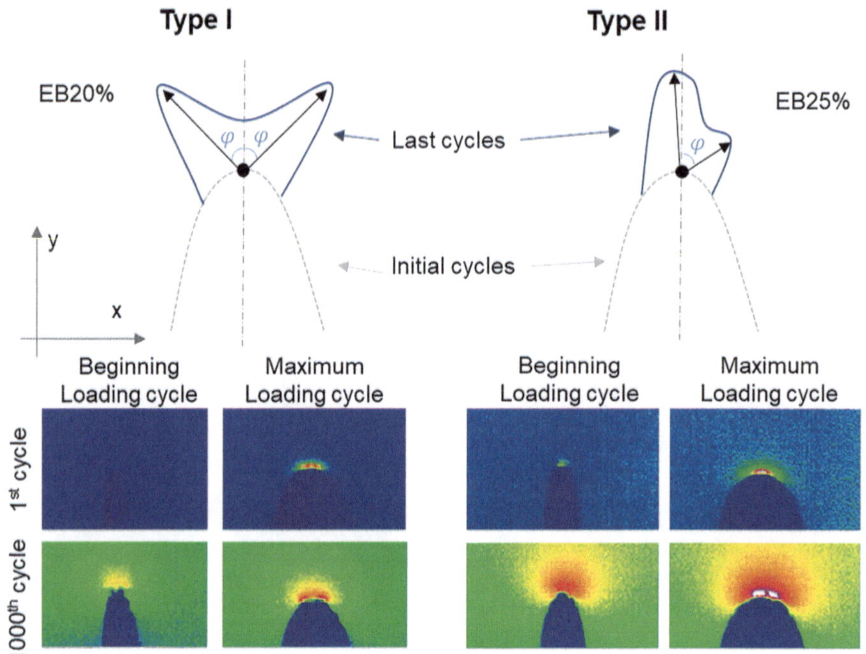

Fig. 13 Schematic representation of the crack contours for the EB 20 (type I) and EB 25 (type II) cases and the recordings of the temperature field 1st and 1,000th cycles, respectively, at the beginning of the loading cycle and at the maximum strain

4.5 Temperature Development During Single and Double Crack Propagation

Crack propagation was recorded initially every 50 cycles in y-direction in the unloaded position, in the case of double cracking for each tip individually, and corrected by

$$da = \frac{da_y}{\cos \varphi} \tag{10}$$

to account for the oblique direction of propagation.

It has already been shown that the highest temperature is present at the crack tip. It correlates for the case of a simple crack with the size of the logarithmic crack growth rate as well as with the von-Mises-strain of the homogeneously stretched area of the specimen [35].

Figure 14 shows the propagation of cracks under different loading regimes together with the changes of the maximum temperature at the crack tip.

Generally, NR20 shows relatively low heat build-up due to small hysteresis. The temperature at the crack tip increases totally by less than 2 K for an amplitude of 25%

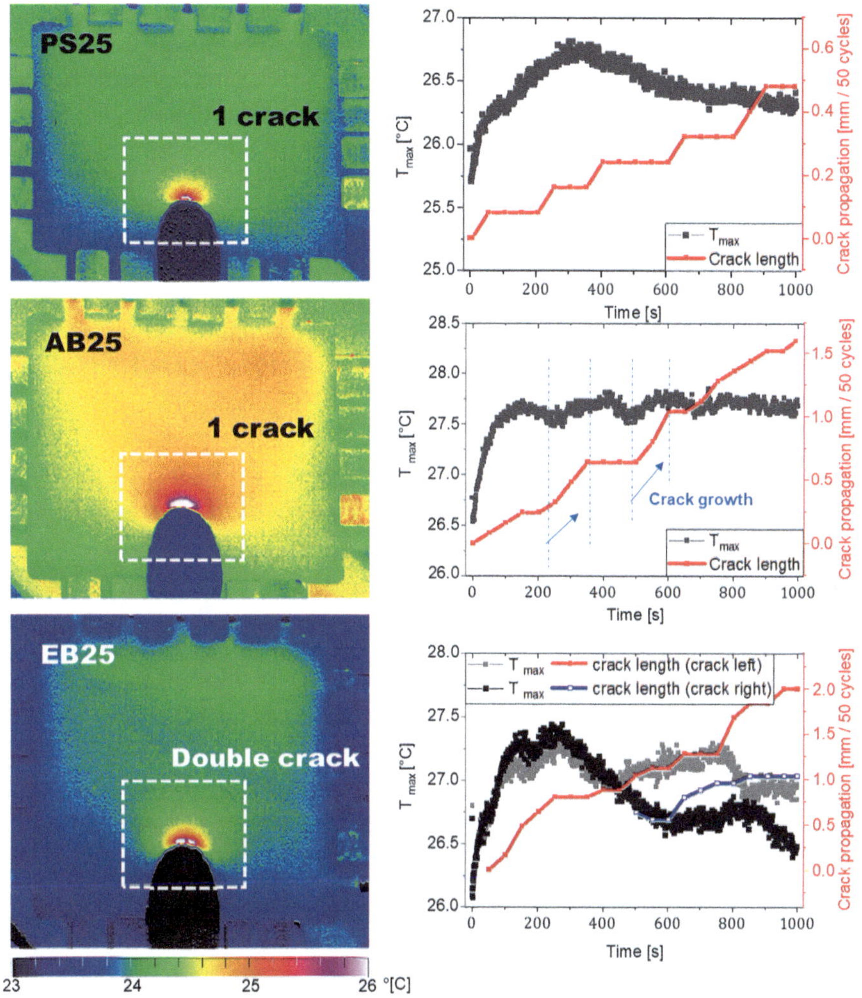

Fig. 14 Temperature fields of NR20 at the last cycle for different loading regimes with $\varepsilon_x = 25\%$ (left), as well as the crack propagation and the maximum temperature at the crack tip region (right) vs. time

over 1,000 cycles at the different loading regimes. The inhomogeneous temperature development (temperature jumps) is independent of the loading condition and was shown for both, the case of a single crack and a branched or double crack. But, generally the spontaneous propagation of a crack was always accompanied with an increase of the temperature at the running crack tip.

For the PS25 loading condition (Fig. 14, top), an almost uniform crack growth pattern can be observed. In this case, the temperature rise and fall are too low to find a clear conclusion about the crack growth. The total temperature variation at the crack tip remains around 1 K.

For AB25, a constant mean crack growth rate and a constant crack tip temperature can be observed (Fig. 14, middle).

When double cracks are formed, the available energy is distributed to both cracks, reducing the crack propagation velocity of the single cracks. Thus, the global crack propagation velocity is approximately half when the second crack is formed in sample EB25 (Fig. 14, bottom). Without crack propagation the temperature at a crack tip drops slightly.

For the analysis of the temperature evolution of NR20, it is not sufficient to analyse the process only as a function of the loading conditions. The investigation shows that several material-dependent factors are responsible for the temperature evolution at the crack tip. Crack propagation proceeds according to different patterns and effects such as: single or double cracks, uniform or irregular crack growth also influence the temperature development.

4.6 *Temperature Distribution Between Individual Crack Tips*

An example of crack growth with 2 crack tips is shown in more detail for loading condition EB25 in Fig. 15.

The crack growth process starts with one crack tip. It can be observed both visually and thermally as a spot of the temperature field. When the dynamic load of approx. 500 cycles is reached, a second temperature spot can be seen. In this case,

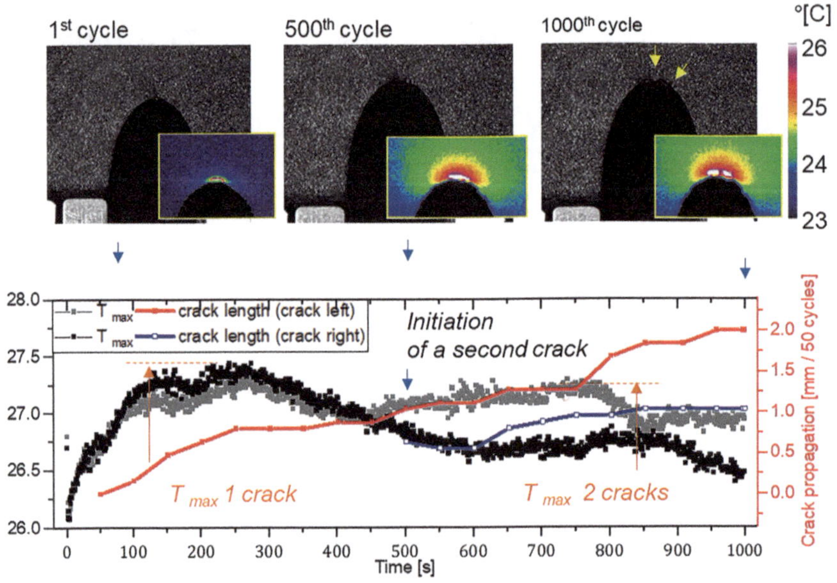

Fig. 15 Double crack formation at loading condition EB25. Development of the temperature at the crack tips, crack propagation and representation of the optical and thermal images at the tips

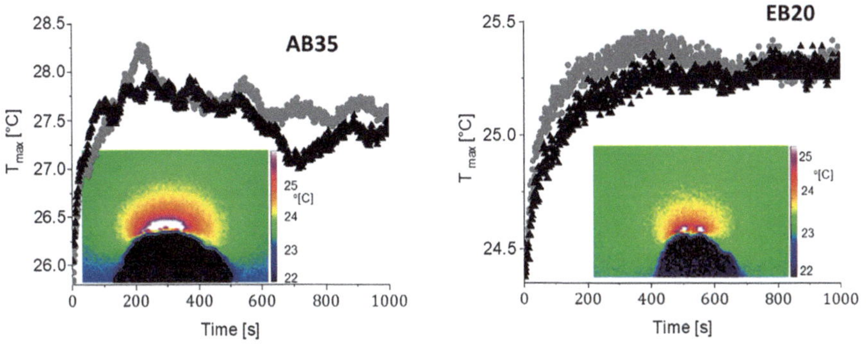

Fig. 16 Temperature development for the two cracks in the AB35 (left) and EB20 (right) loading condition, grey: left crack, black: right crack

it is still difficult to detect the second crack tip optically. As it propagates, the first crack tip cools down. At about cycle 600, the first tip starts to grow again, resulting in a slight increase in temperature. At about cycle 775, the 2nd crack stops, as evidenced by a decreasing temperature. At cycle 850, the temperature of the first crack tip starts to decrease again, while that of the second starts to increase, indicating that the 2nd crack continues to grow. This change is clearly better visible by thermography than in the optical images of the specimen.

At the end of dynamic loading, two crack tips are visible both thermally and optically.

Two other examples are shown in Fig. 16. An erratic temperature development is shown for the AB35 example, while EB20 shows a relatively steady progression for the two propagating cracks.

The studies of the temperature maxima of the notched NR20 samples show that a temperature plateau was not reached in the case of either crack or double crack formation. The continuous dissipative heating of the specimen is superimposed by a stochastic process due to the energy release at the crack tips as well as a corresponding heat conduction into the specimen. In this case, it is not possible to use an exponential fit and define ΔT at the crack tip.

4.7 Crack Growth Rate

Although the crack growth shows growth jumps over the 1,000 cycles (see the Fig. 14), an averaged crack growth rate was estimated via linear approximation.

The cases of double crack formation were considered as a whole process, taking into account from when the two cracks developed and which crack growth type developed. For the loading conditions AB35 and EB25, the crack growth rate was calculated for the initial crack, and after splitting to double crack formation for each crack separately according to Eq. (10) (example in Fig. 17).

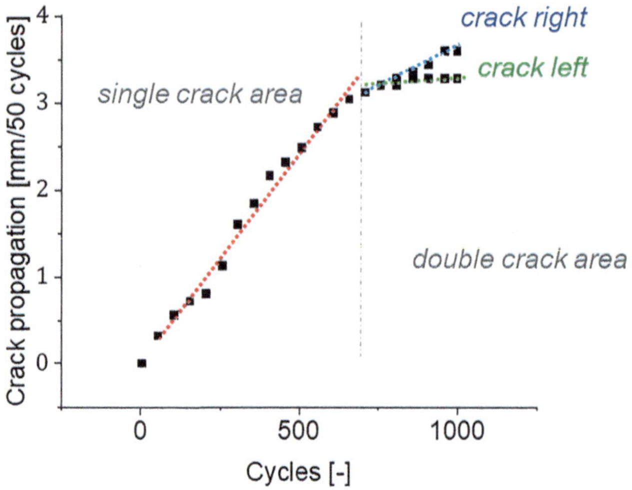

Fig. 17 Crack propagation and linear approximation for the areas with growth of a single crack and double crack for loading condition AB35

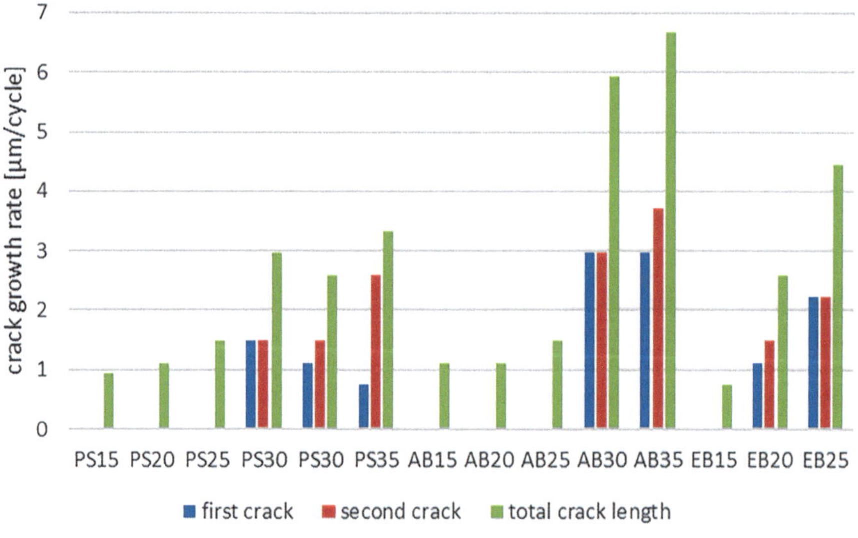

Fig. 18 Crack growth rate of NR20 for different loading conditions and amplitudes (for PS30 two samples were investigated)

An overview about the crack growth rates for all different loading conditions is shown in Fig. 18.

The trend for different loading conditions shows in each case an increase of the crack growth rate with increasing amplitude.

The values of crack growth rate (for a single crack) for PS condition of 15–35% loading and AB condition of 15–25% remain at a comparable level. The average crack growth rate remains below 2 μm/cycle. Higher values are exhibited by the sample AB30. The crack growth rates for double cracks remain below 3 μm/cycle.

4.8 Global Energy and Crack Propagation Under Multiaxial Loading

For the calculation of the energy release rate according to Rivlin & Thomas, an equilibrium curve from the relaxation experiment was taken (Fig. 4), because this approach is relevant for elastic materials.

The tearing energy according to the new method (energy determination by correlations between von-Mises-strain and local sample heating by temperature difference ΔT) was determined by the correlations described in Fig. 7. The correlations for the "Paris Plot" were calculated as

$$\frac{da}{dN} = \Delta w \, h_0, \tag{11}$$

here Δw is the energy density, determined by measuring the von-Mises-strain in the homogeneous stretched area of the specimen; h_0 the initial height of the specimen, da the change in crack length and dN the number of cycles.

The results are shown in Fig. 19.

The results of the classical method (according to Rivlin & Thomas) and the new method (determination of the energy density by von-Mises-strain or local temperature changes) show a good agreement (Fig. 19, left). Closer to each other are the

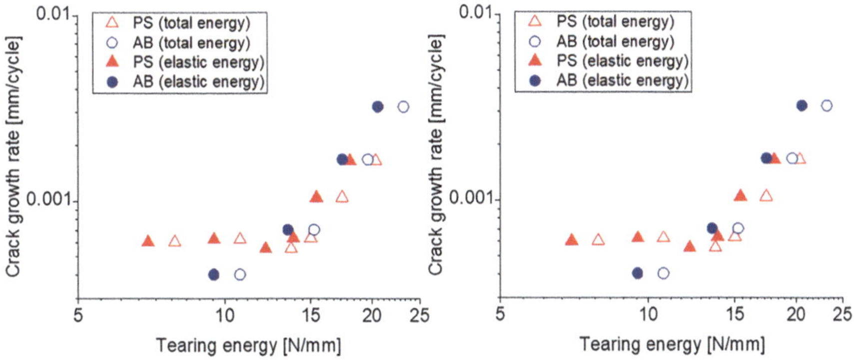

Fig. 19 Left: "Paris plot" for the PS state, calculated according to Rivlin & Thomas and energy determination from the von-Mises-strain, respectively; right: "Paris plot" calculated for PS and AB loading states (with single crack growth) from the von-Mises-strain of a homogeneous stretched area

values of the classical method and calculated values of the elastic curve. This is expected, because during crack propagation elastically stored energy is consumed for crack propagation.

On the right side of Fig. 19, the calculated "Paris plot" is shown according to the new method for PS and AB condition (for the case of a single crack). The trend remains independent of the loading condition and confirms that the relationships between von-Mises-strain of a homogeneous stretched area, ΔT, and Δw determined according to Fig. 7 and Fig. 12 are applicable for the calculation of the tearing energy.

4.9 Tearing Energy for Double Crack Formation

In order to evaluate the crack growth rate and tearing energy in the case of double crack formation, the elastic energy components calculated from a homogeneous stretched area of a specimen were analysed as a function of the actual pattern. The total available elastic energy of the homogeneous stretched area was distributed to both cracks in the case of double crack formation. The results are shown in Fig. 20.

The results confirm that a crack growth process depends on the total available energy and remains independent of the loading condition.

The calculation of global available energy from homogeneous stretched area of a specimen can be determined by means of correlations with von-Mises-strain measurement, or ΔT measurement (shown in Figs. 7 and 12, respectively). The elastic energy contributions calculated by this method can be applied for the analysis of crack growth behaviour in the case of crack doubling or crack deviation.

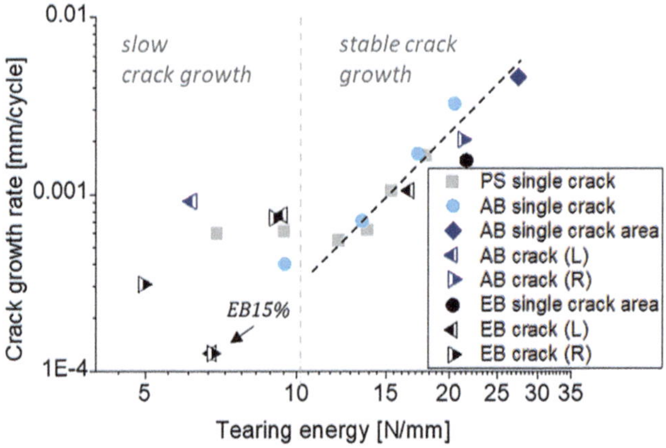

Fig. 20 Tearing energy for all examined states. For double cracking, the areas with one crack or with double cracks (L-crack left, R-crack right) are shown separately

5 Conclusions

The relationships between the von-Mises-strain, dissipative heating and crack propagation were analysed and applied to characterise the material behaviour and the propagation of single or double cracks in filled natural rubber.

The temperature data during dissipative heating a sample, as well as the heat transfer to the environment were described in detail.

The presented combined mechanical and thermal analysis under complex loading conditions shows that for homogeneous deformation and cyclic loading a linear correlation exists between the von-Mises-strain, sample heating, and the dissipated energy in the material. Using a thermal model, it was possible to correlate the energy dissipation with the surface temperature of the sample.

Based on energetic considerations and separating the dissipated energy, the energy-elastic and entropy-elastic components of the internal energy were separated. Thus, the mean ratio of the energy-elastic to total elastic energy of approx. 0.41 was determined for NR20.

When evaluating the crack growth in the case of double cracking in NR20, it was found that the occurrence of the double cracks in the material depends on the loading condition and that the elastic energy stored in the homogeneously stretched specimen area is distributed to the crack tips, considering length and angle of the crack(s). The method describing the relationship between energy dissipation and von-Mises-strain was compared with the classical method (Rivlin & Thomas) and showed good agreement.

The method offers the possibility to analyse the behaviour of crack growth under complex loading conditions independent of the type of loading and whether there is a single or multiple cracks.

Acknowledgements The authors thank Prof. Gert Heinrich for his expert advice. Kristina Eichhorn, Dan Pornhagen, and Dr. Fei Xiang are acknowledged for their assistance with the experiments and Prof. S. Wießner, Dr. Jan. Domurath, Dr. Eric Euchler, Dr. Kai Uhlig for valuable discussions.

References

1. Rivlin RS, Thomas AG (1952) Rupture of rubber. I. Characteristic energy for tearing. J Polym Sci 3:291–318
2. Griffith AA, Taylor GI (1921) The phenomena of rupture and flow solids. Philos Trans R Soc Lond A 221:163–198
3. Mars WV, Fatemi A (2002) A literature survey on fatigue analysis approaches for rubber. Int J Fatigue 24:949–961
4. Gent AN, Lindley PB, Thomas AG (1964) Cut growth and fatigue of rubbers. I. The relationship between cut growth and fatigue. J Appl Polym Sci 8:455–466

5. Schieppati J, Schrittesser B, Wondracek A, Robin S, Holzner A, Pinter G (2018) Impact of temperature on the fatigue and crack growth behavior of rubbers. Procedia Struct Integr 13:642–647
6. Stadlbauer F (2014) Ermüdung und Risswachstum von Elastomerwerkstoffen unter praxisrelevanter Beanspruchung. Dissertation, TU Wien
7. Schieppati J, Schrittesser B, Wondracek A, Robin S, Holzner A, Pinter G (2021) Temperature impact on the mechanical and fatigue behavior of a non-crystallizing rubber. Int J Fatigue 144: 106050
8. Luo W, Li M, Huang Y, Yin B, Hu X (2019) Effect of temperature on the tear fracture and fatigue life of carbon-black-filled rubber. Polymers 11(5):768
9. Brüning K, Schneider K, Heinrich G (2013) In-situ structural characterisation of rubber during deformation and fracture. Fracture mechanics and statistical mechanics of reinforced elastomeric blends. Springer, Berlin
10. Stadlbauer F (2013) Einfluss des Rußfüllgrades von Elastomeren auf Aspekte der Ermüdung in Zug-Druck-Belastung. Kautschuk Gummi Kunststoffe 7–8:37–42
11. Xiang F, Schneider K, Heinrich G (2020) New observations regarding fatigue crack paths and their fracture surfaces in natural rubber: influences of R-ratio and pre-load. Int J Fatigue 135:1–13
12. Schulze A, Heinrich G, Gehde M, Lombardi R (2014) Vorstellung zweier Methoden zur Auswertung von Rissen in Elastomerprüfkörpern. PolyMerTec, Merseburg
13. Stoček R, Kratina O, Ghosh P, Malac J, Mukhopadhyay R (2017) Influence of thermal ageing process on the crack propagation of rubber used for tire application. In: Grellmann W, Langer B (eds) Deformation and fracture behaviour of polymer materials. Springer, Berlin, pp 351–364
14. Marano C, Boggio M, Cazzoni E, Rink M (2014) Fracture phenomenology and toughness of filled natural rubber compounds via the pure shear test specimen. Rubber Chem Technol 87(3): 501–515
15. Stoček R, Stěnička M, Zádrapa P (2020) Future trends in predicting the complex fracture behaviour of rubber materials. In: Hutter K, Müller I (eds) Continuum mechanics and thermodynamics, vol 33. Springer, Berlin, pp 291–305
16. Stoček R, Heinrich G, Gehde M, Kipscholl R (2013) Analysis of dynamic crack propagation in elastomers by simultaneous tensile- and pure-shear-mode testing. In: Grellmann et al (eds) Fracture mechanics and statistical mechanics of reinforced elastomeric blends, LNACM, vol 70, Springer, Berlin, 269–301
17. Stoček R (2012) Dynamische Rissausbreitung in Elastomerwerkstoffen. Dissertation, TU Chemnitz
18. Andreini G, Straffi P, Cotugno S, Gallone G, Polacco G (2013) Crack growth behavior of styrene-butadiene rubber, natural rubber, and polybutadiene rubber compounds: comparison of pure-shear versus strip tensile test. Rubber Chem Technol 86(1):132–145
19. Ghosh P, Stocek R, Gehde M, Mukhopadhyay R, Krishnakumar R (2014) Investigation of fatigue crack growth characteristics of NR/BR blend based tyre tread compounds. Int J Fract 188:9–21
20. Ghosh P, Mukhopadhyay R, Rajasthan K, Stocek R (2016) Durability prediction of NR/BR and NR/SBR blend tread compounds using tear fatigue analyser. KGK 6:53–55
21. Cherepanov GP (1967) Crack propagation in continuous media. PMM 31(3):476–488
22. Rice JA (1968) Path independent integral and the approximate analysis of strain concentration by notches and cracks. J Appl Mech 35:379–386
23. Wunde M, Plagge J, Klüppel M (2019) The role of stress softening in crack propagation of filler reinforced elastomers as evaluated by the J-integral. Eng Fract Mech 214:520–533
24. El Yaagoubi M, Juhre D, Meier J, Kröger N, Alshuth T, Giese U (2018) Lifetime prediction of filled elastomers based on particle distribution and the J-integral evaluation. Int J Fatigue 112: 341–354

25. Wunde M, Klüppel M, Vatterott C, Tschimmel J, Lacayo-Pineda J, Schulze A, Heinrich G (2019) Verbesserung der Laborvorhersagen zum Risswachstum und Verschleiß von LKW-Reifenlaufflächen. Kautschuk Gummi Kunststoffe 10:72–78
26. Nguyen TD, Govindjee S, Klein PA, Gao H (2005) A material force method for inelastic fracture mechanics. J Mech Phys Solids 53:91–121
27. Caimmi F, Calabrò R, Briatico-Vangosa F, Marano C, Rink M (2015) J-integral from full field kinematic data for natural rubber compounds. Strain 51:343–356
28. Calabro R (2013) Mechanical characterization of elastomers under quasi-static and dynamic biaxial loading conditions. Dissertation, Politecnico di Milano
29. Lombardi R, Schneider K, Kipscholl C, Cotugno S, Straffi P, Mensitieri G, Heinrich G (2014) Experimental investigation and constitutive modelling of a filled rubber under dynamic multi-axial loading conditions. Conference paper, RubberCon, Manchester, UK
30. Mai T-T, Urayama K (2021) Biaxial loading effects on strain energy release rate and crack-tip strain field in elastic hydrogels. Macromolecules 54(10):4792–4801
31. Meyer KH, Ferri C (1935) Sur l'élasticité du caoutchouc. Helv Chim Acta 18(1):570–589
32. Treloar LRG (1975) The physics of rubber elasticity. Oxford University Press, New York
33. Dedova S, Schneider K, Heinrich G (2017) Lion A, Johlitz M (eds) Influence of dissipative specimen heating on the tearing energy of elastomers estimated by global and local characterization methods. Constitutive models for rubber X. CRC Press/Balkema, London, pp 219–224
34. Dedova S, Schneider K, Heinrich G (2019) Energy based characterization of fracture and fatigue behaviour of rubber in complex loading conditions. In: Huneau B, Le Cam J-B, Marco J, Verron E (eds) Constitutive models for rubber XI. CRC Press/Balkema, London, pp 363–367
35. Dedova S, Schneider K, Stommel M, Heinrich G (2020) Dissipative heating, fatigue and fracture behaviour of rubber under multiaxial loading. In: Heinrich G, Kipscholl R, Stoček R (eds) Fatigue crack growth in rubber materials. Advances in polymer science, vol 286. Springer, Cham, pp 421–443. https://doi.org/10.1007/12_2020_75
36. Dedova S, Schneider K, Heinrich G (2018) Biaxial testing system for high static and dynamic loading of elastomers and its application. KGK – Kautschuk Gummi Kunststoffe 6:85–90
37. Bergström JS, Boyce MC (1998) Constitutive modeling of the large strain time-dependent behavior of elastomers. J Mech Phys Solids 46:931–954
38. Lion A (1996) A constitutive model for carbon black filled rubber: experimental investigations and mathematical representation. Contin Mech Thermodyn 8:153–169
39. Allen BG, Kirkham MJ, Padget J, Price C (1971) Thermodynamics of rubber elasticity at constant volume. Trans Faraday Soc 67:1278–1292
40. Heinrich G, Kaliske M, Klüppel M, Mark JE, Straube E, Vilgis TA (2003) The thermoelasticity of rubberlike materials and related constitutive laws. J Macromol Sci A 40:87–93
41. Price C, Evans AK, de Candia F (1973) Thermoelastic and thermomechanical studies on natural rubber vulcanized in the swollen state. Polymer 14:338–342
42. Shen M (1969) Internal energy contribution to the elasticity of natural rubber. Macromolecules 2(4):358–364
43. Wolf FP, Allen G (1975) The energetic contribution to rubber elasticity in the range of small uniaxial compression and moderate elongation. Polymer 16:209–217
44. Mohsin MA, Treloar LRG (1987) Thermoelastic measurements of some elastomers under extension and torsion. Polymer 28:1893–1898
45. De Candia F, Amelino L, Price C (1972) Elastic and thermoelastic behaviour of cis-polybutadiene crosslinked in the swollen state. J Polymer Sci Part A-2 10:975–981
46. Wunde M, Klüppel M (2020) Review on the role of phase morphology and energy dissipation around the crack tip during fatigue crack propagation of filler-reinforced elastomer blends. In: Heinruch G, Kipscholl R, Stoček R (eds) Advances in polymer science. Fatigue crack growth in rubber materials. Springer, Berlin, pp 245–271

Adv Polym Sci (2024) 294: 171–192
https://doi.org/10.1007/12_2024_172

Published online: 21 March 2024

Including Temperature Effects in the Theory and Simulation of Problems in Rubber Reinforcement

Reinhard Hentschke

Contents

1 Introduction

Both theory and simulation of polymer materials usually contain temperature as a variable – even though athermal models are not infrequent. Therefore, it is a futile effort to try presenting a comprehensive overview over all temperature-dependent results obtained thus far. The following approach is less ambitious, concentrating on three distinct examples utilizing different simulation strategies to overcome the so-called "scale problem." This should provide the reader with an understanding of how to employ simulations creatively, combined with theory when needed, in the context of reinforced rubber materials.

Reference [1] was published two decades ago but still provides a comprehensive overview over the simulation techniques generally useful for polymer systems. Roughly speaking, there are two types of methods – depending on whether the quantities of interest are related to transport or not. Coefficients of mass and thermal

R. Hentschke (✉)
School of Mathematics and Natural Sciences, Bergische Universität, Wuppertal, Germany
e-mail: hentschk@uni-wuppertal.de

diffusion or viscosity coefficients require simulation methods that contain time explicitly – which generally is Molecular Dynamics (MD). MD simulations follow the time trajectory of the masses based on the Hamiltonian of the system [2, 3]. Usually the masses are the atoms or suitably constructed larger subunits of the polymer. The interaction between the atoms or between the larger subunits is described by phenomenological force fields. In order to avoid atoms and subunits we shall use the term "particles" from hereon. Simulations using particles larger than single atoms are usually referred to as being "coarse grained." If there is no explicit time-dependence underlying the quantities of interest, e.g. phase equilibria, the so-called Monte Carlo (MC) method, which usually means Metropolis Monte Carlo, is an option often preferable to MD [2, 4] (cf. Chapter "Thermodynamics and Molecular Simulation"). MC moves are accepted or rejected, creating a list of properly weighted state points for the calculation of averages, based on ratios of state probabilities which involve the internal energies of the states and attendant types of work. MC can be very efficient since the sought-after quantities are state functions, which means that unphysical MC moves are allowed [2]. An example are simulations of adsorption equilibria during which MC insertion and deletion moves of the adsorbing particles accelerate the calculation significantly. Temperature enters into MD simulations via the motion of the particles and can be extracted using, for instance, the equipartition theorem. In order to adjust temperature in an MD simulation a suitable thermostat has to be employed. "Suitable" means that the thermalized system represents one of the common thermodynamic ensembles. Temperature enters MC simulations directly via the Boltzmann factors weighing the probabilities of the state points. Thus, both methods allow to obtain the temperature dependence of the respective quantities of interest.

However, both methods suffer from the "scale problem" (and this is even more true for the so-called semi-empirical or ab-initio quantum simulation techniques). Using a typical mass m, a typical particle size σ, and a typical interaction particle-particle energy ϵ, one obtains a typical time

$$\tau = \sqrt{\frac{m\,\sigma^2}{\epsilon}}. \tag{1}$$

If m is the mass of an atom or a small molecule, then τ is on the order of picoseconds and σ is in the sub-nanometer range. Rubber generally is a multi-component material containing, for instance, large amounts of filler. The smallest material sample which captures (most of) the macroscopic material properties is close to the mm-range. In addition, technically relevant frequencies can be as low as 1 Hz or less. These are linear dimensions and times/frequencies separated by many orders of magnitude from simulations using chemically realistic force fields. In other words, it is very hard to connect the chemical composition of a material to its macroscopic properties – and possibly predict how the latter change when the material's recipe is altered.

A technique which does not suffer from this predicament is the Finite Element Method (FEM) (e.g., [5]). It numerically solves partial differential equations (PDE) describing macroscopic flow, transport, elastic response, etc. in complex geometries and complex boundary conditions. However, the PDEs contain material coefficients, which must be measured, looked up – or obtained by simulation of the aforementioned type. In the next section we shall discuss the coefficient of thermal diffusion in a strained polymer matrix as one representative example. The scale problem can also be overcome by focusing on suitable laboratory indicators for macroscopic material properties using a combination of theory and simulations of comparatively small systems consisting of a few thousand atoms. An example is the loss tangent. Its peak serves as indicator for "grip" in the case of tire tread rubbers, whereas its behavior for temperatures around 50°C or higher indicates how much rolling resistance to expect. A second example discusses an MD simulation of the interaction between two silica filler particles inside a cross-linked polyisoprene matrix at different temperatures. It is shown how time-temperature superposition can be useful to extrapolate from the high frequencies of atomistic MD to the low frequencies used during mechanical testing. This simulation also shows how a loss mechanism distinctly different from viscous loss, which was predicted theoretically, overtakes the latter's effect on the loss tangent of rubber, tanδ, at high temperatures and therefore contributes significantly and perhaps dominantly to rolling resistance of tires. The study also suggests approaches to reduce rolling resistance by material alterations on the molecular level, i.e. by reinforcing the polymer in the immediate vicinity of the filler particles (e.g., by increasing the cross-link density in this region). Sacrificing a significant amount of the materials chemical identity the scale problem may be overcome by rather massive coarse graining. The smallest unit of the polymer within the network is its Kuhn length and filler particles are modelled via spheres outfitted with simple surface potentials. Rather than presenting an example of this type we choose to focus on very simple simulations of Morse chains. This simple model can be used to study the rupture of polymer chains under stress and allows to obtain simplified failure envelopes. In particular, the influence of temperature on chain failure can be studied. Since the model is so simple it allows a clearer view of the role of the individual polymer strands within a network during failure.

2 Molecular Simulation of Thermal Conductivity Inside a Strained Rubber

Temperature strongly affects the performance parameters of elastomer-based materials. Important examples of such parameters are rolling resistance and grip in the case of an automobile tire. The design of a complex multi-component elastomer-based product will, at some point, involve modelling based on macroscopic material equations solved via finite element codes. These macroscopic equations contain material parameters, like the thermal conductivity and others as well, taken from standard sources. However, inserting fixed parameter values is not always a good method. This is because a material parameter rarely is a mere constant. Continuous advances of computer technology as well as computer simulation methodology nowadays permit the computation of material parameters for variable thermodynamic and mechanical conditions, providing fast and independent access to most material parameters. This approach directly links a parameter of interest to a material's chemical structure and possibly complex external conditions. The simulation method of choice in the case of transport properties is MD.

In the following we concentrate on the thermal conductivity, which in an amorphous polymer material is rather small, i.e. around 0.1 W/mK. Before we discuss one specific example in greater detail, it is useful to provide a brief overview. For a more comprehensive discussion the reader is referred to the review by Algaer and Müller-Plathe [6] on molecular dynamics calculations of the thermal conductivity in molecular liquids, polymers, and carbon nanotubes.

Most simulation studies initially focused on single extended polymer chains [7–11] – in particular polyethylene. An important motivation for this was the experimentally observed increase of thermal conductivity in chain direction, attributed to phonons along the backbone, depending on overall chain orientation or strain. Techniques used in these studies include equilibrium and nonequilibrium molecular dynamics. By comparison, amorphous polymers, e.g., amorphous polystyrene, polycarbonate, polyvinyl chloride, or poly(methyl methacrylate), exhibit a smaller thermal conductivity. Since bulk materials are technically more relevant compared to single chains, it is worth to briefly list some instructive examples. Below we shall discuss one system in detail, where we employ equilibrium statistical mechanics to obtain transport coefficients. However, nonequilibrium methods are perfectly fine as well. This approach was used, for instance, by Lussetti et al. [12], who consider the thermal conductivity in amorphous polyamide-6,6. Using models differing in atomic detail the study finds that the thermal conductivity indeed depends on the number of degrees of freedom. In addition, the thermal conduction is found to be faster parallel to the strained direction. Another application of a nonequilibrium method is described in Rossinsky and Müller-Plathe [13]. Again, the focus is on anisotropy. Here, the material is crystalline polystyrene. The thermal conductance across different interfaces is studied by Ni et al. [14]. The systems are composed of oriented single crystal diamond surfaces and covalently bonded as well as aligned polyethylene chains. In addition, the authors analyze the influence of cross-linking and defects. Using both and comparing nonequilibrium as well as equilibrium MD, Varshney et al. [15] study a cross-linked polymer system, i.e. an epoxy-based thermoset polymer. Enhancement of thermal conductivity of all-atom polyethylene as a function of strain rate is studied by Liu and Yang [16] via MD. The authors find that the thermal conductivity is larger when the strain rate is reduced. Another interesting molecular parameter in this work is molecular weight, which is found to affect the thermal conductivity. The molecular weight dependence is studied more thoroughly by Zhao et al. [17]. These authors use united-atom nonequilibrium molecular dynamics to simulate amorphous polyethylene at ambient conditions varying the chain length from 4 to 1,260. Thermal transport in these systems is dominated by a collision mechanism when the chains are short. For long chains a phonon vibration mechanism dominates. Another example discussing thermal transport affected by boundaries/interfaces is the work of Alaghemandi et al. [18]. The method in this study is reverse nonequilibrium MD applied to polyamide-6,6 oligomers confined by sheets of graphene. Depending on the spacing of the latter, the thermal conductivity parallel to the sheets is enhanced significantly if compared to the bulk [19]. In the aforementioned detailed example we shall focus on bulk polyisoprene. This commercially important polymer has been studied frequently. In He et al. [20] the authors study isoprene oligomer samples of short chains (10 monomer units) without chemical cross-links as well as a quasi-crystalline sample composed of straight chains. The focus is on the anisotropy of the thermal conductivity.

Experimental measurements of the thermal conductivity in elastomer materials began to appear rather early, i.e. in the 1930s [21]. Barnett's work already includes the effect of fillers on the thermal conductivity. Note also that this reference mentions even older work from the 1920s. The early publication of such measurements is not too surprising if one considers the growing use of these elastomers in applications involving material deformation at frequencies where heat generation becomes an issue. Already early in the 1930s Barnett studied the effect of fillers on thermal conductivity [21]. Also noteworthy is the subsequent paper by Dauphinee et al. [22]. These authors discuss the thermal conductivity of elastomers as function of strain and temperature. There are two reviews from this early era, compiling data up until 1966, which can still be read with profit today. The first is Anderson [23] and the second is Carwile and Hoge [24]. The description of experimental techniques as well as detailed measurements of thermal conductivity of carbon black filled rubber can be found in Ref. [25].

In the remainder of this section we present an example taken from Ref. [26]. In this work the authors study the thermal conductivity tensor in a bulk sample of cis-1,4-polyisoprene (PI) rubber. The united-atom force field of the polymer is that of Harmandaris et al. [27]. An important step in the simulation of amorphous polymers is the generation of (cross-linked) chains possessing the correct conformation statistics (cf. Figs. 1 and 2). Here, this is accomplished by polymerization from a liquid of monomers [28]. Initially a number of "starter" monomers are selected. Subsequently a new bond is established between the starter monomer and another free monomer in close proximity. The newly added monomer then becomes the new starter monomer. Depending on the search radius used to find new monomers the algorithm produces polymers with variable dispersity. During the polymerization the density in the system changes and at the end of the polymerization should be close to the experimental value. Additional checks of the final system include the calculation of the thermal expansion coefficient and the calculation of the characteristic ratio (amorphous chains should be ideal). On this basis the authors of [26] study a number of different systems distinguished by polymer molecular weight, different chemical cross-link densities, and temperature. The systems are subjected to volume conserving uniaxial strain of up to 200%. Here, the GROMACS simulation package [29] is used mainly because of its flexibility.

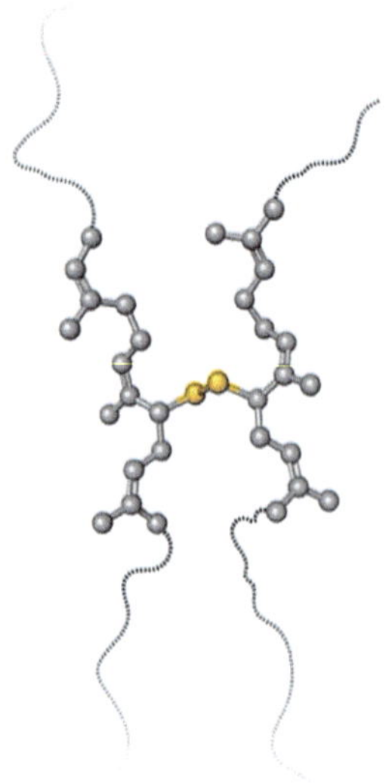

Fig. 1 Type and position of the sulfur bridges

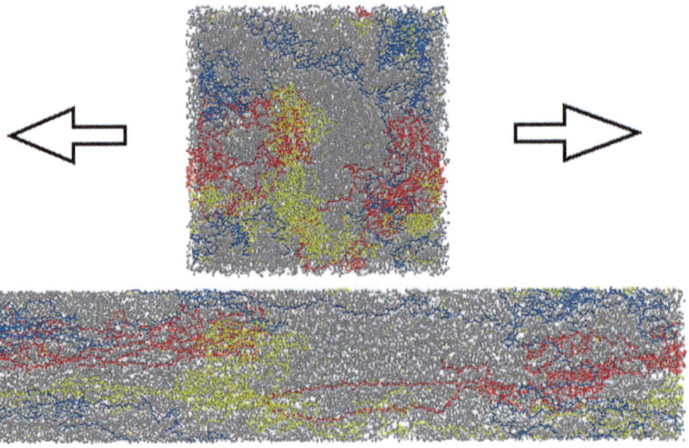

Fig. 2 Example showing one particular system in the undeformed state (top) and at 200% strain (bottom). Selected polymer chains are highlighted in different colors. The system consists of 20 PI chains containing 250 monomers each and 500 sulfur cross-links

The components of the thermal conductivity tensor $\Lambda_{\alpha\beta}$ can be obtained from equilibrium atomistic MD simulations using the following Green-Kubo formula [2, 3], which essentially is an integral over the equilibrium autocorrelation function of the components of the heat current $\vec{J}$:

$$\Lambda_{\alpha\beta} = \frac{V}{k_B T^2} \int_0^\infty dt \left\langle J_\alpha(t) J_\beta(0) \right\rangle. \tag{2}$$

$J_\alpha(t)$ is the instantaneous α-component of the heat current at time t computed via

$$J_\alpha(t) = \frac{1}{V}\sum_{i=1}^{N} v_{\alpha,i}(t)\,\delta e_i(t) + \frac{1}{2V}\sum_{\substack{i,j=1 \\ i\neq j}}^{N} r_{\alpha,ij}(t) f_{\beta,ij}(t) v_{\beta,i}(t)\,. \tag{3}$$

In the first sum on the right side, the quantity $v_{\alpha,\,i}(t)$ is the α-component of the instantaneous velocity of atom i and $\delta e_i(t)$ is the instantaneous energy fluctuation of the same atom. This term is the part of the heat current due to the atoms traveling with a certain velocity while possessing a certain energy deviation relative to the average energy per atom in the system. In the second sum $r_{\alpha,\,ij}$ is the α-component of the vector $\vec{r}_i - \vec{r}_j$, the difference of the position vectors of atoms i and j. $f_{\beta,\,ij}(t)$ is the β-component of the force on atom i due to atom j. Note that here and below we use the summation convention on β. This term describes the interaction contribution to the heat current. The starting point for the derivation of Eq. (2) is the macroscopic transport equation

$$J_\alpha\left(\vec{r},t\right) = -\Lambda_{\alpha\beta}\partial_\beta \delta T\left(\vec{r},t\right) = -V\frac{\Lambda_{\alpha\beta}}{C_V}\partial_\beta \delta e\left(\vec{r},t\right) \tag{4}$$

in conjunction with the continuity equation linking the divergence of the current $\vec{J}\left(\vec{r},t\right)$ with the time derivative of $\delta e\left(\vec{r},t\right)$. Details of the derivation, which itself is quite general and thus applies to other transport coefficients as well, can be found in appendix A of Ref. [26].

Figure 3 shows the diagonal components of the thermal conductivity tensor at $T = 300K$ vs. strain, u. The strain is applied along the z-direction. The experimental data measured along this direction are taken from Tautz [30]. Notice the clearly discernible increase of the thermal conductivity in the long direction in good accord with the experimental data. The bottom panel illustrates that the increase of Λ_{zz} above $u = 1$ is more pronounced when the cross-link density is increased. The thermal conductivity in the orthogonal direction remains quite constant over the entire strain range.

3 Theory and Simulation of the Temperature Dependence of tanδ

There are laboratory indicators of the performance of a rubber product in its intended environment. These indicators are useful in the sense that they can be routinely measured in the laboratory without too much effort. One such laboratory indicator is the loss tangent, i.e. the ratio of the loss modulus μ'' to the storage modulus μ'. The sketch in Fig. 4 shows tanδ versus temperature for an unfilled elastomer and for the same elastomer when it contains a (large) amount of filler. As the sketch shows, the

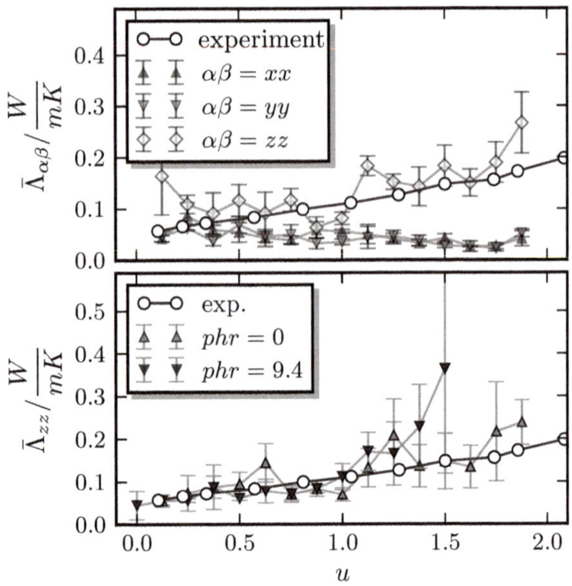

Fig. 3 Top – Diagonal components of the thermal conductivity tensor at $T = 300K$ vs. strain u. The experimental data measured along the long direction are taken from [30]. The number of cross-links is about 100 or about 1.9 phr. Bottom – Thermal conductivity along the stretch direction for systems with very different cross-link concentrations. This figure is taken from Ref. [26]

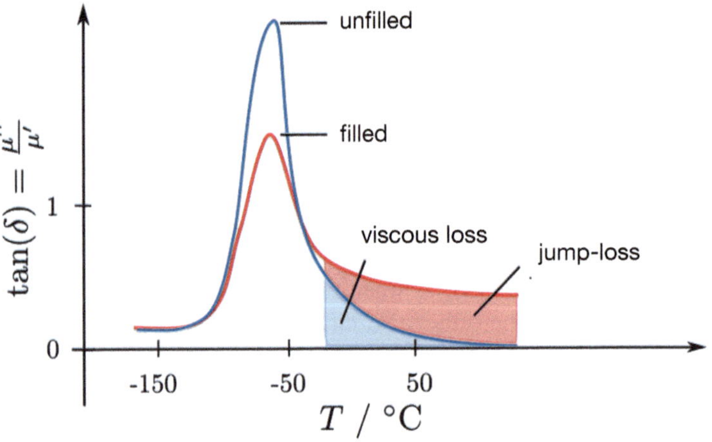

Fig. 4 Sketch of the temperature dependence of the loss tangent of a rubber, the ratio of the loss modulus μ'' to the storage modulus μ', showing the effect of added filler

unfilled elastomer network features a higher peak. Beyond the peak on its high-temperature side, however, a cross-over between the two curves occurs and the filled elastomer produces the higher tanδ. The significance of this is that in the tire industry the loss tangent at temperatures around 0°C is considered a measure for the grip of a tire, whereas the temperature region around 50 to 60°C is a measure for a tire's rolling resistance. Clearly, it is desirable to control this behavior and possibly adjust the material's recipe to "optimize" tanδ.

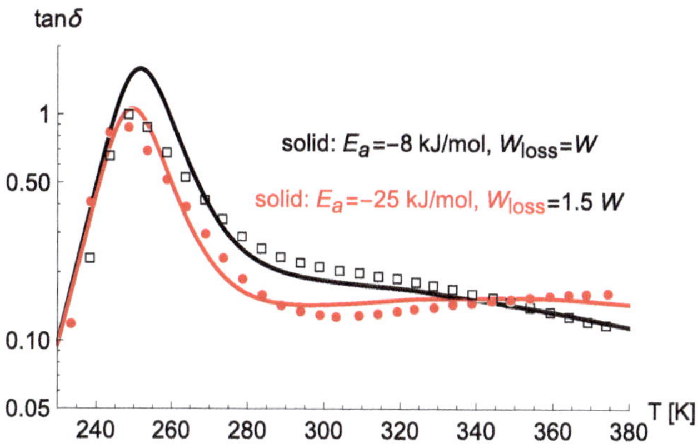

Fig. 5 Experimental tanδ vs. temperature extracted from Fig. 29 in reference [31], obtained for filled (50phr) SSBR Duradene 715, (symbols) compared to Eq. (1) in Ref. [32] (lines). The dynamic strain amplitude is 5%. Hollow squares: filler is carbon black (N234); red solid circles: filler is silica (Hi-Sil 210)

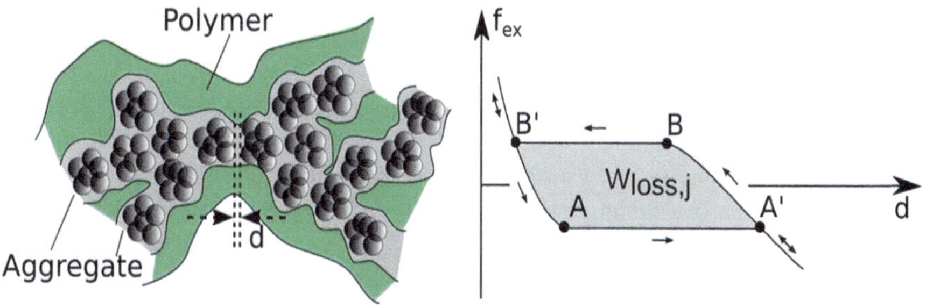

Fig. 6 Left: Cartoon of a filler network branch consisting of aggregates containing (primary) filler particles. For one of the aggregate-to-aggregate contacts the particle-to-particle separation, d, is indicated. Right: d increases and decreases due to an external cyclic force. The combination of the direct particle–particle interaction in conjunction with the elastic restoring force due to the surrounding polymer matrix produces a hysteresis (here the shaded area) $W_{loss,\,j}$ which, together with the viscous loss in the polymer matrix $W_{loss,\,v}$, is the dissipated work W_{loss} in this contact during the strain cycle

Here, we discuss an approach which is a theoretical one complemented by information from atomistic simulations. Figure 5 shows a comparison between the theory in Ref. [32] and experimental data obtained for filled (50phr) SSBR (Duradene 715) when the filler is changed from carbon black to silica. The theory assumes that the filler forms a network in the elastomer matrix (cf. the left sketch in Fig. 6) surrounded by strongly bound rubber (gray area). The quantities $E_a = -$ 8 kJ/mol to $E_a = -25$ kJ/mol are typical physical bond energies (corresponding, for instance, to physisorbed chain segments or hydrogen bonds) in the respective system. The quantity W_{loss} is a measure for the dissipated energy during the cyclic deformation of the aggregate-to-aggregate contacts in a filler network strand. W_{loss} is

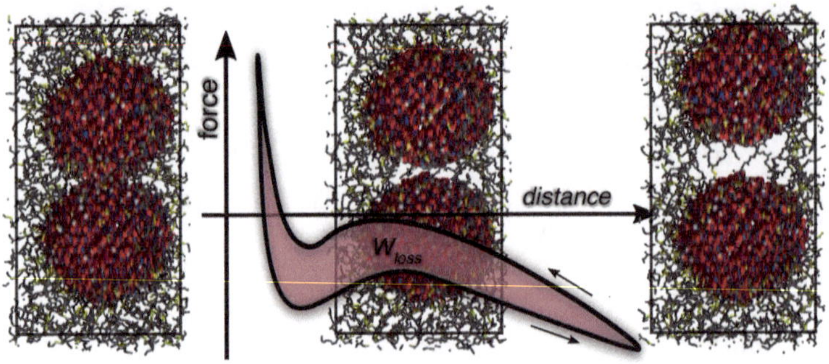

Fig. 7 Cartoon of two silica particles embedded within a cross-linked PI matrix including silane bonding to the particle surfaces. The diagram in the foreground illustrates the energy loss during a cyclic change of the interparticle separation

the sum of the viscous loss caused by the deformation of the polymer in the proximity of the contact plus $W_{\text{loss},\ j}$. The latter arises from the direct particle–particle interaction in conjunction with the elastic restoring force, which leads to a mechanic instability or spontaneous "jump" of the particles and an attendant hysteresis (cf. the right sketch in Fig. 6).

W_{loss} in Fig. 5 was estimated via an atomistic computer simulation [33] depicted partially in Fig. 7. Two silica particles are embedded in a cross-linked PI matrix, which is bound to the particle surfaces via silanes. One of the particles is held fix whereas the other one is moved back and forth by a virtual spring with spring constant k_o. The work done during the spring cycles (cf. the inset in Fig. 7) is recorded. The result is depicted in Fig. 8 for a range of temperatures and two different particle sizes as well as two different spring constants. If the spring constant has a high value (blue data points and curve), only the viscous loss is observed. However, if the moving particle is allowed to respond freely to the interaction with the stationary particle and the surrounding polymer matrix (small spring constant), the aforementioned instability occurs and this causes additional dissipation $W_{\text{loss},\ j}$ (red data and curve). As the temperature is increased, the fractional contribution of $W_{\text{loss},\,j}$ to the total loss W_{loss} increases. Since $W_{\text{loss},\,j}$ is specific to filled elastomers, it explains the high-temperature behavior of tanδ sketched in Fig. 4.

As pointed out in the introduction, atomistic MD simulations are confined to high frequencies. Specifically, the relative motion of the particles corresponds to frequencies in the GHz range. This is shown in Fig. 9 where the simulation data are the red triangles. Using the depicted curve, the high frequencies can be mapped onto the experimental frequency at which the data in Fig. 5 were obtained (here, 10 Hz). While time-temperature superposition works well for unfilled rubber, it works less well for filled rubber. The aforementioned particle–particle and (physical) polymer–particle bonds, for instance, do not obey time-temperature superposition. In other words, the lower of the two temperature axes in Fig. 8 is plagued by this problem. Nevertheless, the function W_{loss} in this figure, in particular its temperature

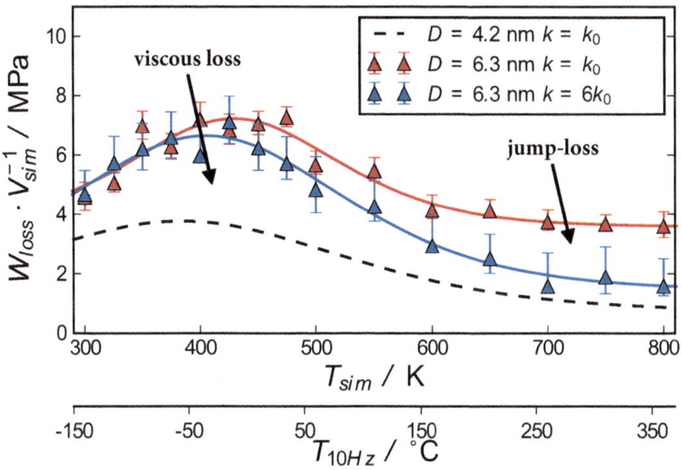

Fig. 8 Effect of different spring constants on the shape of W_{loss} vs. temperature T. The (virtual) spring constant k_o is rather weak and allows the moving particle to respond unimpeded to the forces exerted by the stationary particle and the surrounding cross-linked PI. However, 6 $k_o$is rather rigid, which binds the moving particle to the spring force. The weak spring constant yields the red data, which include the dissipation cause by the mechanism explained in the text, whereas the latter spring constant yields the blue data, whose sole contribution is viscous loss. Note that D is the particle diameter in the simulation

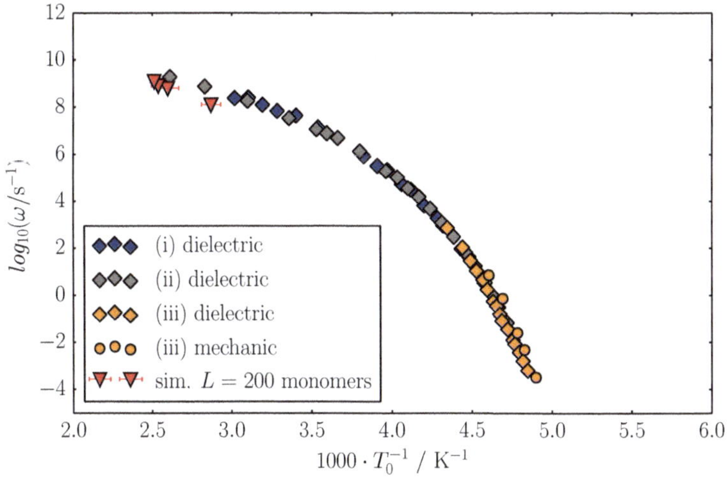

Fig. 9 Frequency vs. inverse temperature of the segment relaxation peak (α-process) for polyisoprene. Experimental molecular weights are between 97.0 and 504.0 kg/mol. This figure is adopted from Ref. [33]

dependence, albeit with adjusted magnitude, is used to obtain the (very reasonable) fits to the experimental tanδ-data in Fig. 5.

4 Failure Envelope of a Morse Chain

According to Wang and Morris [34] (Sect. 6.4.3 in Chapter "Thermodynamics and Molecular Simulation") the mechanism of tensile failure of elastomers is not fully established. They state that tensile failure occurs due to catastrophic tearing by growth of cracks initiated from accidental flaws, microvoids, dewetting, or cavitation from the filler surface. Failure of elastomer materials undoubtedly is an important topic, but predicting failure from theoretical considerations thus far is not possible – quite in line with Wang and Morris' statement that even the underlying mechanism is not well understood. It therefore is sensible to construct theories and simulation schemas aimed not at the detailed description of specific systems but at the understanding of basic mechanisms, including tests of such theories and simulations.

A recent example is Ref. [35]. In this work a theoretical picture describing the tensile strength of elastomers as function of temperature, filler concentration, and particle size as well as cross-link density is developed. However, stress at break always is linked to the strain at break, and this relation could not be established. In the following we want to explore a possible approach to this problem employing a simple simulation model.

Roughly speaking there are two very different scales involved here, i.e. the one on which the energy in the material is distributed and dissipated and the breaking of bonds on the atomic scale. This is why most theoretical attempts (e.g., the early reference by Gul [36] or the recent one by Plagge [37]) at explaining tensile failure in elastomers include Eyring-type expressions for the chain rupture rate k_r, i.e.

$$k_r = \nu\, A \exp\left[-\left(E_b - f\, d\right)/RT\right]\,. \tag{5}$$

The quantity ν is a molecular attempt frequency and A is a factor essentially composed of the partition functions on the two sides of the reaction equation. E_b is a relevant bond dissociation energy. In addition, the chain, due to the externally applied macroscopic stress, experiences a tensile force f. This force is multiplied by the so-called activation length d and reduces the bond dissociation energy.

It is instructive to study the breaking of a single chain segment using a very simple generalization of the freely jointed chain model in which the bonds are breakable, i.e. the bond potential is the Morse potential

$$u_{\text{Morse}}(r) = D\left(1 - \exp\left[-a\left(r - r_o\right)\right]\right)^2 \tag{6}$$

depicted in Fig. 10(a). The quantity D is the depth of the potential minimum at r_o, which is the equilibrium bond length of an isolated bond. The Morse potential is

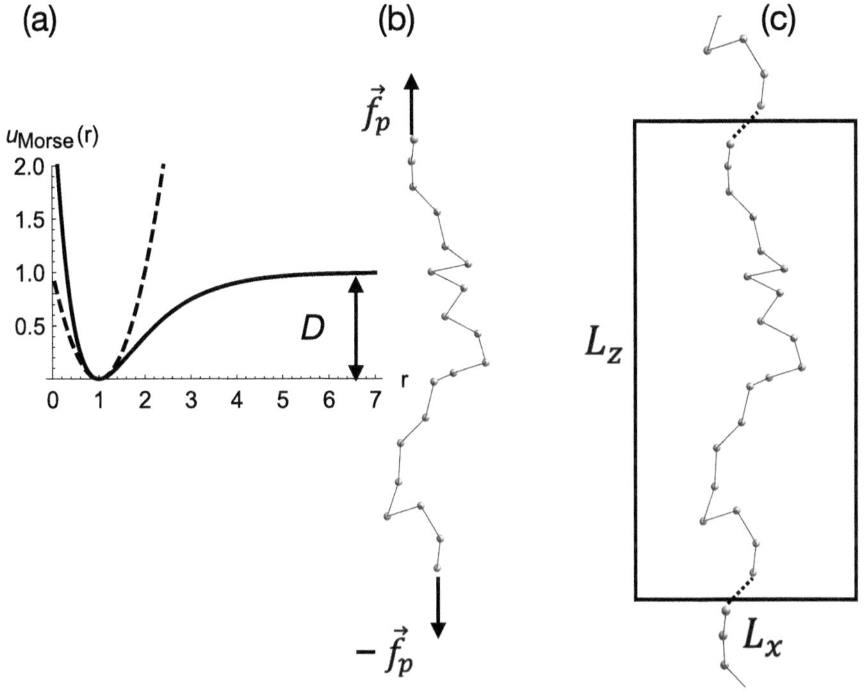

Fig. 10 (**a**) Morse potential ($D = a = r_o = 1$; solid line) including harmonic approximation at r_o. (**b**) Chain of freely jointed Morse bonds. The first and the last bead are pulled in opposite direction by the forces $\vec{f}_p$ and -$\vec{f}_p$, respectively. (**c**) The same chain in a rectangular box with linear dimensions $L_x = L_y$ and L_z. The dotted bonds tie the real chain to its virtual periodic images above and below

harmonic near its bottom, i.e. $u_{\text{Morse}}(r) \approx D\, a^2\, (r - r_o)^2$, but because it possesses a finite depth, the bond will break if the tension force between neighboring beads exceeds $a\, D/2$. There are several reasons for studying such a simple model. One is the activation length. How long is it? Another question is the following. Does the chain segment break when the force which stretches it is dominated by conformation entropy or is significant strain in the individual bonds required? And how does this depend on temperature and possibly chain length, etc.?

The chain is thermalized via Langevin dynamics [38], i.e.

$$\ddot{\vec{r}}_i = \vec{F}_i + \vec{Z}_i - \zeta \dot{\vec{r}}_i. \tag{7}$$

Here, $\vec{r}_i$ is the position of bead i (each bead possesses unit mass), $\vec{F}_i$ is the force on bead i due to neighboring Morse bonds, and $\vec{Z}_i$ is a Gaussian random force on bead i satisfying $\left\langle \vec{Z}_i(t) \cdot \vec{Z}_j(t') \right\rangle = 6\, \zeta\, T\, \delta(t - t')\delta_{ij}$. Here, ζ is a friction parameter and T is the bath's temperature. $\delta(x)$ is the δ-function, whereas $\delta_{ij} = 1$ if $i = j$ and zero otherwise. This in turn will yield an average bead temperature $\left\langle \dot{\vec{r}}_i^{\,2} \right\rangle / 3 = T$. Note that we use units where the Boltzmann constant k_B is equal to one. Langevin dynamics is a variant of MD in which much of the system, here the environment in which the chain moves, is replaced by a heat bath. This method obviously reduces the computational effort. On the other hand, specific and long-range interactions of the chain with its environment are ignored. The method, for instance, does not yield the proper diffusion of a chain in the melt or a solvent due to the neglect of hydrodynamic interactions (e.g., [38]). But for the present purpose it is suitable.

In a first computer experiment we simulate chains with $N = 2$ to $N = 100$ beads at different temperatures. During the simulation a tensile force $\vec{f}_p$ is acting in -z-direction on the first and in z-direction on the last bead of a chain, respectively (cf. Fig. 10(b)). For each temperature we determine the probability $h_{\text{break}}(T)$ of chain dissociation. A chain is considered to be dissociated if at the end of the simulation there exists at least one bond with a length $\geq 5r_o$ (here, we set $r_o = 1$). Other multiples of r_o, e.g. $4r_o$ or $6r_o$, work just as well, i.e. the dissociation criterion is not a sensitive parameter. However, since the activation length d is found to be around $4r_o$ (cf. below), it is reasonable to use a factor multiplying r_o which is (slightly) larger than 4.

Figure 11 (a) shows the probability of chain failure $h_{\text{break}}(T)$ vs. temperature T for $N = 10$ ($a = 1$, $D = 0.6$, $f_p = 0$, and $\zeta = 0.1$). Each simulation run encompasses $2 \cdot 10^6$ time steps of length 0.0025. Each data point in Fig. 11(a) is computed from 10 independent runs at the same temperature. Panel (b) of the same figure shows the temperature T_{break} at which chain rupture occurs (inflection point of the fit to $h_{\text{break}}(T)$) for N between 2 and 100. The error bars indicate the width of the step.

There are two main results here: (i) $T_{\text{break}} \approx 0.1$ when $f_p = 0$, i.e. $D/T_{\text{break}} \approx 10$. This result is not altered significantly by changing a or ζ as indicated. Note that the temperature at which the chains dissociate does depend on the details of the partition functions on the two sides of the reaction as well as on the reaction path. Since the present chain is merely a simple model, one should not expect to find this exact ratio for real polymers. However, quite generally it is observed that the thermal energy $k_B T_{\text{break}}$ is considerably less than what one expects from the mere depths of the bond potentials. (ii) Increasing f_p from 0 to 0.1 reduces T_{break} significantly. The same reduction of T_{break} in comparison with the standard case, i.e. $D = a = 1$, is obtained if $f_p = 0$ but $D \approx 0.6$. This means that in the present case the “activation length” d ≈ 4 (or 4 r_o), which follows via

$$D - f_p d = D' \tag{8}$$

with $D = 1$, f_p=0.1, and $D' = 0.6$. Despite this being a crude argument, it shows that the “activation length” is on the order of perhaps several equilibrium bond lengths and not, for instance, tied to the chain or segment length. The same conclusion was reached by other means in a recent paper studying the fracture of vitrimers [39].

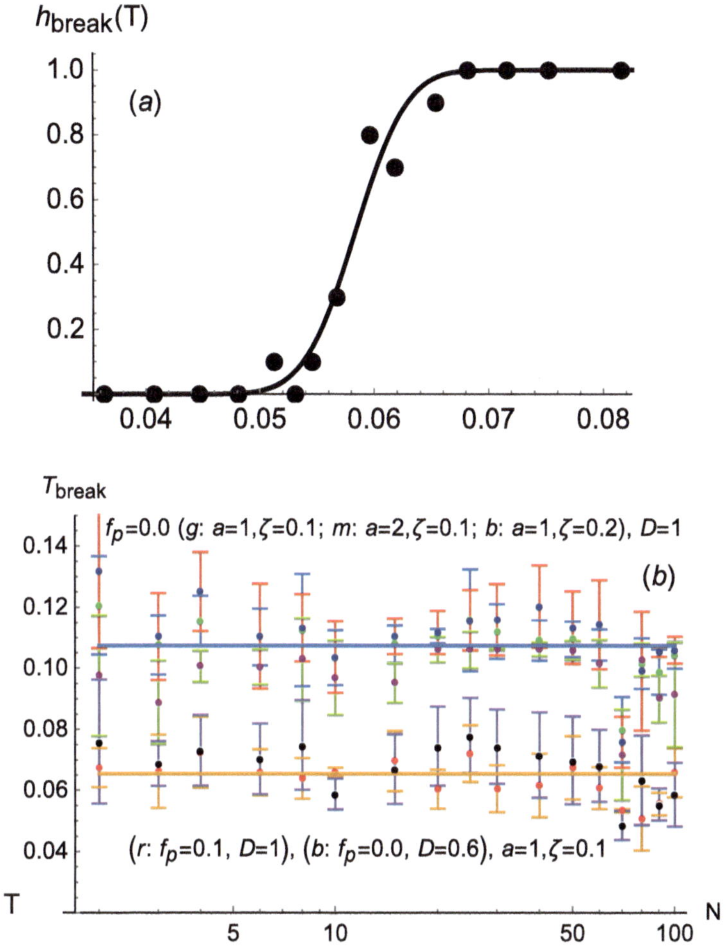

Fig. 11 (**a**) Probability of chain break $h_{\text{break}}(T)$ vs. temperature T for $N = 10$ ($a = 1$, $D = 0.6$, $f_p = 0.1$, and $\zeta = 0.1$). (**b**): Chain rupture temperature T_{break} for N between 2 and 100 for different parameter values. The letters g:, m:, b:, r:, and b: stand for the colors of the dots (green, magenta, blue, red, and black) distinguishing the indicated parameter sets. Horizontal lines are meant to guide the eye. The figure is taken from Ref. [35]

We can modify the setup (cf. Fig. 10(c)) so that it is analogous to the polymer simulation discussed above in the context of thermal conductivity – albeit for a single chain. The strain is given by $\epsilon = \lambda - 1$, where $\lambda = \frac{L_z(t)}{L_z(0)}$, and the stress $\sigma_z(t)$ is given by

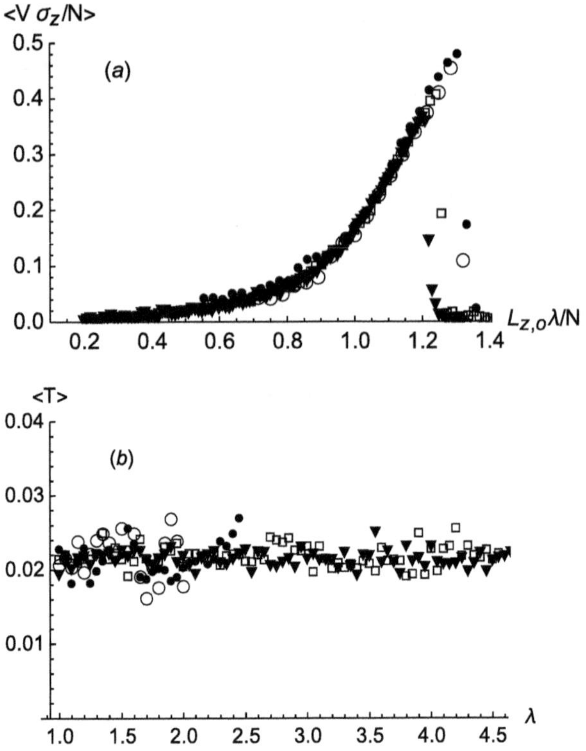

Fig. 12 $\langle V\,\sigma_Z/N\rangle$ vs. $L_{z,\,o}\lambda/N$ (**a**) and $\langle T\rangle$ vs. λ (**b**) for chains consisting of $N = 14$ (open circles), 18 (solid circles), 31 (open squares), and 50 (solid down-triangles) beads. Here, $L_{z,\,o} \equiv L_z(0) = 10$.

$$\frac{V\sigma_z}{N} = \frac{1}{N}\sum_{i=1}^{N-1}\frac{z_{i,i+1}^2}{r_{i,i+1}}\frac{du_{\mathrm{Morse}}}{dr}\bigg|_{r_{i,i+1}} + \frac{1}{N}\frac{z_{1,N}^2}{r_{1,N}}\frac{du_{\mathrm{Morse}}}{dr}\bigg|_{r_{1,N}} - T. \tag{9}$$

The second term accounts for the contribution of the bonds between the actual chain and its periodic images. V is the volume of the simulation box, which is constant during its deformation. The quantity $z_{i,\,i+1}$ is the separation between beads i and $i + 1$ in z-direction, whereas $r_{i,\,i+1}$ is the overall separation of the same two beads. The last term subtracts an ideal gas contribution.

Figure 12 shows $\langle V\,\sigma_Z/N\rangle$ vs. $L_{z,\,o}\lambda/N$ and $\langle T\rangle$ vs. λ for chains consisting of $N = 14$ (open circles), 18 (solid circles) and 31 (open squares) beads. Here, $L_{z,\,o} \equiv L_z(0) = 10$, $T = 0.02$, and the time averages for every data point include 50 units of time at a simulation time step of 0.005. It is sensible to use $\langle V\,\sigma_Z/N\rangle$ instead of $\langle V\,\sigma_Z\rangle$. This assures constant density during deformation. Normalizing chains of different length to equivalent strain by multiplication of λ with $L_{z,\,o}/N$ collapses the curves onto a single master curve. Only the stress at break and the strain at break can vary. Even though Fig. 11(b) shows that breaking occurs at a certain temperature quite independent from the value N, there is considerable scatter. Since the data sets for each N in Fig. 12 are from single simulation runs and the

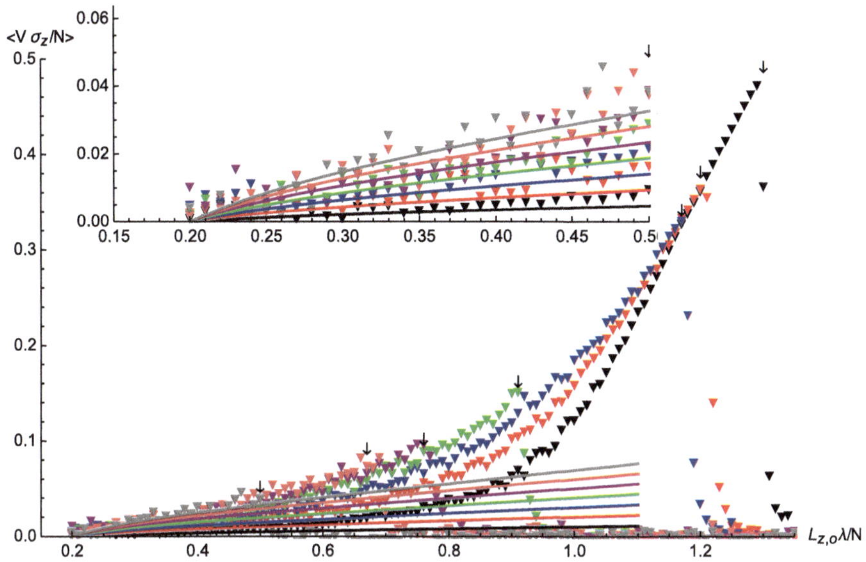

Fig. 13 $\langle V\,\sigma_Z/N\rangle$ vs. $L_{z,\,o}\lambda/N$ for chains consisting of $N = 50$ beads. The solid lines depict the classical result of rubber elasticity at different temperatures. Arrows indicate failure, i.e. chain break, for chains simulated at (from top to bottom) $T = 0.01$ (black), 0.02 (red), 0.03 (blue), 0.04 (green), 0.05 (magenta), 0.06 (pink) and 0.07 (gray)

temperatures also vary along these runs (cf. the right panel in Fig. 12), it is not surprising that tension and extension at break vary.

Along the curve in Fig. 12 we distinguish two strain regions featuring distinct slopes. This can best be understood by comparing the reduced stress-strain curves for different temperature as shown in the next figure, Fig. 13. The lines, whose colors match the color of corresponding data points, represent the classical stress-strain relation of rubber elasticity (Neo-Hooke), i.e.,

$$\frac{V\sigma_{re}}{\nu} = T\left(\lambda - \frac{1}{\lambda^2}\right). \tag{10}$$

The quantity ν is the number of network strands inside the volume V. In particular, $\langle V\,\sigma_Z/N\rangle$ vs. $L_{z,\,o}\lambda/N \equiv x$ must be compared to $T(x - (L_{z,\,o}/N)^3x^{-2})$ vs. x. This shows that the initial increase of the stress with increasing strain is due to the entropic restoring force on the stretched chain. On the other hand, when $\lambda = N/L_{z,\,o}$, which means 1 on the $L_{z,\,o}\lambda/N$-axes in Figs. 12 and 13, the chain is stretched to its full contour length (assuming the bond lengths assume their equilibrium values). Thus for $L_{z,\,o}\lambda/N$ around and above 1 most of the stress is due to the stretching of the Morse bonds, which accounts for the steep increase of the stress-strain curves.

The arrows in Fig. 13 indicate chain breaking for a series of temperature ranging from 0.01 to 0.07. The last value already is close to the thermal breaking of the unstrained chain in Fig. 11. From these points we can construct a failure envelope – originally suggested by Smith [40]. Smith's failure envelope is a plot of log[$\sigma_b T_o/T$], where $\sigma_b$is the stress at break and T_o is an arbitrary reference temperature, versus log [ε_b], where $\varepsilon_b = \lambda_b - 1$ is the strain at break. He observed that the data thus plotted accumulate on a curve which is independent of strain rate and temperature and he concluded that this curve depends only on basic characteristics of the polymer network. Nevertheless, changing the temperature or strain rate shifts the points on the failure envelope, i.e. the fracture point moves clockwise around the envelope, either with rising temperature or with decreasing rate of extension, and limits its predictive power regarding σ_b and ε_b. In addition the failure envelope will be different in the presence of filler and will depend on filler content and cross-link density. Perhaps due to these shortcomings the failure envelope has not received too much attention – even though it is discussed at some length in the recent book by Wang and Morris [34].

Panel (a) in Fig. 14 depicts data taken from Fig. 6.59 in Ref. [34], showing the effects of carbon blacks on failure envelopes of well-vulcanized SBR. Panel (b) contains an analogous plot of the breaking points (indicated for the $N = 50$-chain by arrows in the previous figure). Note that the temperature decreases from left to right starting at $T = 0.07$ in steps of 0.01. The Morse chain is somewhat removed from experimental SB rubber. However, we can, albeit roughly, equate the dissociation temperature $T_{\rm break}$ of about 0.1 in the case of the Morse chain with no force acting on its ends with a thermal degradation temperature of about 450°C for the real polymer (e.g., $T = 0.05$ in model units corresponds to about 90°C). Of course, the temperature range in the model calculation extends down to low temperatures which are unrealistic in comparison with an experimental bulk elastomer. However, provided that the above mapping of the model's temperature to the experimental temperature is reasonable, one may conclude that the relevant breaking occurs in a strain regime which at high temperatures still is dominated by the entropic restoring force but at the lower temperatures is strongly affected by the average strain of the Morse bonds. In any case, there is sufficient similarity between the experimental envelope and the one obtained for the Morse chain to warrant additional consideration.

One might object that already 60 years ago, Bueche has developed theoretical expressions allowing to nicely fit Smith's failure envelope data [41]. Bueche's theory contains a number of adjustable parameters, limiting its predictive power. Perhaps more importantly however, it is a macroscopic theory with significant empirical input. This author thinks that it is desirable to tie the concept of a failure envelope to the molecular structure of the chains. Specifically, it appears that (most of) the shape of the envelope might be deducible from atomistic simulations of isolated polymer chains. There are works where stretching and breaking of isolated realistic chains has been looked at (e.g., [42]). However, this work does not include failure envelopes.

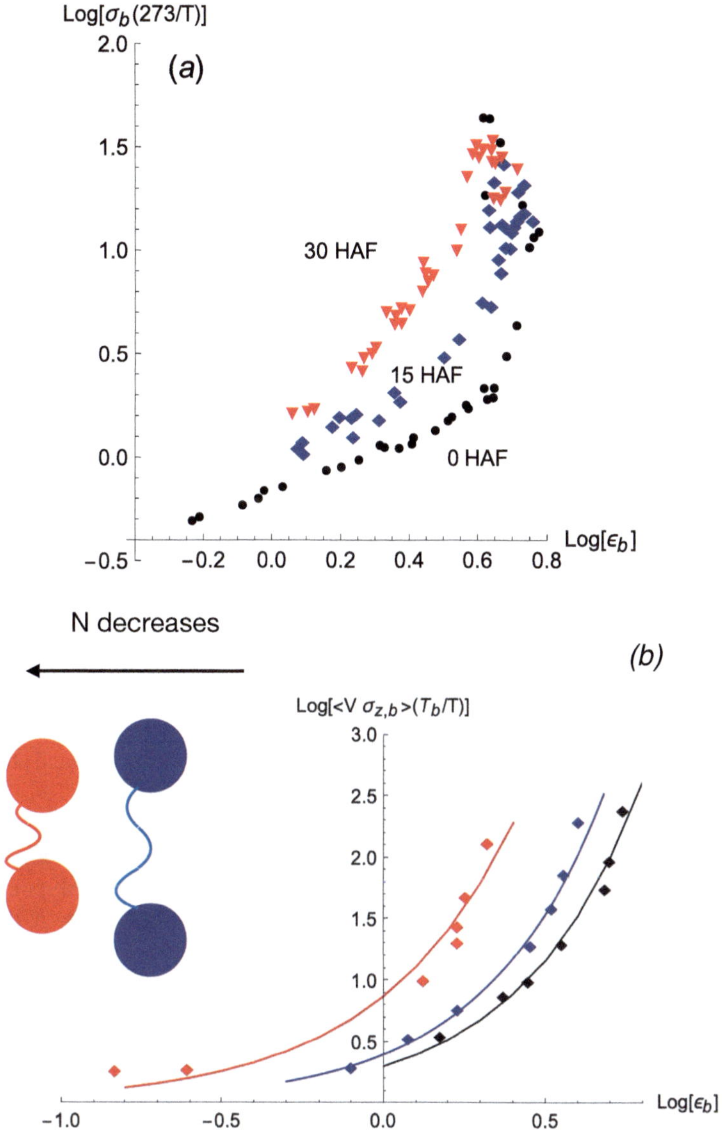

Fig. 14 (**a**) Data taken from Fig. 6.59 in Ref. [34] showing the effects of carbon blacks (HAF) on failure envelopes of well-vulcanized SBR. The unit of the stress at break, σ_b, is MPa. (**b**) Analogous plot based on the stress at break, $\sigma_{z,\,b}$, and the strain at break, ε_b, of Morse chains with different N, i.e. $N = 22$ (red), $N = 37$ (blue), and $N = 50$ (black). The lines are meant to guide the eye. The sketch illustrates the assumption that increasing filler content can be simulated, albeit roughly, by decreasing N

5 Conclusion

Based on three selected, yet quite different, examples this article discussed strategies for extracting information from simulations of rubber with a particular emphasis on temperature effects. Molecular simulation of thermal conductivity inside a strained rubber is one example. Straightforward equilibrium atomistic simulations allow to study thermal conductivity of cross-linked PI and the dependence of the diagonal components in the thermal conductivity tensor on uniaxial strain. Theory and simulation of the temperature dependence of tanδ is very different from this. A central ingredient, the dissipation of energy in a filler–filler contact still is modelled atomistically. However, the result is embedded into a theoretical description of the dynamic moduli. This plus time-temperature superposition yields tanδ over a wide range of temperatures. The model features a novel loss mechanism, which becomes increasingly important with rising temperature. Thus, model and simulation allow predictions regarding the rolling resistance. For instance, if the stiffness of the polymer in the immediate proximity of the fillers can be increased, possibly by a suitable pretreatment of the filler particles, the "jump loss" and the rolling resistance should be reduced. The final example, failure envelope of Morse chains, is again of a different nature. Tensile failure of a rubber material is a complex process. Rather than simulating a complex polymer network it can be instructive to study just one particular chain segment and how it breaks depending on applied force and temperature. In this case the result is a simplified picture of the relation between tensile strength, elongation at break and temperature. In summary, these three examples are meant to acquaint the interested reader with some of the basic concepts and strategies for the application of simulation techniques to rubber materials.

As a final remark it is worth mentioning that in recent work this author and his coworkers have developed and tested a coarse-graining approach to the modelling of the mechanical behavior of filled rubber which is similar to FEM but retains chemical information about the system of interest. The smallest entity is the filler minimum aggregate, whereas the upper spatial limit is in the range of a few micron. The range of accessible frequencies is much enhanced compared to MD simulations. Overall the approach consists of an MC morphology generator modelling the flocculation of filler based on measured surface tensions of the material's components (similar to an idea put forward in Ref. [43]). These morphologies become the input of a subsequent dynamic mechanical analysis, which allows to study the Payne effect depending on strain amplitude, strain frequency, and composition [44, 45]. However, the model does not include explicit temperature dependence aside from the temperature dependence of the measured surface tensions and thus relies on the (approximate) validity of time-temperature superposition.

Acknowledgments Much of the work described here has benefited from discussions I have had over the years with partners in industry, especially Drs. Ali Karimi-Varzaneh and Frank Fleck from Continental Reifen Deutschland GmbH, Hannover.

References

1. Kotelyanskii M, Theodorou DN (2004) Simulation methods for polymers. Marcel Dekker, New York
2. Hoover WG (1991) Computational statistical mechanics. Elsevier, Amsterdam
3. Heyes DM (1998) The liquid state – applications of molecular simulations. Wiley, New York
4. Hentschke R (2022) Thermodynamics.2nd edn. Springer, Heidelberg
5. Chandrupatla TR, Belegundu AD (2002) Introduction to finite elements in engineering.3rd edn. Pearson Prentice Hall, Delhi
6. Algaer EA, Müller-Plathe F (2012) Molecular dynamics calculations of the thermal conductivity of molecular liquids, polymers, and carbon nanotubes. Soft Mater 10:42–80
7. Choy CL, Luk WH, Chen FC (1978) Thermal conductivity of highly oriented polyethylene. Polymer 19:155–162
8. Wang J-S, Li B (2004) Intriguing heat conduction of a chain with transverse motions. Phys Rev Lett 92:074302
9. Henry A, Chen G (2008) High thermal conductivity of single polyethylene chains using molecular dynamics simulations. Phys Rev Lett 101:235502
10. Hu G-J, Cao B-Y, Li Y-W (2014) Thermal conduction in a single polyethylene chain using molecular dynamics simulations. Chin Phys Lett 31:086501
11. Lin Y-P, Zhang M-Y, Gao Y-F, Mei L-Y, Fu Y-Z, Liu Y-Q (2014) Effect of stretching on the thermal conductivity of single polyethylene chains by molecular dynamics simulations. Acta Polym Sin 6:789–793
12. Lussetti E, Terao T, Müller-Plathe F (2007) Nonequilibrium molecular dynamics calculation of the thermal conductivity of amorphous polyamide-6,6. J Phys Chem B 111:11516–11523
13. Rossinsky E, Müller-Plathe F (2009) Anisotropy of the thermal conductivity in a crystalline polymer: reverse nonequilibrium molecular dynamics simulation of the δ phase of syndiotactic polystyrene. J Chem Phys 130:134905
14. Ni B, Watanabe T, Phillpot SR (2009) Thermal transport in polyethylene and at polyethylene-diamond interfaces investigated using molecular dynamics simulation. J Phys Condens Matter 21:084219
15. Varshney V, Patnaik SS, Roy AK, Farmer BL (2009) Heat transport in epoxy networks: a molecular dynamics study. Polymer 50:3378–3385
16. Liu J, Yang R (2010) Tuning the thermal conductivity of polymers with mechanical strains. Phys Rev B 81:174122
17. Zhao J, Jiang J-W, Wei N, Zhang Y, Rabczuk T (2013) Thermal conductivity dependence on chain length in amorphous polymers. J Appl Phys 113:184304
18. Alaghemandi M, Garib-Zahedi MR, Spohr E, Böhm MC (2012) Thermal conductivity of polyamide-6,6 in the vicinity of charged and uncharged graphene layers: a molecular dynamics analysis. J Phys Chem C 116:14115–14122
19. Eslami H, Mohammadzadeh L, Mehdipour N (2012) Anisotropic heat transport in nanoconfined polyamhde-6,6 oligomers: atomistic reverse nonequilibrium molecular dynamics simulation. J Chem Phys 136:104901
20. He Y, Ma L-X, Tang Y-Z, Wang Z-P, Li W, Kukulka D (2015) Thermal conductivity of natural rubber using molecular dynamics simulation. J Nanosci Nanotechnol 15:3244–3248
21. Barnett CE (1934) Thermal properties of rubber compounds I. Thermal conductivity of rubber and rubber compounding materials. Ind Eng Chem 26:303–306
22. Dauphinee TM, Ivey DG, Smith HD (1950) The thermal conductivity of elastomers under stretch and at low temperatures. Can J Res 28a:596–615
23. Anderson DR (1966) Thermal conductivity of polymers. Chem Rev 66:677–690
24. Carwile LCK, Hoge HJ (1966) Thermal conductivity of soft vulcanized natural rubber: selected values. Rubber Chem Technol 39:126–131
25. Hands D, Horsfall F (1977) The thermal diffusivity and conductivity of natural rubber compounds, vol 50. Rubber Chemistry and Technology, pp 253–265

26. Engelmann S, Meyer J, Hentschke R (2017) Computer simulation of thermal conductivity in vulcanized polyisoprene at variable strain and temperature. Phys Rev B 96:054110
27. Harmandaris VA, Doxastakis M, Mavrantzas VG, Theodorou DN (2002) Detailed molecular dynamics simulation of the self-diffusion of n-alkane and cis-1,4 polyisoprene oligomer melts. J Chem Phys 116:436–446
28. Hager J, Hentschke R, Hojdis NW, Karimi-Varzaneh HA (2015) Computer simulation of particle-particle interaction in a model polymer nanocomposite. Macromolecules 48:9039–9049
29. Pronk S, Páll S, Schulz R, Larsson P, Bjelkmarand P, Apostolov R, Shirtsand MR, Smith JC, Kasson MP, van der Spoel D, Hess B, Lindahl E (2013) GROMACS 4.5: a high-throughput and highly parallel open source molecular simulation toolkit. Bioinformatics 29:845–854
30. Tautz H (1960) Das Verhalten der Wärmeleitfähigkeit von Kautschuk bei der Dehnung. Kolloid-Zeitschrift 174:128–133
31. Wang M-J (1998) Effect of polymer-filler and filler-filler interactions on dynamic properties of filled vulcanizates, vol 71. Rubber Chemistry and Technology, pp 520–589
32. Hentschke R (2018) Macroscopic mechanical properties of elastomer nano-composites via molecular and analytical modelling. Soft Materials 16:31
33. Meyer J, Hentschke R, Hager J, Hojdis NW, Karimi-Varzaneh HA (2017) A nano-mechanical instability as primary contribution to rolling resistance. Sci Rep 7:11275
34. Wang M-J, Morris M (2021) Rubber reinforcement with particulate fillers. Hanser Verlag, Cincinnati
35. Hentschke R (2022) Tensile strength of rubber described via the formation and rupture of load-bearing polymer chains. Phys Rev E 106:014505
36. Gul V (1961) Mechanism of rupture of high polymers. Rubber Chem Technol 34:101–118
37. Plagge J (2022) Mastering of filled rubber strength beyond WLF: competition of temperature, time, crack deflection and bond breaking. Polymers 14:765
38. Doi M (1996) Introduction to polymer physics. Clarendon Press, Oxford
39. Song Z, Shen T, Vernerey FJ, Cai S (2021) Force-dependent bond dissociation explains the rate-dependent fracture of vitrimers. Soft Matter 17:6669
40. Smith TL (1963) Ultimate tensile properties of elastomers. I. Characterization by a time and temperature independent failure envelope. J Poly Sci A 1:3597–3613
41. Bueche F, Halpin JC (1964) Molecular theory for the tensile strength of gum elastomers. J Appl Phys 35:36
42. Hanson DE, Martin RL (2009) How far can a rubber molecule stretch before breaking? Ab initio study of tensile elasticity and failure in single-molecule polyisoprene and polybutadiene. J Chem Phys 130:064903
43. Stöckelhuber KW, Wießner S, Das A, Heinrich G (2017) Filler flocculation in polymers – a simplified model derived from thermodynamics and game theory. Soft Matter 13:3701–3709
44. Viktorova M, Hentschke R, Fleck F, Prange C, Karimi-Varzaneh HA (2020) Mesoscopic model for the simulation of dynamic mechanical properties of filled elastomers: model construction and parameterization. Applied Polymer Materials 2:5521–5532
45. Viktorova M, Hentschke R, Fleck F, Taherian F, Karimi-Varzaneh HA (2022) A mesoscopic model for the simulation of dynamic mechanical properties of filled elastomers: filled binary polymer blends. Comput Mater Sci 212:111597

Adv Polym Sci (2024) 294: 193–216
https://doi.org/10.1007/12_2024_168

Published online: 13 February 2024

Deformation-Induced Structure and Temperature Evolution of Natural Rubber Vulcanizates Characterized by Advanced In Situ Experiments

E. Euchler, K. Schneider, A. K. Sambale, M. Schwartzkopf, and S. Wießner

Contents

E. Euchler (✉)
Research Division Elastomers, Leibniz-Institut für Polymerforschung Dresden e.V., Institute of Polymer Materials, Dresden, Germany
e-mail: euchler-eric@ipfdd.de

K. Schneider and A. K. Sambale
Department Materials Engineering, Leibniz-Institut für Polymerforschung Dresden e.V., Institute of Polymer Materials, Dresden, Germany

M. Schwartzkopf
Deutsches Elektronen-Synchrotron, Hamburg, Germany

S. Wießner
Research Division Elastomers, Leibniz-Institut für Polymerforschung Dresden e.V., Institute of Polymer Materials, Dresden, Germany

Chair of Elastomeric Materials, Institute of Materials Science, TUD Dresden University of Technology, Dresden, Germany

Abstract Natural rubber (NR) vulcanizates show deformation-induced structure evolution, e.g., crystallization, explaining the outstanding mechanical performance compared to other elastomers. In this study, combined in situ experiments were realized by synchronizing wide-angle X-ray scattering (WAXS) and infrared (IR) thermography on uniaxially deformed NR vulcanizates reinforced with different carbon black content. Thus, new experimental results were generated that characterize the structural evolution under deformation by a crystallization index and temperature change as a function of mechanical stress and stretch. Owing to the coupled methods, the following findings can be summarized: (i) deformation-induced crystallization is primarily controlled by the mechanical stress; (ii) at cyclic loading, hysteresis occurs and crystallites melt within a narrow stress range; (iii) IR thermography allows to follow crystallization as a function of deformation; (iv) the onset of crystallization causes a rapid increase in temperature, however, only a slight increase in stress; (v) deformation-induced crystallization shows a saturation plateau, which hardly changes with ongoing deformation; (vi) surface temperatures show a certain fluctuation that may indicate a non-uniform crystallite distribution.

Keyword Coupled methods · Crystallization · Energy balance · Infrared thermography · Mechanical loading · Natural rubber · Structure and temperature evolution · X-ray scattering

1 Deformation-Induced Structure Evolution in Natural Rubber

Technical applications for elastomers range from tires and conveying belts to membranes and seals. In particular, elastomers based on natural rubber (NR) are characterized by their comparatively high resilience [1]. Another important aspect is the use of NR as a bio-based polymer: NR gum is produced from the milky sap of rubber plants, most commonly *Hevea brasiliensis* (latex). The ecological footprint can be further reduced by optimizing the service life of products based on NR vulcanizates.

From a micromechanical perspective, NR shows outstanding performance due to the ability to undergo so-called strain-induced crystallization (SIC) [2, 3]. In that case, stretched polymer chains strongly align and orient preferentially in the direction of the principal strain, resulting in domains of short-range order forming fibrillar crystallites within the polymer matrix [4]. As a result, the mechanical behavior of NR vulcanizates is significantly different compared to non-crystallizable elastomers, which can be attributed to the formation of crystalline areas [5]. However, the exact descriptions of the mechanisms of reinforcement by SIC on a molecular scale have not yet been clarified. Already known is the effect of additional intermolecular interactions between the chains in the crystallite. In addition, the density is locally increased. According to Flory's theory, crystallites can act as highly functional cross-links [6]. In [7], however, it is considered decisive that mechanical energy can be temporarily stored in crystallites. In [8], a strengthening effect similar to particle reinforcement was observed – e.g., a stopping of cracks in crystalline regions. However, the high deformations can also promote damage mechanisms, e.g., local growth of intrinsic defects – cavitation [9, 10], and ultimately lead to the formation of macroscopic cracks [11]. This results in an interplay of mechanisms that locally increase or decrease the load-bearing capacity of NR and has an enormous influence on material behavior, especially concerning crack growth [12]. As a result, complex crack deflection and branching phenomena can occur influencing the failure as well as wear behavior and, thus, SIC has a remarkable impact on the lifetime performance of NR-based vulcanizates [13–15].

Deformation-induced crystallization is a reversible mechanism based on the orientation of aligned and oriented polymer chains. At stretch levels of $\lambda \gtrsim 2...5$, crystallites with longitudinal dimensions of about 10 to 20 nm in the stretching direction are formed [16, 17]. The size of crystallites is almost independent of λ but is affected by the filler content and crosslink density of the elastomer [5, 16, 18]. For cyclic deformation, the crystalline regions vanish during unloading, whereby the stretch level for melting last crystallites is smaller than the onset stretch of SIC. Melting of crystallites is an endothermal process leading to a reduction of surface temperature in the case of a priori deformed NR vulcanizates.

By crystallization, SIC leads to the formation of additional X-ray-sensitive scatterers in the elastomer. Therefore, for a qualitative and quantitative analysis, wide-angle X-ray scattering (WAXS) can be used [19–21]. Such experiments can be performed with laboratory or synchrotron radiation, while the latter offers extremely high spatial and temporal resolution due to its high coherence, small beam size, high flux, and high brilliance [22]. Synchrotron experiments can provide in situ data in resolutions of several nanometers and scanning times of a few milliseconds, which is necessary to detect nanoscale crystallites that are formed in about 20 ms [23]. Obtained WAXS data are used to examine the crystalline structure by the presence of diffraction peaks indicating SIC. Especially from the dominating diffraction peaks (1 2 0) and (2 0 0), intensities can be extracted to calculate a SIC index, which is a value considering the intensities of the crystalline and the amorphous phase. Among other factors, the SIC index is influenced by deformation rate, stretch amplitude, temperature of the test environment, molecular weight, and network architecture of the elastomer [5, 20].

Infrared (IR) thermography is another powerful method to study the effect of SIC by measuring and visualizing changes in surface temperature due to the heat release of energy associated with the ordering of polymer chains as an exothermic process [6, 24]. Additionally to SIC-attributed heat generation, dissipative heat generation due to the viscoelastic behavior of elastomers is overlaying the obtained temperature profile. However, a drastic increase in energy associated with the ordering of polymer chains during SIC can generate localized heat, leading to significant temperature variations on the surface [25]. IR thermography can capture and quantify such temperature changes, providing insights into the occurrence and distribution of crystallization [26, 27]. Absolut temperature values are not required for such analysis, already relative temperature changes under deformation may be used as an indicator for the onset of SIC. For elastomers without SIC, e.g., styrene-butadiene rubber (SBR), only viscoelastic dissipation leads to a moderate increase in temperature under deformation.

In this study, X-ray scattering and IR thermography are coupled to complement the mechanical characterization of NR vulcanizates and provide insights into the evolution of structural changes associated with SIC. Such unique in situ experiments assist in investigating correlations between the crystallization, temperature variations, and mechanical response under deformation. The results provide valuable insights into the phenomena of structure evolution in elastomers and soft matter, in general. The investigations collect an outstanding data set to improve the description of deformation-induced crystallization and, in the future, to develop suitable numerical approaches for modeling and simulation.

2 Experimental

2.1 Material and Specimens

In this study, unfilled natural rubber (NR00), as well as NR reinforced by 20 or 50 phr (parts per hundred rubber) carbon black (NR20 and NR50), have been used to produce vulcanizates. NR, type *Standard Malaysian Rubber* (SMR 10), was mixed with the respective amount of CB, type *Corax*® N330, 3.0 phr zinc oxide, 2.0 phr stearic acid, 1.0 phr antioxidant/antiozonant, type N-(1,3-dimethylbutyl)-N′-phenyl-p-phenylenediamine (6PPD), 1.5 phr sulfur and 1.5 phr vulcanization accelerator, and type N-cyclohexylbenzothiazol-2-sulfenamid (CBS). The mixing was performed at DIK Testing Company ("DIK Prüfgesellschaft mbH", Germany) using a two-roll mill at 60°C [28].

The vulcanization was performed at a heating press at 150°C, while the vulcanization time was estimated by rheometrical analysis using a Rubber Process Analyzer, type *SIS V-50* (TA Instruments, US) to be 18.36 min. A special type of sample – similar to pure-shear geometry containing bulges for proper fixing in the testing device – was produced [14]. From these samples, specimens in strip-shaped geometry with a length, l_0, of 10.0 mm, width, w_0, in the range of 3 to 5 mm and thickness, a_0, of 1.1 mm were punched out. For vulcanized NR, equilibrium swelling tests in toluene were done. According to the Flory-Rehner equation, apparent crosslinking densities were determined for NR00, NR20, and NR50 to be 1.28×10^4 mol/cm^3, 2.46×10^4 mol/cm^3, and 4.39×10^4 mol/cm^3, respectively [28].

2.2 *In Situ Experiments*

2.2.1 Mechanical Characterization

For in situ experiments, the mechanical testing was realized using a customized stretching device for quasi-static deformation consisting of two-step motors and a 100N load cell [29]. The displacement, l, was estimated with respect to the distance of specimen clamps. In this study, a cyclic uniaxial deformation was applied as shown in Fig. 1. The deformation speed was set to 300 mm/min. The deformation regime can be separated into three phases: (i) and (ii) first and second loading-unloading block with elevated stretch maxima and constant minima; (iii) third loading-unloading block with decreasing stretch minima and constant maxima. Since the ultimate properties [28], stress and stretch at break, are different for NR00, NR20, and NR50 (Table 1), the applied deformation regimes vary in terms of maximum displacement. The three regimes for each material are illustrated in different colors in Fig. 1.

Nominal stress, σ, and stretch, λ, were calculated according to Eqs. (1) and (2) considering the initial specimen dimensions: width, w_0, thickness, a_0, and parallel length, l_0. Stretch is defined as the ratio between the current length, l, and initial length of a specimen, l_0. Since the stretch calculation refers to the change in clamp distance, estimated maximum or SIC onset values may vary from known literature values, where displacement is precisely measured by extensometers. All measurements were realized at 23°C and 33% humidity.

$$\sigma = \frac{F}{w_0 \cdot a_0} \tag{1}$$

$$\lambda = \frac{l}{l_0} \tag{2}$$

2.2.2 X-Ray Scattering

The unique stretching device is fully implemented into the beamline instrumentation at the *Micro- and Nanofocus X-ray Scattering* (MiNaXS) beamline P03 at PETRAIII light source of *Deutsches Elektronen-Synchrotron* (DESY), Hamburg, Germany [30]. Details on technical features and possible applications of the experimental equipment are published elsewhere [29]. To analyze deformation-induced structure formation, wide-angle X-ray scattering (WAXS) has been applied. For WAXS, a focused rectangular beam of 22 μm × 32 μm size, horizontal and vertical, respectively, was used. The applied energy was 11.65 keV and the wavelength was 0.106 nm. Scattered photons were captured by a 2D detector, type LAMBDA 3 M (X-Spectrum GmbH, Germany), consisting of four LAMBDA 750k modules with each 516 × 1,556 pixels with a size of 55 μm × 55 μm arranged in a Γ-shape

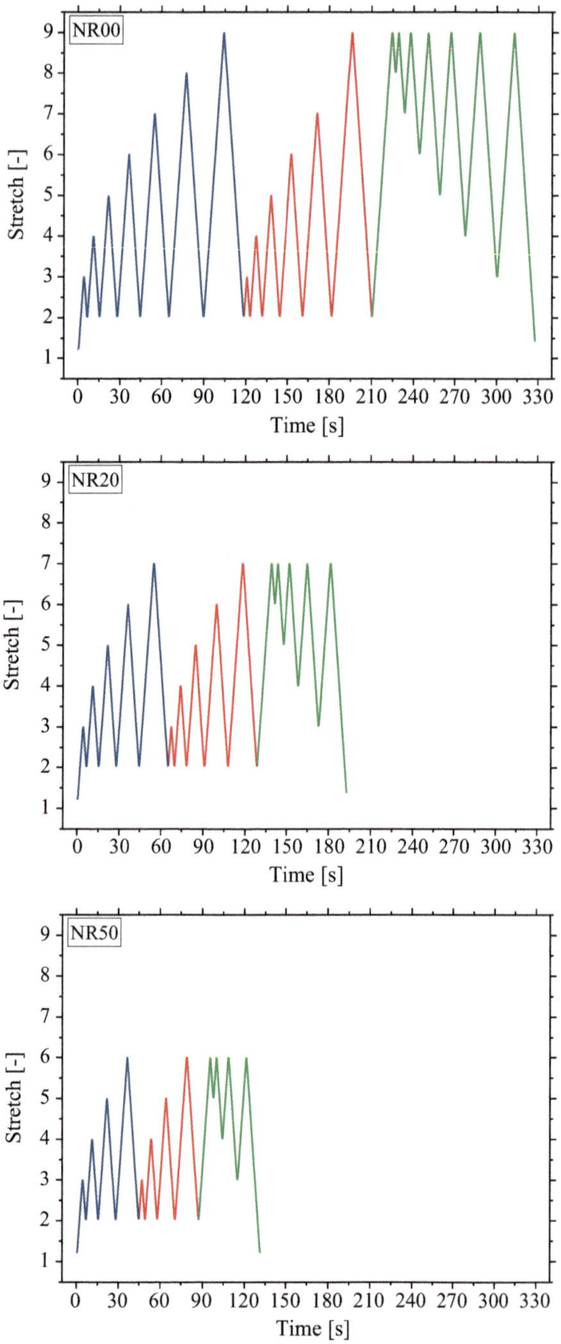

Fig. 1 Different deformation regimes: Stretch as a function of time for NR00, NR20, and NR50

Table 1 Ultimate properties of NR00, NR20, and NR50 [28]

	NR00	NR20	NR50
Stretch at break [−]	8.1	6.7	5.6
Stress at break [MPa]	14.3	29.3	30.7

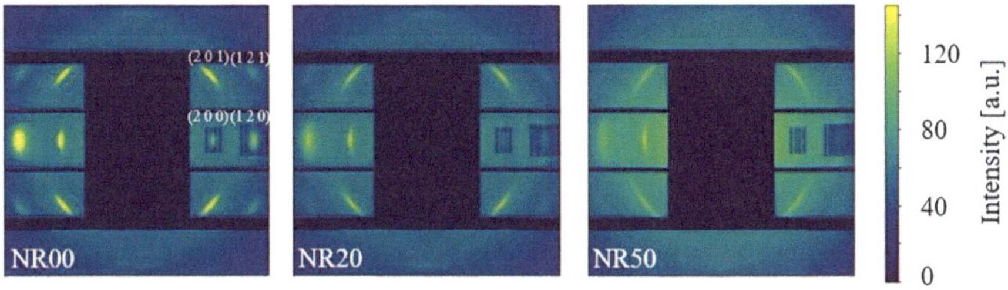

Fig. 2 Typical harmonized WAXS patterns of highly stretched NR vulcanizates showing characteristic diffraction peaks (200), (120), (201), and (121) for NR00, NR20, and NR50

geometry. The sample-detector distance was set to 265 mm. A beamstop with a photodiode was used to avoid radiation damage from high-intensity X-ray beams and to measure the beam intensity after transmission through the specimen. The additionally registered small-angle X-ray scattering (SAXS) was not evaluated within this study. The stretching device was arranged perpendicular to the beam. For in situ experiments, an exposure period of 0.200 s was applied, while the exposure time was 0.199 s. The remaining time of 0.001 s was required for data read-out.

Obtained WAXS patterns were corrected concerning background and specimen thickness. Assuming fiber symmetry, the WAXS patterns were harmonized by mirroring with respect to the tensile axis as well as the equatorial direction to fill the blind regions between the modules of the detector. Typical post-processed WAXS patterns of highly stretched NR vulcanizates are shown in Fig. 2. Typical WAXS patterns of the highest deformation of the first loading-unloading block (Fig. 1) are shown for NR00, NR20, and NR50. The predominant (2 0 0) and (1 2 0) peaks originate from the scattering of lattice planes aligned normally to the stretching direction. The rectangular areas nearby the diffraction peaks, highlighted on the right-hand side of each WAXS pattern, were used to determine the SIC index. When comparing the scattering patterns, diffraction peaks become broader and less sharp as the content of CB increases.

In this study, the equatorial diffraction peak (2 0 0), as one of the main peaks, was used to characterize the degree of crystallization by estimating a SIC index. In this case, the SIC index is a relative quantity of an averaged peak intensity with respect to the intensity of the amorphous halo at a certain deformation. For this purpose, the mean value of the scattering intensity was determined in a rectangular area before and behind the diffraction peak and its mean value was in turn subtracted from the mean value of the intensity of a rectangular area around the peak. The rectangular areas selected for the peaks are shown in Fig. 2.

2.2.3 Infrared Thermography

A commercial infrared (IR) thermo-camera, type *ImageIR 8300* (Infratec GmbH, Germany), with resolution of 640 × 512 pixels and a spectral range of 2.0 to 5.7 μm was used to monitor temperature changes during the deformation of the NR vulcanizates. Using a mirror with a hole in the experimental setup, the IR thermo-camera was arranged to the center position of the NR specimens and captured IR images with a frame rate of 20 Hz simultaneously with mechanical and X-ray data.

Post-processing was performed using the *IRBIS®3* Software (Infratec GmbH, Germany). Nominally, temperature differences of 0.025 K can be detected within the calibrated temperature range, which was 10 to 100°C, for this study. Since IR measurements are very sensitive in terms of evaluating absolute temperature values, the estimation of the material-, specimen-, and boundary condition-depending emissivity coefficient, ε_{IR}, is a challenging issue. ε_{IR} is a measure of the fraction of heat being emitted relative to that energy emitted by a thermally black surface. In this study, $\varepsilon_{IR} = \text{const.}$ was assumed and the effects of the material and the experimental setup, i.e., using a mirror, on ε_{IR} are neglected, because instead of absolute values, the temperature changes were estimated considering a constant surrounding temperature, T_{surr}. The estimated temperature changes can be considered as independent values to quantify deformation-induced structural changes. Especially, the onset of deformation and SIC can be estimated by the progress of heat generation. Figure 3 shows representative IR images for NR00, NR20, and NR50 captured at the same deformation level as considered for Fig. 2.

To determine the surface temperatures of NR specimens, two surface elements were analyzed, one around the beam position and one slightly offset from this position. Both surface elements have a size of 20 × 20 pixels, which corresponds to 0.4 mm × 0.4 mm and are therefore significantly larger than the crystallite domains. Within these areas, measured temperature values fluctuated with an average of ±0.5 K. Although, the X-ray beam introduces heat to the NR specimens, the differences in temperatures between the analyzed areas near and far from the beam position are negligible.

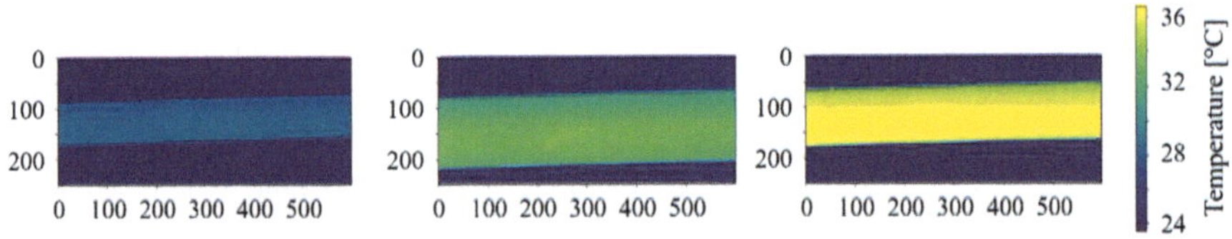

Fig. 3 Typical IR images of highly stretched NR vulcanizates at the highest deformation of the first loading-unloading block (Fig. 1) are shown for NR00, NR20, and NR50. The axes scaling is given in pixels

2.2.4 Coupling of Methods

The experimental challenge of this study is the coupling of different methods to investigate deformation-induced events in NR vulcanizates. Besides the advantages of the individual methods, the performance of in situ experiments requires some adjustments as well as synchronization of scattering experiments, mechanical measurements, and thermography. Figure 4 shows a scheme of the applied experimental setup at the PETRAIII beamline P03. However, simultaneously performed SAXS measurements, for example to study a deformation-induced defect formation and evolution, is not part of this paper.

To avoid material degradation due to radiation damage during the comparable long time of cyclic deformation, which can be several minutes of beam radiation, the mode of specimen loading was adopted. In this case, the deformation was defined to be asymmetric: While the global deformation speed was constant, the speed distribution was set to a ratio of 60:40 concerning the movement of the specimen's clamps. As a result, the X-ray beam constantly illuminates neighboring areas and beam-induced changes can be significantly suppressed.

In general, the mechanical data were synchronized with recorded WAXS patterns. For this purpose, the analog output signals of the load sensor were fed into the IO cards of the *Raspberry Pi Logic Controller* (PiLC) module, which is part of the *Tango Control System* of the beamline. IR data were synchronized with this dataset via logged timestamps. For analysis, *Python*-based evaluation routines were used [29, 30].

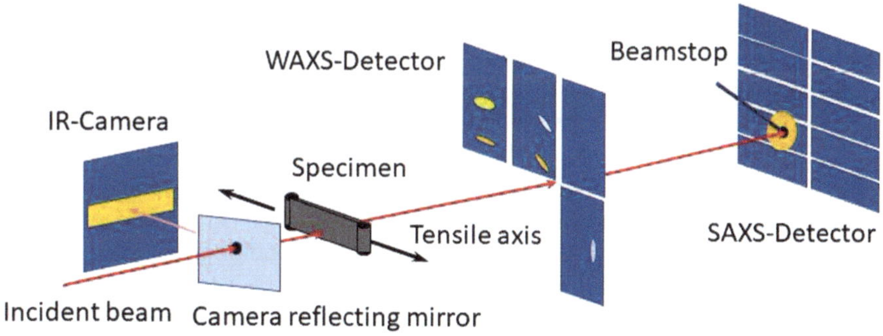

Fig. 4 Scheme of the experimental setup used at PETRAIII beamline P03 for in situ experiments to investigate deformation-induced structural changes in NR vulcanizates

3 Experimental Results

3.1 Stress-Stretch Behavior

Figure 5 shows the stress-stretch curves for NR00, NR20, and NR50. Comparing the first and second loading-unloading blocks (blue & red curves), the stress softening under cyclic deformation due to the Mullins effect [31] can be seen. The third loading-unloading block (green) allows the investigation of the hysteresis behavior of the specimens with decreasing stretch minimum (Fig. 1). Worth noting, the obtained stretch levels are based on estimating the change in clamp distance and not directly from specimen deformation. Therefore, the estimated stretch is slightly increased as the specimen is partially pulled out of the clamps under high load. As a consequence, the determined stretch levels are probably slightly higher compared to literature values, where the stretch was determined by optical analysis.

3.2 Stress, SIC Index, and Temperature Change as a Function of Time

The temporal evolvement of stress, SIC index, and temperature change of NR00, NR20, and NR50 is shown in Figs. 6, 7, and 8. In all cases, the material response depends on the deformation regimes characterized by the three loading-unloading blocks indicated by different colors, see also Fig. 1.

First, with increasing CB content, the strength and stiffness of the NR vulcanizates as well as the hysteresis loops are increasing. Structure evolution due to SIC can be discussed by an increase in the slope in the stress-time curves.

Second, the behavior of the SIC index as a function of time illustrates the results obtained by WAXS. Due to the asymmetric speed distribution of the step motors of the stretching device, the X-ray beam illuminates different specimen positions leading to a certain fluctuation in the WAXS data. In the first loading-unloading block, local deformations are high enough to induce the alignment and orientation of polymer chains to form crystalline structures. After this conditioning, the response is different for the second loading-unloading block.

In some cases, there is no or suppressed SIC, which is reflected in reduced peak heights. Interestingly, the data achieved from the third loading-unloading block express that a certain unloading is required before crystallites melt. If the applied stretch minimum is too high, the SIC index will remain at values above zero. This behavior has consequences for the interpretation of the service lifetime performance of NR-based elastomer parts: if the cyclic deformation does not fall below a certain stress or stretch, the stiffening of SIC remains on a considerable level and, thus, acts as a reinforcing factor for the subsequent loading.

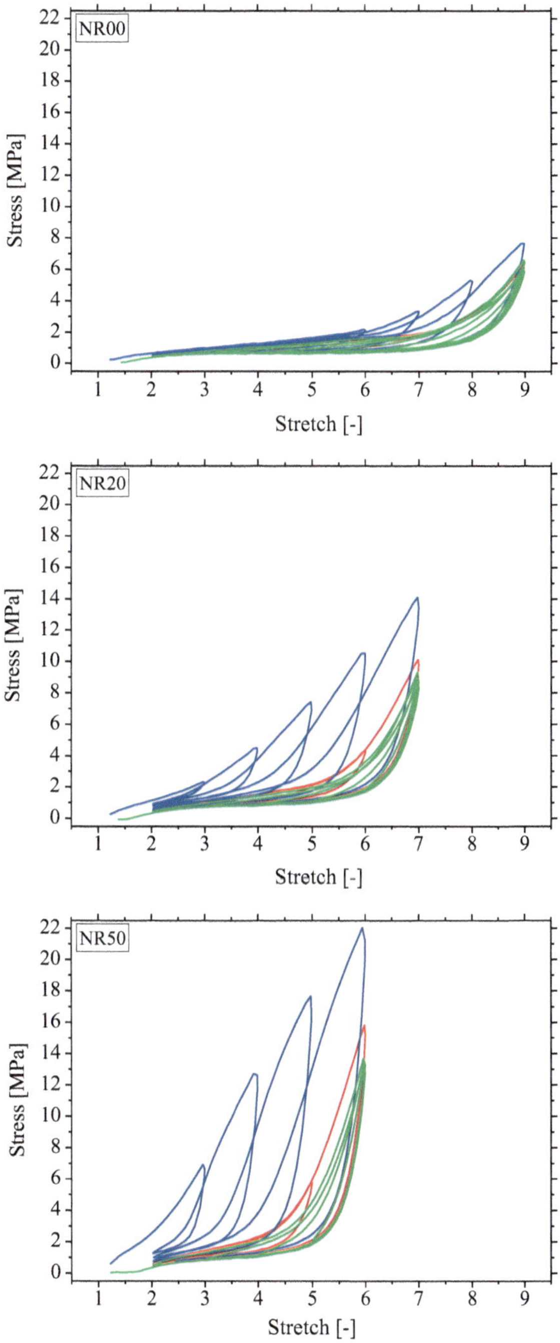

Fig. 5 Stress-stretch behavior of NR00, NR20, and NR50. Different colors represent the different loading-unloading blocks according to Fig. 1

Fig. 6 Stress as a function of time for NR00, NR20, and NR50

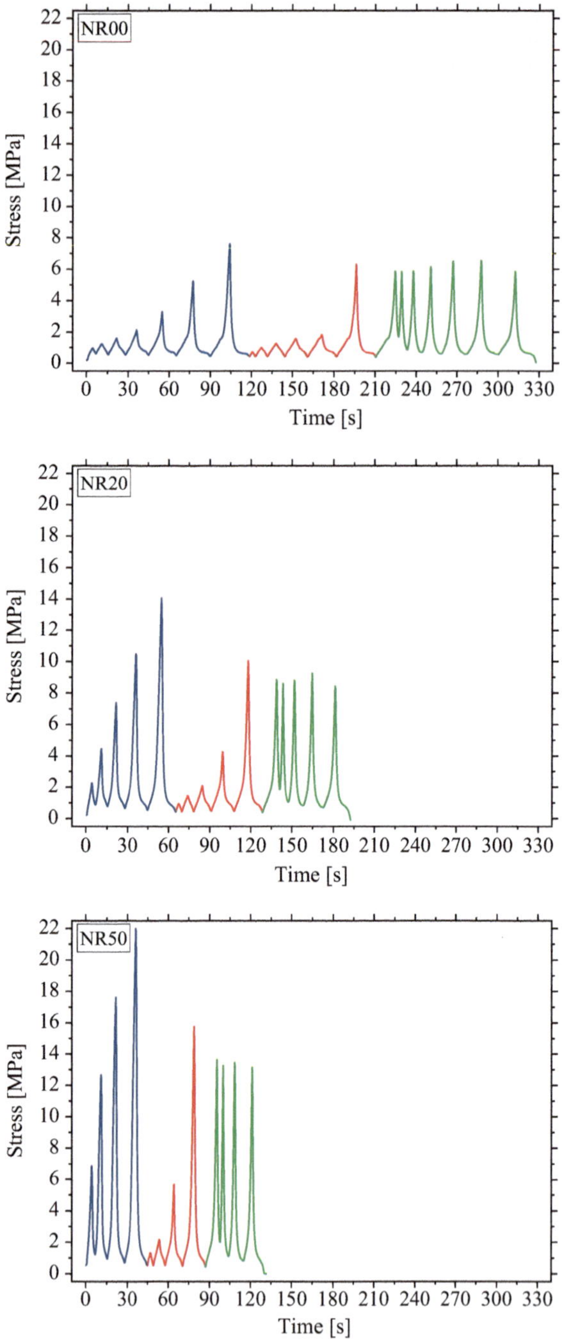

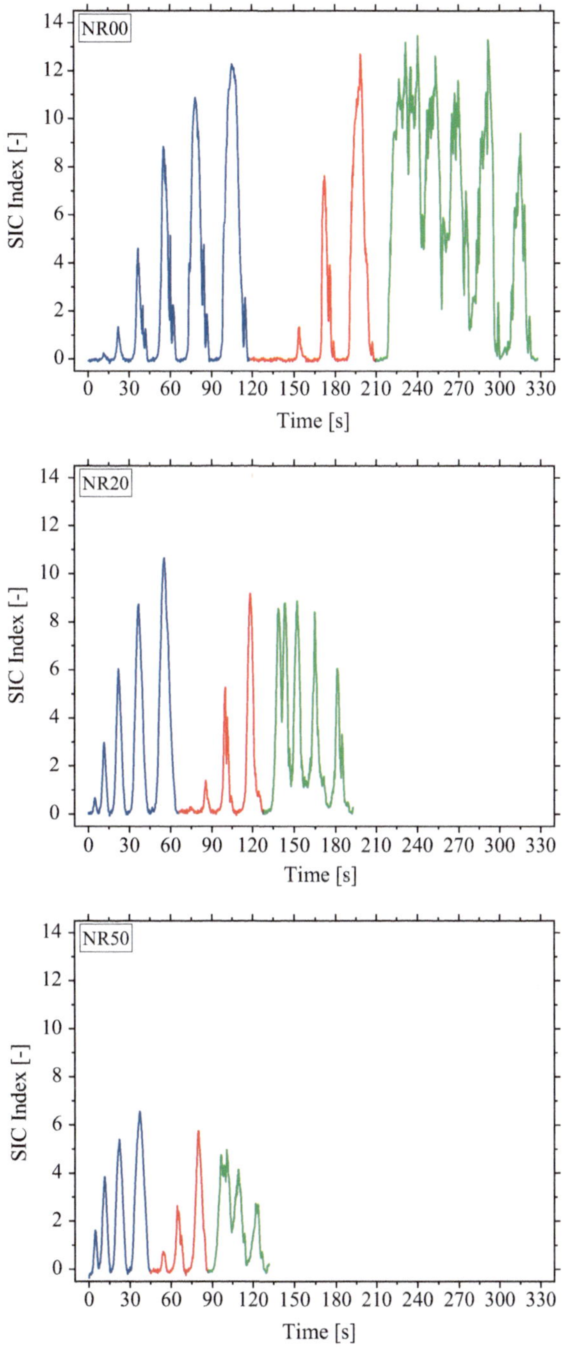

Fig. 7 SIC index as a function of time for NR00, NR20, and NR50

Fig. 8 Temperature change as a function of time for NR00, NR20, and NR50. The curve of NR00 is not available for the total duration. Horizontal lines shall guide the eyes to zero Kelvin

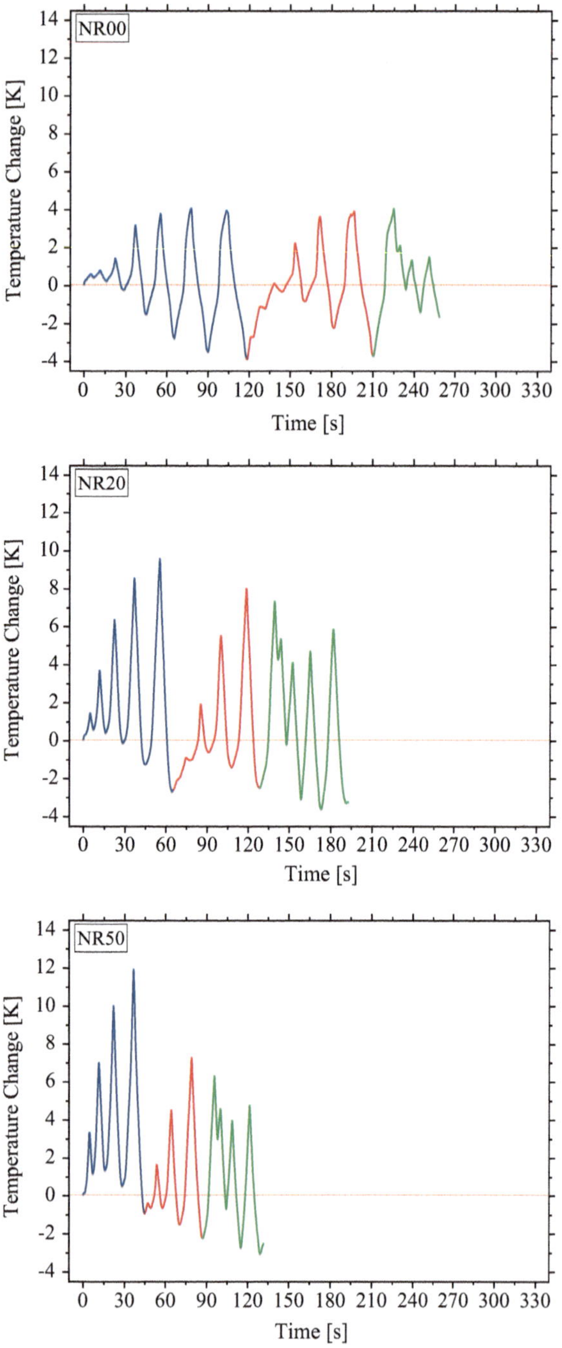

Third, similar to the stress response, the temperature change as a function of time is characterized by a remarkable inflection point indicating the onset of SIC. However, the contribution of SIC is superimposed by dissipative heat generation.

For all materials, NR00, NR20, and NR50, the melting of crystallites reduces the surface temperature; however, the effect is less dominant for CB-reinforced vulcanizates. The fluctuations in the SIC index and temperature change curves indicate a slightly inhomogeneous distribution of crystallites along the specimen, which can be recognized by the small shift in the beam position with respect to the specimen applied to avoid beam damage.

3.3 *Estimation of Crystallization Onset*

Representatively, Fig. 9 shows stress, SIC index, and temperature change as a function of time for the last cycle of the second loading-unloading block. For better comparability, the time for the corresponding cycle duration starts at zero. The dashed lines indicate the SIC onset estimated by the corresponding experimental data. Thereby, the temperature increase occurs first followed by the rise in the SIC index and the increase in stress occurs with a small delay. The results suggest a complex structure evolution where (i) a certain heat generation acts as a precursor for the crystallization, which (ii) progressively changes the elastomer structure until a sufficient crystalline fraction is formed and, thus, (iii) affects the mechanical response as shown by a remarkable stiffening. The results highlight, that SIC can be monitored indirectly by different experimental methods, while all give a certain contribution to better understanding its complexity. Considering the experimental data shown in Fig. 9, the values of crystallization onset stretch are quantified in Table 2.

The onset values were determined considering the first deviation from the linear behavior of each data curve. Worth noting, that the SIC onset values may differ from the literature ones [5, 7, 8, 32], because of the methodological approach to consider clamp distance for stretch calculation instead of measuring the deformation fields directly on the specimen.

3.4 *SIC Index and Temperature Change as a Function of Stress*

In the literature [5, 7, 8, 32], deformation-induced crystallization in NR vulcanizates is more or less exclusively attributed to stretch, λ, or strain ε, with $\lambda = \varepsilon + 1$. Hereby, stretch is a quantity representing the extensibility of an NR specimen, but not the stress-carrying capability of the material itself. From the mechanical point of view, the onset of SIC occurs most probably at the same stress level. This effect is

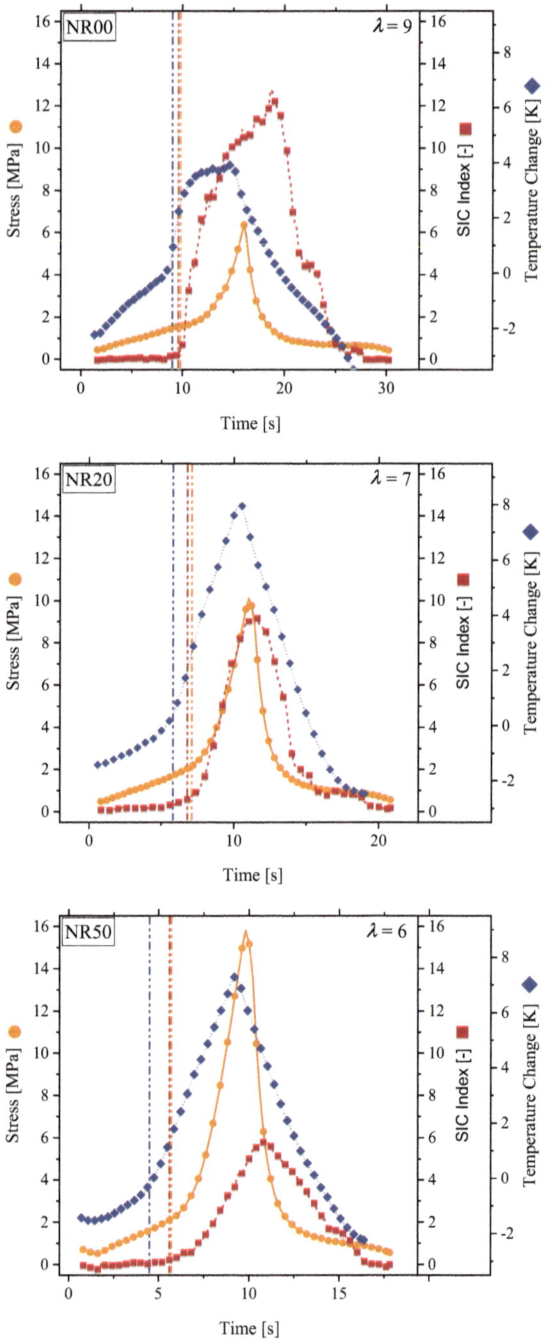

Fig. 9 Stress (*yellow filled circle*), SIC index (*red filled square*), temperature change (*blue filled diamond*) as a function of time – for better comparability, the X-axes are scaled differently to show a full cycle on the width of the abscissa. Estimated onsets of crystallization for NR00, NR20, and NR50 are indicated by corresponding vertical lines

Table 2 SIC onset stretch of NR00, NR20, and NR50 estimated according to Figs. 6, 7, and 8

λ_{SIC} [−]	NR00	NR20	NR50
Mechanical testing	6.0	5.0	4.0
WAXS	5.9	4.9	3.9
IR thermography	5.6	4.4	3.4

indicated by a shift in the onset of SIC to higher stretch values after mechanical softening under cyclic loading. The results in Fig. 10 support this assumption and suggest that the SIC shall be related to *stress* as the triggering quantity [19].

With increasing number of cycles, the material response is softer and, thus, NR specimens have to be stretched to larger levels to reach the required stress value (Fig. 10). When reaching a certain stress value – approximately 1.6 MPa, 1.8 MPa and 2.9 MPa for NR00, NR20, and NR50, respectively – the SIC index increases rapidly suggesting a strong crystallinity growth. However, elevating deformation and ongoing stress increase lead to a saturation in the SIC index. The steady-state plateau is almost constant at the first phase of unloading. Suddenly, the SIC index decreases dramatically representing the melting of crystallites – approximately at 0.7 MPa, 1.0 MPa, and 1.4 MPa for NR00, NR20, and NR50, respectively. Interestingly, temperature change-stress curves show similar hysteresis curves suggesting a strong correlation between crystallization and heat generation under applied mechanical stress due to uniaxial deformation [25–27].

4 Discussion

Due to the strip-shaped specimen geometry, a uniaxial tensile deformation can be assumed. In the parallel section of the specimen, the following energetic consideration should therefore apply. For each cycle, the loading paths transfer mechanical energy into each volume element of the system with an introduced energy density:

$$w_{\text{mech}} = \int \sigma \, d\varepsilon \tag{3}$$

The mechanical energy is not only stored elastically but also leads to an increase in the temperature of the specimen. Due to the temperature difference between the specimen and the surrounding, T_{surr}, heat – or thermal energy – is released into the environment or absorbed from there. For a volume element $dV = dx \cdot dy \cdot d$ at the specimen's thickness $a(t)$, the following applies

$$\frac{\delta Q}{dV} = \delta q = -2 \int \frac{\alpha \, (T(t) - T_{\text{surr}})}{a(t)} dt \tag{4}$$

where α represents the heat transfer coefficient and T is the measured surface temperature. The factor "2" considers that heat is emitted from both sides of a flat

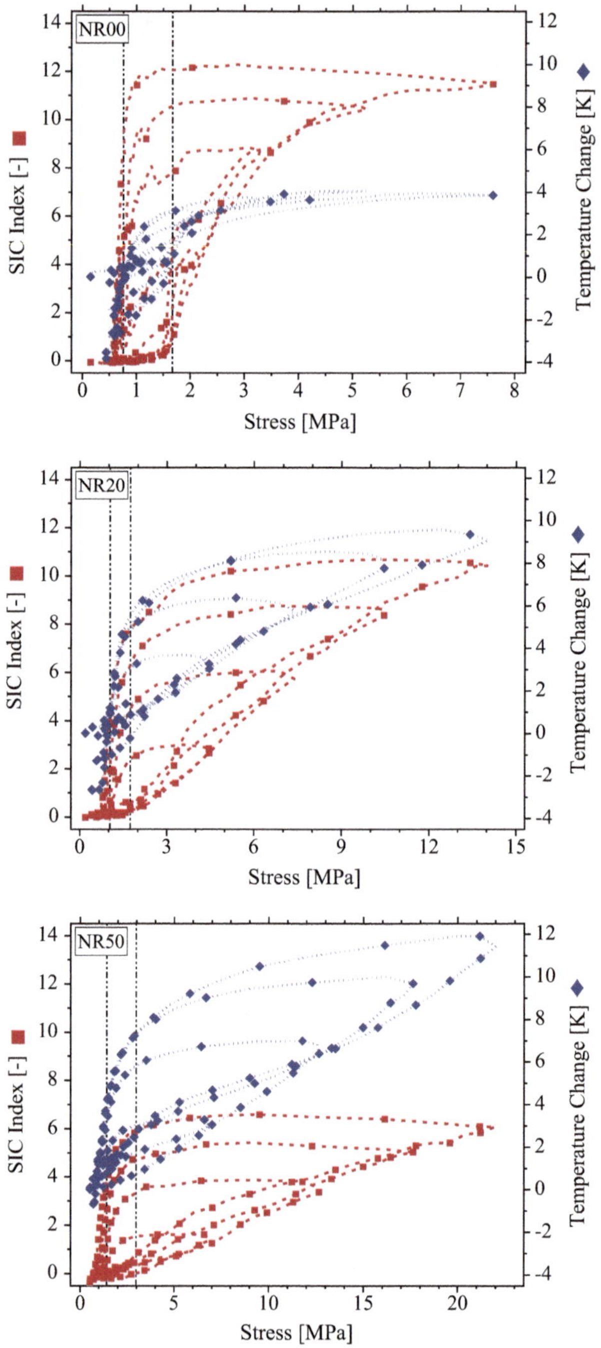

Fig. 10 SIC index (*red filled square*) and temperature change (*blue filled diamond*) as a function of stress for NR00, NR20, and NR50 – for better visualization, the X-axes are scaled differently. Vertical lines to guide the eyes for the onset of crystallite formation and melting

specimen, i.e., $A_1 = l(t) \cdot w(t)$. The heat emission of the two smaller specimen sides, i.e., $A_2 = a(t) \cdot w(t)$, is neglectable, especially due to the ongoing thickness reduction under tension. According to the first law of thermodynamics, the difference between the two quantities is equal to the change in the internal energy of the volume element of the specimen, with

$$w_{\text{mech}} + \delta q = du \tag{5}$$

In general, there are two parts of the internal energy. The thermal one is represented by $du_{\text{therm}} = \rho\, c_p\, dT$, where ρ is the material density and c_p is the specific heat capacity. Additionally, the mechanical part stores energy reversibly, analogous to a stretched spring, without the stress-stretch behavior for the case of unloading being known in the specific case. For SIC, the heat of crystallization, which can be expressed by $du_{\text{SIC}} = \Delta H_{\text{cr}}\, \rho\, dc_{\text{SIC}}$, is added to the system, which leads to a further increase in the thermal part of the internal energy. Here, ΔH_{cr} describes the enthalpy of crystallization of the crystallinity achievable via SIC, normalized to the crystallinity fraction c_{SIC} as a representative of the whole crystallinity within a specimen.

$$\begin{aligned} w_{\text{mech}} + \delta q = & \int \sigma\, d\varepsilon - 2 \int \frac{\alpha\, (T(t) - T_{\text{surr}})}{a(t)}\, dt \\ & = \Delta u = \rho\, c_p\, \Delta T + \Delta u_{\text{mech}} + \Delta u_{\text{SIC}} \end{aligned} \tag{6}$$

At individual points in the shown time charts (Fig. 8), the missing parameters of Eq. (6) can be estimated by considering certain boundary conditions. For example, assuming that the stress at the stretch minimum is close to zero and there is no crystallinity, i.e., $\Delta u_{\text{mech}} = 0$ and $\Delta u_{\text{SIC}} = 0$, the following assumption holds

$$w_{\text{mech}} + \delta q = \rho\, c_p\, \Delta T \tag{7}$$

Using known quantities of T_{surr}, ρ, c_p and assuming that these parameters are constant over the experimental time, the heat transfer coefficient, α, can be calculated. Table 3 summarizes the applied parameters for energy calculations.

In Fig. 11, the energy contributions as a function of time, dw_{mech}, δq, and du_{therm}, show that the main part of dw_{mech} is finally converted to heat and transferred to the surroundings – represented by δq. After the three loading-unloading blocks, a rather small difference remains in the thermal part of internal energy, $du_{\text{therm}} = \rho\, c_p\, dT$. In

Table 3 Parameters to calculate the different energy parts

Parameter	NR00	NR20	NR50
α [mW/(K mm^2)]	0.04...0.09	0.04...0.06	0.03...0.09
ρ [g/cm^3]	0.97	1.04	1.14
c_p [J/(g K)]	0.97	0.90	0.83
T_{surr} [K]	297.65	297.40	297.75

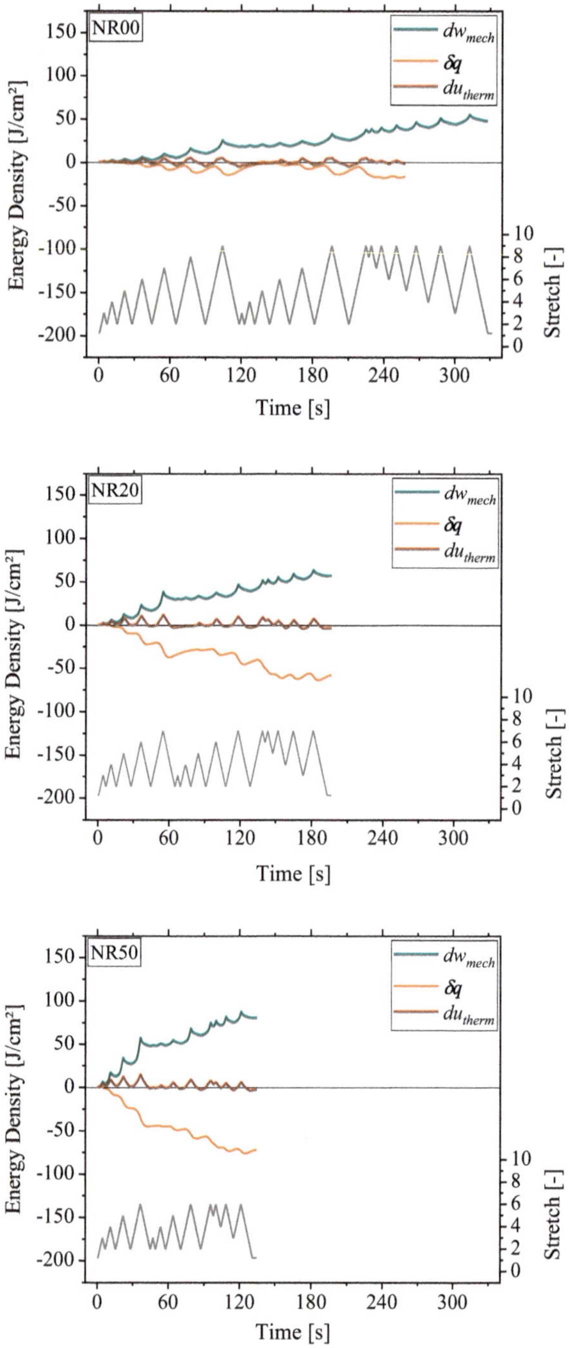

Fig. 11 Energy contributions dw_{mech}, δq, and du_{therm} as a function of time for NR00, NR20, and NR50

SIC, heat is released when aligned and oriented polymer chains interact by crystallization, whereas no additional elastic energy is stored. Besides WAXS experiments, SIC can be monitored and described by IR thermography. However, to distinguish between the different parts of the internal energy, X-ray scattering measurements are mandatory.

After correcting for the heat dissipation to the environment, δq, and considering the change in the thermal proportion of the internal energy, du_{therm}, the mechanically stored proportion of the internal energy, dw_{mech}, can be determined from the energy balance, taking into account the crystallization enthalpy of NR, $\Delta H_{\mathrm{cr,\ NR}}$. This is initially of the same order of magnitude as the thermally stored proportion but increases in relation to the thermal proportion as crystallization begins. This fact can be understood according to the higher energy elasticity of the semi-crystalline NR compared to the considerable proportion of entropy elasticity of the amorphous NR.

5 Conclusion

Owing to in situ experiments at a synchrotron beamline, new insights into structural changes and mechanical response could be achieved for NR vulcanizates, both unfilled and carbon black (CB) reinforced ones. Besides wide-angle X-ray scattering (WAXS) measurements, deformation-induced structural changes in NR can also be monitored using infrared (IR) thermography coupled with mechanical testing. The performed in situ experiments provide a unique database to better understand the process of SIC under uniaxial, cyclic deformation. The results can be condensed to the following:

- WAXS and IR thermography are suitable methods to monitor deformation-induced structure evolution.
- Crystallization onset can be estimated by IR thermography, WAXS, and stress-stretch curves.
- SIC seems to be stress-controlled; as a consequence, the onset stretch is shifted due to stress softening under cyclic deformation, e.g., Mullins effect.
- SIC leads to heat generation while melting of crystallites requires heat.
- During loading, SIC reaches a saturation, where further deformation does not change the SIC fraction anymore.
- Fluctuations in IR and WAXS data suggest an inhomogeneous distribution of crystallites during deformation and along the specimen.
- Energy elasticity of semi-crystalline NR is higher compared to the considerable proportion of entropy elasticity of amorphous NR.

Furthermore, the time-synchronized data allow a comparison of the SIC onsets determined by the three different experimental methods. The observed differences in this regard are understood to represent the successive events of SIC, which are (i) polymer chain alignment and first heat generation, (ii) followed by the formation of first crystallites, which, (iii) affect the macroscopic mechanical response when a certain SIC fraction is reached.

Acknowledgments The authors thank the *Micro- and Nanofocus X-ray Scattering* (MiNaXS) beamline P03 at PETRAIII of *Deutsches Elektronen-Synchrotron* (DESY), Hamburg, Germany, for approving beamtime. The research was partly founded by the *Deutsche Forschungsgemeinschaft* (DFG, German Research Foundation) – Projektnummer 517438497.

References

1. Mars WV, Fatemi A (2004) Factors that affect the fatigue life of rubber: a literature survey. Rubber Chem Technol 77(3):391–412
2. Katz JR (1925) Röntgenspektrographische Untersuchungen am gedehnten Kautschuk und ihre mögliche Bedeutung für das Problem der Dehnungseigenschaften dieser Substanz. Naturwissenschaften 13:410–416
3. Toki S, Hsiao BS (2003) Nature of strain-induced structures in natural and synthetic rubbers under stretching. Macromolecules 36(16):5915–5917
4. Andrews EH, Gee G (1964) Crystalline morphology in thin films of natural rubber II. Crystallization under strain. Proc R Soc Lond Ser A Math Phys Sci 277(1371):562–570
5. Candau N, Laghmach R, Chazeau L, Chenal J-M, Gauthier C, Biben T, Munch E (2014) Strain-induced crystallization of natural rubber and cross-link densities heterogeneities. Macromolecules 47(16):5815–5824
6. Flory PJ (1949) Thermodynamics of crystallization in high polymers. IV. A theory of crystalline states and fusion in polymers, copolymers, and their mixtures with diluents. J Chem Phys 17: 223–240
7. Le Cam JB (2017) Energy storage due to strain-induced crystallization in natural rubber: the physical origin of the mechanical hysteresis. Polymer 127:166–173
8. Saintier N, Cailletaud G, Piques R (2011) Cyclic loadings and crystallization of natural rubber: an explanation of fatigue crack propagation reinforcement under a positive loading ratio. Mater Sci Eng A 528(3):1078–1086
9. Gent AN, Lindley PB (1959) Internal rupture of bonded rubber cylinders in tension. Proc R Soc London Ser A 249(1257):195–205
10. Euchler E, Bernhardt R, Schneider K, Heinrich G, Wießner S, Tada T (2020) In situ dilatometry and X-ray microtomography study on the formation and growth of cavities in unfilled styrene-butadiene-rubber vulcanizates subjected to constrained tensile deformation. Polymer 187: 122086
11. Lake GJ, Thomas AG, Lawrence CC (1992) Effects of hydrostatic pressure on crack growth in elastomers. Polymer 33(19):4069–4074
12. Le Cam JB (2010) A review of volume changes in rubbers: the effect of stretching. Rubber Chem Technol 83(3):247–269
13. Xiang F, Schneider K, Heinrich G (2020) New observations regarding fatigue crack paths and their fracture surfaces in natural rubber: influences of R-ratio and pre-load. Int J Fatigue 135: 105508
14. Schneider K, Xiang F, Mishra D, Heinrich G (2023) Fatigue crack propagation of silica and carbon black filled natural rubber at elevated temperatures. Int J Fatigue 177:107968
15. Osumi R, Yasui T, Tanaka R, Mai T-T, Takagi H, Shimizu N, Tsunoda K, Sakurai S, Urayama K (2022) Impact of strain-induced crystallization on fast crack growth in stretched cis-1,4-polyisoprene rubber. ACS Macro Lett 11(6):747–752

16. Poompradub S, Tosaka M, Kohjiya S, Ikeda Y, Toki S, Sics I, Hsiao BS (2005) Mechanism of strain-induced crystallization in filled and unfilled natural rubber vulcanizates. J Appl Phys 97(10):103529
17. Albouy P-A, Sotta P (2020) Draw ratio at the onset of strain-induced crystallization in cross-linked natural rubber. Macromolecules 53(3):992–1000
18. Yamano M, Yamamoto Y, Saito T, Kawahara S (2021) Preparation and characterization of vulcanized natural rubber with high stereoregularity. Polymer 235:124271
19. Trabelsi S, Albouy PA, Rault J (2002) Stress-induced crystallization around a crack tip in natural rubber. Macromolecules 35(27):10054–10061
20. Brüning K (2014) In-situ structure characterization of elastomers during deformation and fracture. Springer, Cham
21. Albouy P-A, Vieyres A, Pérez-Aparicio R, Sanséau O, Sotta P (2014) The impact of strain-induced crystallization on strain during mechanical cycling of cross-linked natural rubber. Polymer 55(16):4022–4031
22. Stribeck N (2007) X-ray scattering of soft matter. Springer, Berlin
23. Albouy P-A, Guillier G, Petermann D, Vieyres A, Sanseau O, Sotta P (2012) A stroboscopic X-ray apparatus for the study of the kinetics of strain-induced crystallization in natural rubber. Polymer 53(15):3313–3324
24. Khiêm VN, Le Cam JB, Charlès S, Itskov M (2022) Thermodynamics of strain-induced crystallization in filled natural rubber under uni- and biaxial loadings, part I: complete energetic characterization and crystallinity evaluation. J Mech Phys Solids 159:104701
25. Plagge J, Klüppel M (2018) Determining strain-induced crystallization of natural rubber composites by combined thermography and stress-strain measurements. Polym Test 66:87–93
26. Zybell L, Domurath J, Schneider K (2017) Thermomechanics of strain-induced crystallization in natural rubber under cyclic loading. Proc Appl Math Mech 17(1):493–494
27. Le Cam J-B, Albouy P-A, Charlès S (2020) Comparison between x-ray diffraction and quantitative surface calorimetry based on infrared thermography to evaluate strain-induced crystallinity in natural rubber. Rev Sci Instrum 91:044902
28. Xiang F (2022) The role of strain-induced crystallization and cavitation on the mechanism of crack splitting and crack deviation in natural rubber. Dissertation, TUD Dresden University of Technology, Dresden, Germany
29. Euchler E, Sambale AK, Schneider K, Uhlig K, Boldt R, Stommel M, Stribeck A, Schwartzkopf M, Rothkirch A, Roth SV (2022) Beamline-implemented stretching devices for in situ X-ray scattering experiments. J Phys Conf Ser 2380(1):012109
30. Buffet A, Rothkirch A, Dohrmann R, Korstgens V, Abul Kashem MM, Perlich J, Herzog G, Schwartzkopf M, Gehrke R, Muller-Buschbaum P, Roth SV (2012) P03, the microfocus and nanofocus X-ray scattering (MiNaXS) beamline of the PETRA III storage ring: the microfocus endstation. J Synchrotron Radiat 19:647–653
31. Mullins L (1948) Effect of stretching on the properties of rubber. Rubber Chem Technol 21(2): 281–300
32. Chenal J-M, Gauthier C, Chazeau L, Guy L, Bomal Y (2007) Parameters governing strain induced crystallization in filled natural rubber. Polymer 48(23):6893–6901

Adv Polym Sci (2024) 294: 217–250
https://doi.org/10.1007/12_2024_170

Published online: 8 March 2024

Theoretical and Experimental Characterization of Heat Transfer in Polymers and Elastomers: A Review

C. Breitkopf

Contents

Abstract Thermal conductivities represent one of the most important properties of polymer and elastomer materials that determine their final use in industry and research. Therefore, there exists a great interest to understand the inherent structure-property relations between the molecular structure and the thermal conductivity. The review summarizes approaches to calculate thermal conductivities ranging from macroscopic to microscopic level, whereas it is generally accepted that

C. Breitkopf (✉)
TU Dresden, Faculty of Mechanical Engineering, Chair of Thermodynamics, Dresden, Germany
e-mail: cornelia.breitkopf@tu-dresden.de

an appropriate design of new materials with optimized thermal properties is only accessible via knowledge-based bottom-up approaches. Both approaches – macroscopically and microscopically based – are presented in retrospect with their historical development. Supplementary to the theoretical view elaborated experimental methods are described with the focus on elastomers.

Keywords Elastomers · Experiments · Flash method · Green-Kubo · Hot wire method · Molecular dynamics · Phonon transport · Polymers · Theory · Thermal conductivity

1 Introduction

Polymers and elastomers have an extensive range of properties, which make them advantageous for numerous industrial and daily life applications. Properties like chemical inertness, light densities, easy and still low costs in production enable a wide application range. The thermal conductivity κ is one of the most important properties, which determines the final usability of a polymer or elastomer material. Table 1 shows thermal conductivities for one representative example, e.g. polyethylene based on different preparation techniques, which emphasizes the strong structure-property correlation for thermal conductivities of the final modified material.

The thermal conductivity as a property physically relates the local heat flux across a surface to a temperature gradient at this point perpendicular to the surface. Therefore, polymers as well as elastomers and their composites possess certain thermal properties depending on their manufacturing. Pure polymers and also elastomers appear mainly as thermally resistive substances showing low thermal conductivities κ ($\kappa = 0.1$–0.5 W/m/K) in their amorphous state [2], see also Table 2. Thus, applications in thermal management are often limited [3]. An exception would be some modern applications like thermal barrier coatings or thermoelectrics, which need low thermal conductivities to impede heat transfer and to maintain high-temperature differences in the material [4].

To increase the application readiness, higher thermal conductivities, e.g. an improved heat dissipation, would be of high interest. Polymers showing higher κ-values above 10 W/m/K could indeed get competitive to common materials as for example in LED devices [5]. Molecular simulations on individual polymer chains indicated and predicted higher thermal conductivities [6], thus opening new applications in materials science and semiconductor industry for polymer nanofibers and

Table 1 Thermal conductivities κ [W/m K] at 293 K for polyethylene polymers based on different preparation conditions according to Zhang et al. [1]

Polymer	κ [W/m K]
Crystalline polyethylene nanofibers	90 (150 K)
Polyethylene films	62
High-density polyethylene nanofibres	9
High-density polyethylene/boron nitride	3.57

Table 2 Thermal conductivities κ [W/m K] for selected elastomers and polymers according to Mehra et al. [25] (see references for measurements at certain temperatures in there)

Elastomer	Temperature [K]	κ [W/m K]
Polybutadiene (extrusion grade)	293	0.220
Poly(butadiene-styrene) copolymer (SBR)		
23.5% styrene content		
Pure gum vulcanizate	293	0.190–0.230
Carbon black-filled vulcanizate	293	0.300
Polyisobutylene	293	0.130
Polystyrene	273	0.105
	373	0.128
	473	0.130
	573	0.140
	673	0.160
Polyurethane	293	0.310
Polyacrylonitrile-butadiene-styrene (ABS)	413	0.330

composites. Also, new developments in science and research open the access to new materials as interactive polymer and elastomer composites [7, 8], which might have improved thermal properties. Intensive research is done in this field as for example the research training group of the DFG at TU Dresden [9].

An equally large application branch as thermal barriers with a high innovation potential are heat transfer problems in engineering applications like cooling or refrigeration. Here, polymers are nowadays also attractive materials for use as heat exchangers. Cevallos et al. [10] described the history and future use of polymer-based heat exchangers in a review article. A new trend in research and possible application is the active control of heat flow through a material-based thermal management [11] as well as exotic aspects like thermal rectification [1, 12, 13]. In general, for cooling or refrigeration two directions of development are of interest. Either first create a design for low thermal conductivities to minimize heat dissipation or secondly, to increase the thermal conductivity to enable functional materials like electronics to be heat resistant. As materials get more complex, a knowledge-based understanding of heat transport and its detailed transport measurements is mandatory. Moreover, the thermal conductivity has a complex interplay with other properties and/or influencing factors like temperature, density, crystallinity of the phase, appearance of grain boundaries, and material impurities. For more information, the reader is referred to dos Santos et al. [14]. Consequently, all new materials and composites must be adapted to their needs without any stress reaction according to temperature changes like shrinking or delamination. To further control the thermal conductivity, for semicrystalline and crystalline polymers the latent heat storage may be an option, which would open also new applications.

Further applications depend on the ability of an efficient thermal management as for instance for electronic systems [15], soft robotics, artificial skin, wearable electronics, batteries, thermal insulation, and thermoelectrics [16], and thermal insulators [17].

Especially crosslinked elastomers are meanwhile basis of materials for use in transportation, electronics, construction, and sporting goods industry [18]. Examples are given by Chen et al. [19] for the investigation of the heat dissipation capacity of conducting polymers and elastomers for stabilizing cell-phone covers or encapsulation of solar cells or as in situ heat sink in LED.

With the growth of the semiconductor industry, a tremendous interest appeared to investigate the thermal transport in solid silicon materials. Interestingly, semiconductors also have crystalline and amorphous parts similar to polymers. Thus, intensive research that has been done on heat transport in semiconductors can be also transferred in their basics to polymers and elastomers [20]. Besides the thermal transport also the view to electronic properties and transport may support the understanding of polymers, of course this is valid vice versa. Understanding detailed transport mechanisms of electronic excitations in conjugated polymers is then key to advancing organic optoelectronics [21].

Rubbers are elastomers, which can be reinforced by additional materials to form valuable composites that possess improved properties with respect to applications. Properties of pure rubbers may be tuned by adding fillers or additives depending on the final application. Functional rubbers for specific applications (e.g. electronics) are often filled with carbon nanotubes (CNT), and those composites have been studied extensively [22–24]. Moreover, carbon black and also silica fillers reinforce the tire rubbers as well as technical rubbers, which is generally an important business with millions of tons of production per year. A literature summary with respect to other polymer-filler combinations and the ability to open numerous new applications can be found in Henry [26].

An improvement of elastomer properties with respect to potential new applications is possible via the knowledge-based design of appropriate composites. However, as will be discussed later, the accompanied increase of interfaces and boundaries in the new composite material may not always be beneficial on the final thermal transport and thus the thermal conductivity.

Numerous studies with respect to individual aspects of thermal conductivity for polymers exist; however, comprehensive review articles including elastomers and their composites are rather limited. A few review articles were published as for instance Zhang B. et al. [1], Koh [4], Zhang T. et al. [17], Campaioli [21], Mehra et al. [25] (although the article suffers partly from a poor writing style), Henry [26], and Anderson [27].

More general textbooks, handbooks, and monographs with basic background in the physical theory of heat transport can be found here [28–38]. Herein, also details for the fundamental theory can be found, where parts of them will be introduced and discussed in Chapter "Effects of Temperature on Uniaxial and Multiaxial Fatigue of Natural Rubber Under Relaxing and Non-relaxing Loadings". The textbook by Bird et al. [36] together with its valuable historical notes for the origin of heat transport equations is one of the rare textbooks, which treat comprehensively both, the microscopic (statistic) as well as the macroscopic (engineering) view accompanied with examples, however, lacks the microscopic concept of phonon transport. The handbook by Mark [3] describes physical properties, mainly electrical conduction and within its Chapter "Characterization of Crack Propagation in Natural Rubber

Under Complex Loading Conditions Using Thermographic Methods" gives a macroscopic overview for thermal properties and tables for thermal conductivities with respect to specific applications.

A summary table with thermal conductivities for numerous polymers showing a high chemical variety is given in Mehra et al. [25] (pages 32–37) giving an overview for nylons, polycarbonates, polyesters, polyethers, polyketones, epoxides, halogenated olefin polymers, hydrocarbon polymers, polyimides, phenolic resins, polysaccharides, polysiloxanes, polysulphides and polysulfones, polyurethanes, and vinyl polymers. Table 2 summarizes thermal conductivities of selected elastomers and polymers from this reference emphasizing the low values of thermal conductivities of elastomers.

Many industrially used polymers are amorphous or at least semicrystalline. Among them elastomers are prominent representatives. Formulated in very general terms, the microscopic structure is essentially responsible for the fact that elastomers have a very low thermal conductivity. Long polymer chains with a high aspect ratio cause strong bending and curvature leading both to highly disordered and entangled amorphous structures, which hinder heat transport on the molecular level (details see in Chapter "Effects of Temperature on Uniaxial and Multiaxial Fatigue of Natural Rubber Under Relaxing and Non-relaxing Loadings").

Another important aspect is that the thermal conductivity κ for amorphous thermoplastics as a function of temperature differs from the temperature dependence of κ for crystalline polymers [27]. For amorphous polymers κ is the same below the glass transition or increases with temperature. Reason for that special behaviour is the high segmental mobility for amorphous polymers. This increase in κ by structural changes was shown by Eiermann [39], who investigated the thermal conductivity parallel and perpendicular to the direction of stress and found an anisotropy for κ (details see also Chapter "Effects of Temperature on Uniaxial and Multiaxial Fatigue of Natural Rubber Under Relaxing and Non-relaxing Loadings"). dos Santos et al. [14] analysed as well amorphous thermoplastics within the range of phonon heat conduction theory used for ceramic materials [40] and also described this anisotropy of the thermal conductivity.

Based on the works of Eiermann [39], Yamamoto [41] studied the influence of structural effects like additional intermolecular bonding schemes on κ for polystyrene and PMMA. Here, rather than discussing the impact of the change of molecular weight through cross-linking like it is typical for elastomers, the impact of the relative number of van-der-Waals bonds was put in the focus. For elastomers in general, the thermal conductivity of crosslinked materials increases with increasing degree of crosslinking. This conceptual idea for elastomers was transferred to polymers like PMMA. Several empirical formulas were derived describing (i) the thermal conductivity for crosslinked polymers via a resistance approach, (ii) the thermal conductivity for amorphous copolymers, and (iii) the thermal conduction and heat capacities through van-der-Waals bonds. Theoretical explanations had been made incorporating as for instance the Grüneisen equation making use of the Debye temperature [42] or the Tarasov expression for the heat capacity [43], which was further also related to the sound velocity. Details are explained in Chapter "Effects of

Temperature on Uniaxial and Multiaxial Fatigue of Natural Rubber Under Relaxing and Non-relaxing Loadings".

Rubbers show a correspondingly low thermal conductivity (see Table 2) as the strength of interatomic binding forces is typically low for organic carbon chains [27]. A comparison is made by Rehner [44] for amorphous rubbers to be similar to liquids, and attempts have been made to find expressions based on statistical thermodynamics for liquids. Quite often, especially in the beginning of polymer and elastomer research, the thermal conduction behaviour of amorphous polymers was described analogues to amorphous materials like glass [64].

Material design options to manipulate thermal conductivities depend on the final application. Either lowering κ by introducing some disorder see Henry [26], however engineering higher thermal conductivities needs quite a lot of efforts as is necessary for composites. For polymer composites, adding fillers of high conductivity leads to an increase in κ (see Chapter "Effects of Temperature on Uniaxial and Multiaxial Fatigue of Natural Rubber Under Relaxing and Non-relaxing Loadings").

The variety of polymer structures and their applications have led to the use and development of numerous theoretical applications on both, macroscopic and microscopic levels, e.g. either continuum-based or molecular-based. Some of the most important methods will be described in the following highlighting concepts and ideas to describe and improve thermal conductivities by using structure-property relations.

2 Macroscopic and Microscopic Heat Transport in Polymers: An Overview

The chapter will give a general as well as a short historical review with respect to heat transfer in solid materials or polymers, respectively. This comprises crystalline as well as amorphous polymers and elastomers. Since historically the greatest focus has been on crystalline structures, most literature sources often start with the description of crystalline materials and do not explicitly differentiate between special features in theory for crystalline and amorphous polymers. Certain technical terms as for instance Debye model, phonons or mechanism will be presented always with slightly different focus of explanation depending on the reference.

2.1 General Introduction

In short, in this chapter, the macroscopic heat transport is described first followed by more detailed parts of microscopic heat transport accompanied by selected examples.

The transfer of heat as exchange of energy is an ubiquitous phenomenon, which is not limited to polymers. Therefore some general aspects are summarized here. First, the macroscopic heat transport is described. Heat transfer is defined by Fourier's law of heat conduction [28, 36, 45], which represents the classical macroscopic view (see Eq. 1).

$$\vec{j}\,(x, y, z) = -\kappa \cdot \nabla T = -\kappa \cdot \text{grad}T \tag{1}$$

where $\vec{j}$ is the heat flux, κ the thermal conductivity, T the temperature in K, and $\nabla$is the Nabla operator, which is applied to the temperature in all spatial directions x, y, and z. The unit used for κ is W/m/K. Fourier's law relates the heat flux $\vec{j}$, which equals the transfer rate per unit area in a particular direction x, y, or z to the temperature gradient in that direction. The law applies to all states of matter like gases, liquids, and solids as long as a continuum approach is used. The negative sign indicates that heat flows against the gradient of temperature from the hot to the cold end. The thermal resistivity, which is not explicitly treated here, is the reciprocal of the thermal conductivity.

Classically in the macroscopic view, heat transport is subdivided into conduction, convection, and radiation depending on the involved media [4, 28]. For polymers and elastomers as solid materials, conduction is the main process to be recognized in heat transfer processes. An engineering-related view to all heat transfer aspects on a phenomenological level can be found in the comprehensive textbooks of Nellis and Klein [28], Bird et al. [36], and Beek et al. [45].

It is generally accepted that today's modern applications, however, require a change of the scientific perspective to heat transport. While engineering applications are mainly performed based on a continuum approach, this does not apply to problems in the nanoscale. As long as solids, liquids, and gases behave and are seen as continuum, e.g. the distances for interactions between microscale energy carriers are small compared to the dimensions of interest, then Fourier's law is to be applied without any risk. A summary of common resistance formulae for continuum approaches for plane wall, cylindrical shell, spherical shell, convection, and radiation can be found in Nellis and Klein [28]. Especially, one criterion in particular to decide between a continuum or a microscale model is the check of the dimensionless Knudsen number. It relates the characteristic length scales of energy carriers to the materials problem length scale. For values much less than unity, the continuum theory is sufficient [28].

Current requirements for modern materials, however, can only be met with a strong focus on the microscale side of heat transfer processes. The basic theory for this approach was already developed more than 100 years ago by Debye [46] and Peierls [33], and allows in combination with high computational power nowadays a detailed view of heat transfer in nanostructures, where characteristic lengths of nanostructures need to be taken into account for a correct representation of the thermal conductivity.

From the point of view of physics, molecular heat transport is one of the molecular transport phenomena, e.g. transport of energy, mass or momentum. A similar structure of their mathematical description is common to all of them. The transport equations for energy, mass or momentum are always derived from an

Table 3 Transport coefficients

Transport coefficient	Transport	Law
Thermal conductivity κ	Thermal energy	Fourier
Diffusion coefficient D	Mass	Fick
Viscosity η	Momentum	Newton

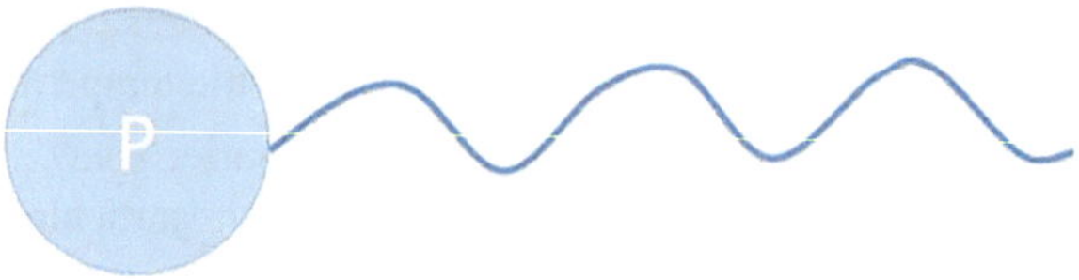

Fig. 1 Schematic representation of the wave-particle nature of phonons according to Mehra et al. [25]

adequate transport coefficient, e.g. thermal conductivity, diffusion constant and viscosity, respectively related to the change of temperature, concentration or velocity, respectively [29]. A summary of these basics is given in Table 3.

2.2 *Phonon Transport in Solid Materials*

Based on the early work of Debye from 1911 "Theorie der spezifischen Wärme von Kristallen" and also Peierls in the 1920s, heat conduction in dielectric crystals was described by the concept of mean free path of thermal waves (quanta) called phonons, which was summarized by Anderson [27] in a review paper.

A phonon is the quantum mechanical description of an elementary vibrational motion in which a lattice of atoms or molecules uniformly oscillates at a single frequency [47]. Phonons are also called quasiparticles [48]. Mehra et al. [25] describe phonons as quantized energy of lattice vibration or form of heat energy and suggest a representation of the wave-particle nature of phonons, which is reproduced in Fig. 1. The tail is intended to illustrate the wave nature of a phonon, whereas the head represents the particle nature.

Basic physical principles of describing the thermal conductivity, which in their original approaches date back to Peierls [49], are comprehensively explained in Chen [72], where the phonon transport and the resulting thermal conductivity is described by the specific heat c, the phonon group velocity ν, and the relaxation time τ between two phonon collisions, see Eq. 2. This equation will appear throughout the following text in different notations depending on the reference source or textbook and its interpretation or focus.

$$\kappa \sim c\, v^2 \tau \tag{2}$$

For polymer materials the phonon transport is governed by the so-called phonon group velocity, the mean free path length, and the specific heat [25].

Moreover, phonons are scattered on their way through the crystal bulk. Reasons for scattering are crystal imperfections thus influencing the direction, momentum, or

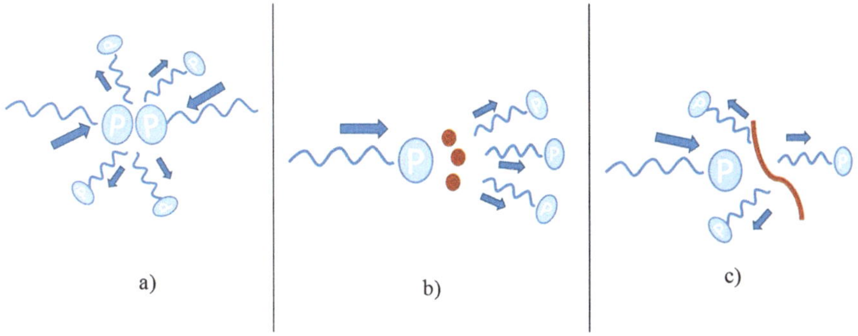

Fig. 2 Schematic representation of scattering pathways for phonons. (**a**) Phonon-phonon scattering, (**b**) phonon-impurity scattering, (**c**) phonon-boundary scattering. Adopted from Mehra et al. [25]

energy of the phonons leading to a decrease of κ [25]. In Fig. 2 three basic scattering pathways are illustrated.

Henry [26] provides a qualitative view to Eq. 2 highlighting the linear or quadratic dependencies via the group velocities of the thermal conductivity. The group velocity is seen as one of the major influencing factors to achieve certain thermal conductivities. It is emphasized that the effect of the velocity on the thermal conductivity is basically related to the elastic modulus of the material. The modulus itself is further strongly related to the underlying molecular structure. As carbon-carbon bonds are the typical bonds in polymers, they possess the highest phonon velocities. Therefore, linear and crystalline or semicrystalline polymers, which have carbon backbones are potentially suitable for high thermal conductivities. However, in amorphous polymers due to disorder in the solid structure, the thermal conductivity is significantly decreased by several scattering processes.

Phonons are transported as well across interfaces or boundaries of the polymer chains. This interfacial transport has a high impact at low temperatures because the mean free path length of phonons is the longest there. For nanostructures where the mean free path of the phonons approaches the length scale of the material, the thermal transport deviates from the Fourier law [25, 50].

Phonon scattering at the interfaces of layers, solids, or crystallites is one main reason that limits the thermal conductivity of polymer composites [25]. In (crystalline) composite materials, phonon scattering is mainly attributed to interfacial thermal resistances which arise from an acoustic mismatch at the filler-matrix interface [25, 26, 51]. The overall phonon pathways sum up to an integral effect, which can be finally measured as thermal conductivity. Figure 3 relates the phonon transport to the macroscopic view according to Fourier and shows a schematic representation of phonon transport.

Phonon-phonon collisions, which may occur, are complex and have to be considered for the overall heat transport. Such collisions are distinguished to as a normal or as Umklapp mechanism [72]. For normal phonon-phonon collisions, energy as

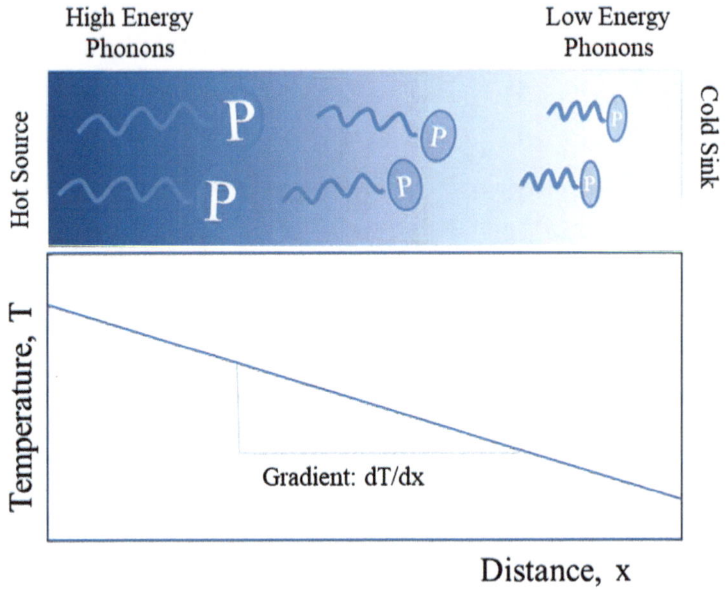

Fig. 3 Temperature gradient as measured macroscopically and related phonon transport. Adopted according to Mehra et al. [25]

well as momentum are conserved, for the Umklapp mechanism, the energy is conserved only.

The relaxation time (see Eq. 2) as the average time interval between two successive collisions is related to the mean free path length – for more information the reader is referred to the following sources (Koh [4] and Zhang [1] and general textbooks of solid state physics [32, 34, 36]). Relaxation times for phonon scattering are on the order of picoseconds, thus usually electrons relax much faster by phonon scattering than by photon scattering, which is the order of nanoseconds [52].

A discussion of macroscopic properties as the glass transition temperature and its relation to microscopic heat transport can be found in Eiermann [53]. Amorphous high polymer materials show a maximum for the glass transition temperature, like seen for natural rubbers, polyester-urethane and others. In the article, Eiermann developed a model for the heat transport in amorphous materials based on the preliminary works of Debye (see above). According to Debye, heat is transported in solids by elastic waves, which undergo scattering at inhomogeneities of the solid and thus disperse intensity along the pathway. The distance for the lowering of intensity is called the mean free path length of the waves or, according to the particle view, the phonons. The thermal conductivity is thus defined as:

$$\kappa = B\, \varrho\, c\, v_{\text{elastic}}\, l \qquad (3)$$

where B is a pure number in the order of magnitude one, $v_{elastic}$ is the velocity of elastic waves, and l is the mean free path of elastic waves or particles.

2.3 Structural View on Microscopic Heat Transport

The mean free path length is for amorphous solids, according to Debye, independent of the temperature as scattering paths predominate [53]. For amorphous silicate, the mean free path was shown to be in the range of atomic distances [54]. Thus, the value of the mean free path was set by Eiermann equal to the distance of two neighbour atoms. A coupling of this approach with the macroscopic view to heat transport gave him also the opportunity to measure heat resistances similar to electrical resistances. Samples need to fulfil for that measurement, that the sample size resembles the value for the mean free path length. For this approach, amorphous materials are seen as a network of elemental heat resistances. The article then translates this into a practical approach relating it to the mass of the sample, the speed of sound, and the heat capacity [53]. Eiermann also analysed the temperature dependence of the glass transition for various polymers and elastomers. It was shown that the strength of bonds (represented by spring constants) varied according to the kind of the bonding, e.g. covalent ("Hauptvalenzen") or van-der-Waals bonds. Covalent bonds show a low resistance to heat transport while van-der-Waals bonds are characterized by high heat resistances. The thermal conductivities of amorphous polymers took values between amorphous glasses and liquids and showed a maximum at their glass transition temperatures. Eiermann [42] investigated also the anisotropy of κ of polymers depending on the stress applied. While κ was increased in the direction of stress, it decreased perpendicular to the stress. It was concluded that this is due to improved transport pathways along the covalent main chain valences and van-der-Waals bindings for the direction of the stress. It was not attributed to crystallization as for this reason also an increase of the heat conductivity in the other direction of stress would have been observed.

2.4 Phonon Scattering Processes

The background theory of phonon scattering mechanisms for crystalline substances with respect to phonons as bosons, and their description by the Debye model as well its improvement by the Bose-Einstein-model can be found in the textbook by Hertel [55].

For solid materials, there are two contributions, namely from electrons and phonons, which sum up to the total thermal conductivity for non-magnetic solid materials [1].

$$\kappa = \kappa_{\text{electron}} + \kappa_{\text{phonon}} \tag{4}$$

Phonons in amorphous materials as polymers or elastomers have a mean free path less than 10 nm [56]. The transport takes place along the axis (chains in polymers) similar to C–C bonds in graphite and diamond [57], and thus determines the value of the thermal conductivity. As learned from carbon-based materials, the dimensionality of the material has an impact on κ [57]. Single-layer graphenes had a larger thermal conductivity compared to its bulk lattice. The reduction of scattering channels was discussed as the reason for this behaviour, and the effect was traced back to symmetry constraints [59] as well as the removal of inter-layer coupling via strong anharmonic van-der-Waals forces [60]. The thermal transport in 1D-model-polymers showed diminished van-der-Waals forces [61].

Phonon scattering in crystals is divided into three basic types of scattering (i) phonon-phonon scattering, (ii) phonon-impurity scattering, and (iii) phonon-boundary scattering [62]. All three phonon transport modes are described in detail in Mehra et al. [25] (see Fig. 2).

Phonon scattering is a temperature-dependent phenomenon as temperature has a significant influence on the mean free path length. Thus, the phonon scattering mechanisms are dominant at different temperature regimes. This is also prominent when looking to phonon group velocities. Single polymer chains for instance show a low scattering and provide an effective path for phonon transport. For a single model polymer chain, the phonon group velocity reaches 16,000 m/s (see in [63]). Such structures have flexible bonds thus resulting in a high elastic modulus in contrast to bulk polymers [26]. In contrast to single polymer chains, bulk polymers do show a significantly reduced thermal conductivity due to increased scattering [26] caused by disordered structures and entanglements in the bulk [64, 65]. In general, thin diameters in materials enhance phonon transport [25] thus indicating a strong dependence on the dimensionality of the material.

The transformation of this basic knowledge into recipes for managing polymer properties with respect to enhanced phonon transport can be also found in Mehra et al. [25].

The main phonon scattering pathway in amorphous polymers is structural scattering, which can be easily understood. Amorphous polymers are characterized by a random curvature, e.g. numerous bends in a chain thus reducing the maximum mean free path for phonons. As such, amorphous polymers resemble amorphous inorganic materials like glass according to the fundamental article of Klemens [66]. In this pioneering work, he already described 1951 the statistical equilibrium behaviour for phonons and calculated κ for quartz glass. His work is based on an even earlier fundamental work of Peierls (1929), Landau and Rumer (1937) and Pomeranchuk (1941), details see in [66]. The maximum value of κ for amorphous polymers without any additives possesses a similar course like the volumetric specific heat [26]. Klemens states in his publication that crystalline and amorphous solids may be treated similarly. According to him, each atom has a fixed equilibrium position and its thermal vibrations can thus be resolved into travelling waves [66]. Those normal modes are no longer plane waves for amorphous structure, which lacks regularity. When now analysing the displacements, the interchange of energy between the waves has to be recognized. Those interactions have been called structure scattering [66]. More details are given also to the interaction pattern for longitudinal phonons, which contain non-conserving momentum processes (Umklapp) as well as a description of κ for glass with respect to low temperatures.

The so-called Klemens' approach [66] distinguishes two basic contributions to κ arising from the conservation of the momentum. The first contribution is the so-called normal one κ_{I} conserving the momentum, whereas the so-called Umklapp contribution κ_{II} is not momentum conserving. Based on the early work of Peierls, Landau and Pomeranchuk, he derived mathematical expressions (see Eqs. 5–7) for two contributions κ_{I} and κ_{II} by additionally employing the Debye theory of specific heat of solids.

$$\kappa\,(T) = \kappa_I(T) + \kappa_{II}(T) \tag{5}$$

$$\kappa_I(T) = \frac{1}{3}\int_0^{\infty} \frac{\partial}{\partial T}\left(E(k,T)_I\right) \upsilon_I \,\frac{l_I^{'}\, l_I}{l_I^{'} + l_I}\, dk \tag{6}$$

$$\kappa_{II}(T) = \frac{1}{3}\int_0^{\omega_1} \left(\frac{\partial}{\partial T}\left(E(k,T)_{\text{eff}}\right)\left[\frac{l_{II}^{'}\, l_2}{l_{II}^{'} + l_2} + \frac{l_{II}^{'}\,(\omega)}{1 + \frac{l_2}{l_{II}^{'}}}\right] dk\right.$$
$$+ \frac{1}{3}\int_0^{\infty} \frac{\partial}{\partial T}\left(E(k,T)_{\text{eff}}\right) \upsilon_{II}\, l_{II}^{'}\, dk \tag{7}$$

where l is the relaxation length for single modes (details for contributions depending on conservation of momentum or not, see paper), k is the normal mode of the wave number, ν is the group velocity of phonon-wave, E is the energy content per unit volume and w is the frequency. Using this approach [26, 72], the thermal conductivities for the low and moderate temperature range could be estimated for quartz glass. Most amorphous polymers, like PMMA or PET, show a similar typical curve trend for both values of κ_{I} and κ_{II} and result in an average thermal conductivity of maximal 0.2 W/m K [26, 74].

Polymers and elastomers have different kinds of bonding thus causing different bond interactions. In their chains, they are primarily covalent while between chains van-der-Waals interactions dominate. Covalent bonds are stronger compared to van-der-Waals interactions inducing different contributions to the phonon transport. Moreover, ordered structures grant an improved phonon transport, which is further enhanced via a perfect low dimensionality. 1D-structures compared to 3D-structures and oriented chains in contrast to chaotic structures have better phonon transport as scattering processes are diminished [25, 67]. In general, polymer chains with an appropriate elastic modulus, given mainly via linear chains and less branching have optimal phonon transport properties [25, 68].

Extending this structural view to polymers or elastomers, which have crystalline and amorphous parts in their structures, the phonon transport contributes to a different amount to the heat conductivity. Ordered structures and alignments are positive factors with respect to an increased thermal conductivity. Therefore in semicrystalline polymers, the higher the crystalline parts the higher the heat conductivity [25, 69, 70]. Fillers on carbon-basis, like graphene or CNTs promote crystallization thus increasing the thermal conductivity finally [71]. Most crystalline polymers are actually semi-crystalline [56], e.g. crystalline and amorphous regions influence the overall behaviour of the polymer and thus also the heat conductivity.

Again back to the original work of Debye and extending the interpretation from crystalline to amorphous materials, polymers and elastomers are dielectrics, thus electron contribution to the thermal conductivity is negligible. Phonons as quasiparticles are the main heat carriers for polymers and elastomers. The thermal conductivity based on phonons can be calculated according to the Debye equation [27, 28, 72, 73], see eq. 8.

$$\kappa_{\mathrm{phonon}} = \frac{1}{3}\, c_{\mathrm{v}}\, v_{\mathrm{g}}\, l \tag{8}$$

The volumetric specific heat c_{v} is a function of the phonon frequency, polarization, and temperature. The group velocity v_{g} of phonons describes how fast phonons propagate in a material and depends via phonon dispersion on the phonon frequency and its polarization [4]. The mean free path l describes the average distance, which phonons can travel without being effected by any type of scattering. For amorphous polymers, l is an extremely small constant (i.e. a few angstroms) due to phonon scattering from numerous defects, leading to a very low thermal conductivity of polymers. Koh [4] illustrates in detail the heat conduction by phonons in crystalline dielectrics, which is characterized by contributions of different polarizations (longitudinal or transverse) accompanied by a broad distribution of frequencies and lifetime of phonons.

Applying the theoretical basics as seen in Eq. 8, Hansen and Ho [74] showed that for linear amorphous high polymers κ depends on the molecular weight and the degree of crystallinity. The dependence on the molecular weight was described by a theoretical model, which comprised polymers of low and high molecular weights. At low molecular weights κ was found to be proportional to the molecular weight, at higher molecular weights κ tends to a nearly constant value [74, 90]. Moreover, the authors stated that the thermal conductivity for semicrystalline polymers (here linear polyethylene) was found to be in accordance with the phonon concept in dielectric crystals and that the results prove that κ can be calculated by recognizing thermal conductivities of all fractions of the material, e.g. amorphous and crystalline, via the per cent crystallinity concept [74]. Eiermann [53] classified via its theoretical considerations, the thermal conductivity for amorphous polymers in between inorganic glasses and liquids and derived an expression which describes the decrease of κ with increasing molecular weight.

2.5 Phonon Behaviour and Transport

Scattering processes in crystalline materials are caused by impurities, anharmonicities, and grain boundaries. The mean free path and the group velocity are mutually correlated, see details in [4]. Phonon transport is of course also temperature dependent. Heat is transferred at high temperatures mainly by so-called long-wavelength acoustic phonons [4].

Phonons are formed in general when atoms vibrate. Transferring this idea to an ideal harmonic solid, an infinite thermal conductivity would result [25]. The mean free path length of a phonon is always a compromise between different scattering modes and may be calculated according to Matthiessen's rule [25].

Matthiessen's rule states that the overall phonon scattering rate v results from the sum of individual scattering rates v_i (e.g. phonon-impurity, phonon-electron, phonon-boundary, phonon-phonon) as seen in Eq. 9 [1]:

$$\frac{1}{\nu} = \frac{1}{\nu_{\text{phonon - impurity}}} + \frac{1}{\nu_{\text{phonon - electron}}} + \frac{1}{\nu_{\text{phonon - boundary}}} + \frac{1}{\nu_{\text{phonon - phonon}}} \tag{9}$$

In dielectric materials such as polymers, contributions from electrons are negligible. The understanding of phonon transport is somehow incomplete as there is a lack of a direct measure of mean-free paths of phonons [4] or the definition of phonon wave vector and group velocity [1]. An accurate calculation of phonon scattering is part of extensive research. A more detailed view can be found in Zang et al. [1] or in textbooks [32–34] where the mean free path is correlated to the phonon relaxation time. Moreover, Zhang et al. [1] discuss in detail the influence of phonon dispersion, which may be described by empirical potential functions or ab initio calculations and is highly dependent on the number of atoms in the model. Low-frequency phonons play a preferential role in heat conduction. The relaxation time is still to be seen in relation to phonon-material interactions thus opening the wide range of scattering phenomena caused by vacancies, impurities, dislocation, and grain boundaries as stated above [75].

In addition to the well-known diffusive and ballistic thermal transport regimes, thermal phonons can also propagate hydrodynamically as a collective motion when the momentum-conserving phonon-phonon scattering (normal process) is dominant [76]. In this phonon hydrodynamic regime, many unique transport behaviours have been demonstrated so far, such as second sound, the Poiseuille flow, ultrahigh thermal conductivity, and its strong size effects. More recently, hydrodynamic phonon transports in new systems have received extensive research attention. For instance, hydrodynamic phonon transport and second sound were reported in two-dimensional materials and graphite [76].

Using general equations as already given above may sometimes be problematic for amorphous polymers. One way out of this was provided by the description of polymers by a new master equation. This was done in an application study regarding optoelectronics [21].

An earlier theoretical framework on heat conduction in amorphous materials was first proposed by Einstein and later refined by Slack, see in Wingert et al. [20]. The theory by Slack [77] is based on the assumption of independent or non-correlated vibrations among neighbouring atoms or lattices, which is in contrast to that of extended waves in crystals. Heat conduction is described by a "random walk" of these independent oscillators, each at a characteristic frequency, the Einstein frequency. Slack formulated this problem by equating the minimum mean free path to be the same as the wavelength of the phonon. The thermal conductivity obtained from this model is known as the "minimum thermal conductivity". In 1992, Cahill et al. [78] further extended this model by dividing the sample into regions of definite size, with velocity being the low-frequency speed of sound (i.e. Debye model) [20].

2.6 Quantum Mechanics View and Bond Types

Heat conduction in crystalline solids is generally described with the phonon concept as already described before. Phonons as vehicle to describe heat transport should be recognized as "particles", which are generated at the hot end of a sample and will be conducted to the cold end. Similar to photons, the concept of wave-particle duality is also applied, which describes the particle phonon as a superposition of waves, which are localized in a certain space. Also similarly to photons, phonons are classified as bosons, which do not follow the Pauli-principle. The statistical view to phonons is the basis for analysing the heat capacity of solids, which then is the leading to the value of κ [30].

As already mentioned phonons are waves of oscillatory atomic vibrational energy. Depending on their polarization (longitudinal or transverse) they transport different amounts of heat and possess a broad range of frequencies and lifetimes [4].

Phonon-phonon interactions have to be distinguished with respect to the solid material itself, e.g. crystalline or amorphous. Pathways of phonon transport depend on the availability of translational symmetry, thus phonon wave vectors and group velocities are not well defined. Some authors therefore suggest phonon-hopping processes [79]. In Zhang et al. [1], the references with respect to this field (e.g. Refs [37–40]. therein) are not correctly cited! However, the concept is successfully adopted for amorphous silicon referring to semiconductor applications [80].

In general, the energy transfer proceeds via three modes: (i) energy transfer related to molecular motion, (ii) energy exchange by bonded interactions, e.g. covalent bonds, and (iii) energy exchange by non-bonded interactions, e.g. van-der-Waals forces [25, 81]. All modes influence the microscale phonon transport differently. The flexibility of the backbone chain, visible via the segmental motion is an important factor. Rigid backbones and a π-conjugation in the polymer chain rather suppress segmental rotations and thus large phonon group velocities [17]. Moreover, the structural conditions also prevent phase transitions as well as reduce disorder scattering thus enabling large phonon group velocities. The relation between the nanoscale molecular structure and the resulting property of thermal conductivity is clearly shown here.

The structure of the backbone based on covalent bonding, which can be increased via crosslinking on covalent bonding, is one prerequisite for high thermal conductivities. The thermal resistance of a structure characterized by covalent bonds is ten times smaller compared to only interchain van-der-Waals forces [53]. Via crosslinking new chemical bonds are generated thus enhancing the thermal conductivity. This effect is, however, ambiguous as newly created bonds increase in general the thermal conductivity, whereas branching is counter-productive with respect to the alignment of chains [82]. This could be shown very impressively via molecular dynamics simulations on polyethylene. The fibres showed a decrease in κ for crosslinking densities above 10% (see in [81]). The design of adequate linkers in the chain network offers another option to generate improved thermal conductivities. Strong linkers between chains have been found to have a positive effect on κ [83]; however, more bulkier groups decreased vice versa κ due to phonon scattering caused by segmental rotations [17].

Nanoscale structural features provided by non-bonding forces such as van-der-Waals or hydrogen bonds also contribute to the final κ. Especially a hydrogen-bonded network has a higher impact on κ due to the different energetics for the different bond types. In general, for crystalline polymers κ increased significantly due to hydrogen bonding by creating new phonon transport paths [84]. Similar to the impact of additional crosslinking bonds in covalently bonded networks, introducing hydrogen bonds however increases the number of ambiguous effects. Hydrogen bonds are strongly structure-dependent and a tuning of bond strength may be possible via the appropriate choice of the involved molecules, thus determining the strength of the hydrogen bond. Further influencing factors with respect to hydrogen bonds may be the density of these bonds throughout the structure, its ability to couple chains, and its dispersion in the bulk material. In general, hydrogen bonds open the ability of generating a higher structural order thus reducing phonon scattering [85]. A high amount of hydrogen bonds was found to be very positive on final thermal conductivities [83].

With regard to the structural features of amorphous polymers, the possibilities of structural influence become smaller. For amorphous polymers, structural scattering is much easier and delivers some structure-property-relations to interpret low thermal conductivities. Amorphous polymers have in general no ordered structures, and show a low alignment combined with an irregular structure thus opening a lot of scattering paths. The orientation of chains is not ordered, it is rather randomly with numerous curvatures in the polymer chains. Phonons are exposed to high tortuosities. This decreased propagation of vibrational waves is causally responsible for very short mean free path lengths, and thus low thermal conductivities [86]. In general, the mean free path length in amorphous polymers is lower than 10 nm [56].

2.7 Characterization and Structure-Property Relations

A quite simple approach based on X-ray results for semicrystalline polymers to propose macroscopic properties (stress-strain behaviour, compression modulus, shear modulus, elastic properties) was proposed very early by Kausch-Blecken von Schmeling [87]. Simple resistance models have been set-up to derive empiric equations for the thermal conductivity, and the linear isothermal compressibility for differently shaped isotropic and anisotropic phases. A new model for amorphous polymers was developed. Polymers of interest have been PC, PMMA, PVC, and PS.

One important aspect in describing the thermal conductivity for amorphous polymers, which has to be recognized, is the appearance of a glass transition temperature. The mobility of the polymer chains changes significantly when reaching the glass transition temperature. A high mobility of the chains at higher temperatures decreases the elastic modulus. In case of amorphous polymers, this changed mobility, however, leads finally to a constant κ or a maximum of κ around the glass transition temperature. The thermal conductivity variations registered around the glass transition temperature of amorphous polymers can be explained by considering the reduction in the phonon velocities above this glass transition temperature [14]. Another interesting effect was observed for semicrystalline polymers, which showed anomalous behaviour around the glass transition [14].

The anomalous behaviour of κ for polymers was already earlier described for amorphous polymers and elastomers by Knappe et al. [88]. Measurements on various samples like polystyrene, polymethylmethacrylate, polyvinylchloride, and polycarbonate revealed that the thermal conductivity in the direction of stretch is greater and perpendicular to this direction is lower than in the un-stretched isotropic state. The degree of anisotropy depended on the degree of stretch and on the molecular structure of the polymeric chain [88]. There exists a close relation between the anisotropy of thermal conductivity and the thermal expansivity. From measurements on polyvinylchloride, it followed that the anisotropy of the thermal conductivity remained rather unchanged when heating the polymer to temperatures above the glass-transition temperature in the stretched state. Therefore, it may be expected that also strained elastomers will have an anisotropy in their thermal conductivity. Measurements on several strained elastomers in this direction perpendicular to the strain had been done therefore, which showed that strained elastomers with strong polar side groups have a considerable amount of anisotropy, while aliphatic polymers like natural rubber and polyisobutylene have no remarkable effects [88, 89].

Further structure-property relations can be found. The molecular weight is one important influencing factor. The thermal conductivity increases when the molecular weight is increasing [90]. A theoretical model, based on early research of Überreiter and Otto-Laupenmühlen from 1953, who first described the dependency of κ on the molecular weight [91], turns out the improved propagation of vibrational energy along the chain, which is reduced when the linearity is decreased by branching [26]. The influence of the molecular weight is also discussed in detail in relation to scattering processes. Phonons are scattered due to boundary scattering [26, 27, 56], which occurs at the ends of the chains. Chen [72] has described that cumulative effect via the Matthiessen rule (see Sect. 2.5 in Chapter "Effects of Temperature on Uniaxial and Multiaxial Fatigue of Natural Rubber Under Relaxing and Non-relaxing Loadings") as the combination of parallel scattering effects, e.g. intramolecular, intermolecular, structural, and boundary scattering, which do show all characteristic relaxation time.

To describe all parameters, which have an influence on thermal conductivity, several theoretical approaches appeared over the last 100 years and reflect the strength of statistical thermodynamics as well as theoretical chemistry. One has to distinguish general approaches, which differ in their theoretical depth as either first-principles based or classical molecular dynamics. In combination with Green functions, non-equilibrium approaches, a wide variety of approaches exist. One selected approach, namely Green-Kubo, will be introduced in short in Chapter "Identification of the Part of the Hysteresis that is Converted into Heat" along with examples for the calculation of thermal conductivities for elastomers.

3 Green-Kubo Approach for Determination of Thermal Conductivities for Elastomers

The search for new elastomer materials with high thermal conductivities, and thus new potential industrial applications significantly profits from approaches, which relate structure and property on a molecular level. Those knowledge-based theoretical models enable a prediction of fundamental material properties based on computer-designed structures, which may not yet have been synthesized or are difficult to measure.

In general, the different levels of theory give rise to model approaches ranging from ab initio, semi-empirical to empirical models. Hereunder, the Green-Kubo approach is one excellent method to simulate transport properties. An overview to Green-Kubo but also to Equilibrium Molecular Dynamics – EMD, Non-Equilibrium Dynamics –NEMD, Effective Medium Theory –EMT, and Boltzmann Transport Equation – BTE can be found in Breitkopf [92], Henry [26] or Mehra et al. [25]. Only, the Green-Kubo method will be discussed here in detail, first with a short introduction in basic details followed by examples for crosslinking effects on thermal conductivities in polybutadiene or silicone rubbers, and finally the influence of randomly sulphur-crosslinked polybutadiene on the glass transition temperatures of those model elastomers.

3.1 Introduction to Green-Kubo Approach

To give just a short introduction, the Green-Kubo method is basically an equilibrium technique based on the fluctuation-dissipation theorem and was derived accordingly via the linear response theory [67]. Based on considerations of Drude [93] for the transport of electrons in metals applying kinetic theory, Kubo formulated later this linear response formalism which was initially also related to electrical transport, see also in Ventra [94].

These early ideas of Drude have been merged by Kubo [95] and Green [96] and led to the naming of this theoretical approach: Green-Kubo method. The Green-Kubo equations deliver exact mathematical expressions for different transport coefficients such as thermal conductivity, viscosity or diffusion coefficient as integrals of time correlation functions. In a small limited field area, all transport coefficients of interest are expected to show a linear dependence. The fluxes (here the heat flux for the thermal conductivity) are linear to the special applied field (here: temperature). The heat flux and the temperature are conjugate variables to each other. This relation between a thermodynamic force and its conjugate thermodynamic flux is called a linear constitutive relation [97]. Green and Kubo found exact expressions for these linear coefficients, and showed that the resulting linear coefficients are proportional to the time dependence of equilibrium fluctuations of the conjugate flux.

It has to be stated that the simulation of amorphous polymers compared to crystalline materials is extremely computationally demanding [16]. This statement can be also completely applied to elastomers. Large systems with variable chain conformations require long equilibrium times in the computer simulations and may influence the interpretation of the final outcome, see Henry [26].

The Green-Kubo approach with respect to heat transport provides a relation between κ as the transport coefficient to the integral over the time correlation function of the heat flux, see Eqs. 10–12. Due to the discretization of time steps in molecular dynamics, the integrals can be rewritten as sum [98]. The property temperature itself in molecular dynamics simulations is further available via the equipartition theorem.

$$\kappa \sim \int_0^\infty < j_Q(t) \times j_Q(0) > dt \tag{10}$$

$$\kappa = \frac{V}{3k_B T^2} \int_0^\infty < j_Q(t) \times j_Q(0) > dt \tag{11}$$

$$j_Q = \frac{1}{V}\frac{d}{dt}\left(\sum r_i\, e_i\right) \tag{12}$$

Equation 11 relates the thermal transport coefficient to equilibrium properties (see also in Evans and Morris [99]). Moreover, in their Chapter "Kinetics of Self-heat Build-up in Carbon Black Filled Natural Rubber Caused by Cyclic Mechanical Loading" a comprehensive overview is given for all Green-Kubo relations of Navier-Stokes transport coefficients as for instance self-diffusion, thermal conductivity, shear viscosity, and bulk viscosity thus showing the computational strength of this approach.

The quantity e_i in Eq. 12 represents the total energy of the model polymer, and is calculated from the overall kinetic and potential energy in the molecule on a classical mechanical basis, e.g. sum of potential and kinetic energy.

The equations summarize different nanoscale contributions to reflect the macroscale heat transfer. First, the convectional part, e.g. a diffusive contribution based on the atom movement, whereas the second part describes the intermolecular contribution or per-atom stress tensor. For details, the reader may be referred to as Hoover et al [100] and Zwanzig [101].

Modern computer systems allow a fast calculation of the autocorrelation function. To receive an adequate autocorrelation function, different single simulation steps applying specific ensembles have to be processed one after the other with care using appropriate thermostats and barostats. First, a NPT-ensemble is applied to equilibrate the model system and correct densities for a given temperature and pressure. Afterwards, a NVT ensemble is applied to derive average values for κ. Details with respect to this sequence of simulation steps can be found in Vasilev, Lorenz, and Breitkopf [102–104] as well as Alamfard, Lorenz, and Breitkopf [105].

3.2 *Thermal Conductivities and Glass Transition Temperatures for Elastomers via Green-Kubo Calculations*

The Green-Kubo method was applied to calculate thermal conductivities for polybutadiene crosslinked with sulphur [105]. Those simple model elastomer structures have been merely a starting point of a number of calculations to ensure the applicability of the used approach for this kind of substance. Having this as a reference, which confirmed to be able to reproduce experimental values for thermal conductivities, allows to extend the calculations on new elastomer structures of interest, which have not yet been prepared. Thus, Green-Kubo is expected to predict new elastomers as well as elastomer composites, see [9].

Simulations have been done using the LAMPS software and a united-atom force field. The molecular structure of polybutadiene was modified by random crosslinking with sulphur, thus generating model systems with variable amounts of sulphur bridges, see Fig. 4a, b. The sulphur bridges always consisted here of two sulphur atoms only.

The calculations of various crosslinked polybutadiene model structures showed that a uniform distribution of sulphur (20% sulphur) led to very low thermal conductivities of 0.182 W/m K, whereas a random distribution of sulphur in the model structure resulted for a comparable sulphur content of 20% to a thermal conductivity of 0.242 W/m K, thus confirming by this comparison, that real crosslinked structures may preferably possess a random distribution of crosslinkers.

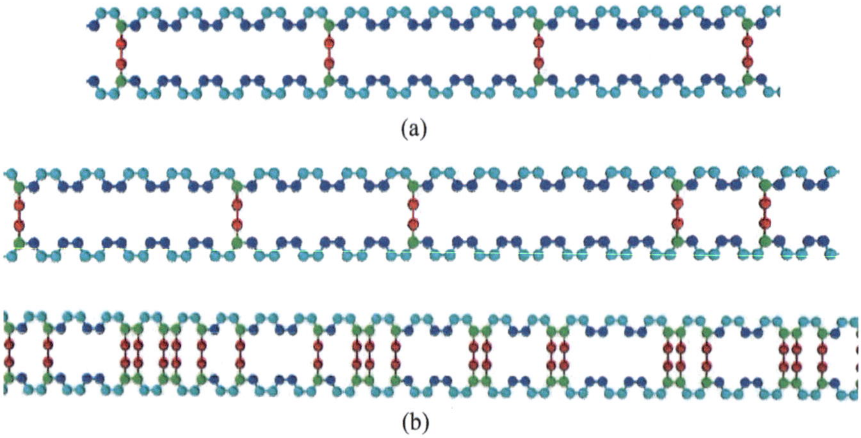

Fig. 4 Simulated model structures for sulphur-crosslinked polybutadiene (sulphur atoms shown in red). Taken from Alamfard et al. [105] (**a**) Polybutadiene model structure with a uniform distribution of sulphur bridges. (**b**) Two polybutadiene model structures with random distribution of sulphur bridges and different sulphur contents

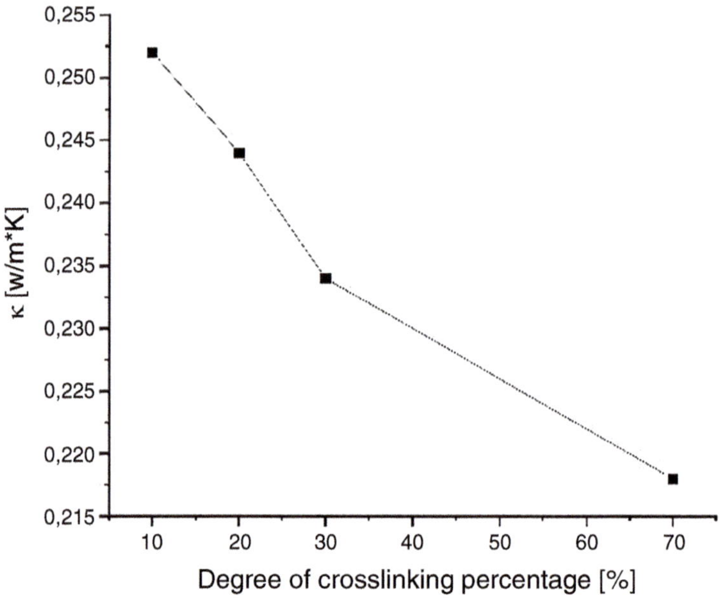

Fig. 5 Calculated thermal conductivities κ [W/m K] for model polybutadienes with random sulphur-crosslinks and different amounts of sulphur contents. Reproduced from Alamfard et al. [105]

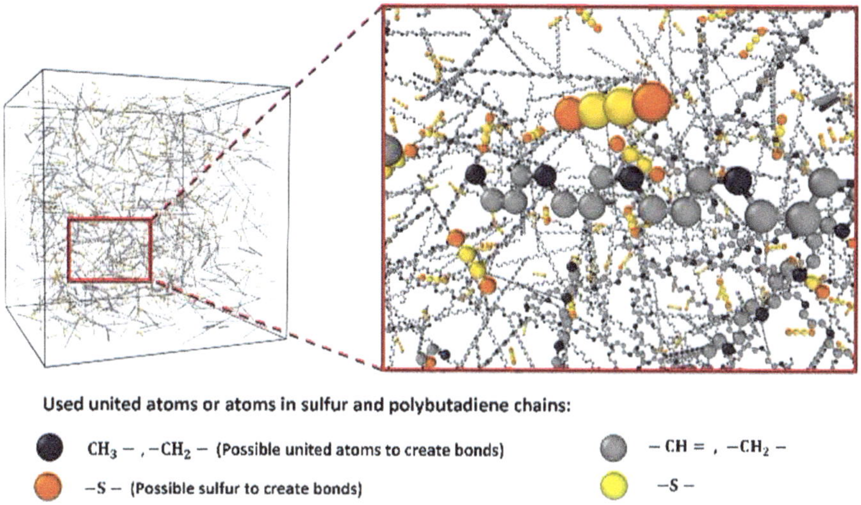

Fig. 6 Polybutadiene model structure with variable length of sulphur bridges, here four sulphur atoms. Taken from Alamfard et al. [107]

Further results for a random distribution of sulphur in polybutadiene can be found in Fig. 5.

With increasing sulphur content in the model structures, which may even be beyond real sulphur contents in vulcanized polybutadiene, the resulting thermal conductivities decreased as the periodicity along the chains is disturbed by the sulphur bridges. Compared to pure polybutadiene, which has a thermal conductivity of 0.17 W/m K (measured [104]), the effect of crosslinking leading to an increased thermal conductivity for sulphur crosslinked polybutadiene is clearly visible.

Green-Kubo was also successfully applied to silicone rubbers, where model structures with a variable degree of crosslinking ranging from 0% to 14% crosslinking have been investigated [102]. Silicone rubbers are expected to have good performance properties for soft robotics. The type of crosslinking was chosen according to Kanan et al. [106]. Calculated values for these structures varied from 0.13 W/m K for the pure silicone rubber to 0.16 W/m K for crosslinked silicone rubbers showing a value close to experimental thermal conductivities of 0.17 W/m K (see in Vasilev et al. [102]).

Recently, the Green-Kubo method was applied to polybutadiene model structures having more than two sulphur bridging atoms, see Fig. 6. For those structures also the structure-property relations with respect to the resulting glass transition temperatures were investigated. Modelling details as for instance an automatic bond creation to generate random structures are given in Alamfard et al. [107].

The first effect of longer sulphur bridges was seen as an increase in the thermal conductivities. Moreover, for model structures with a constant degree of crosslinking but variable sulphur chain length, the glass transition temperature was found to increase from model structures having one sulphur atom as a bridge (179°C) to model structures having longer bridges up to eight sulphur atoms (216°C). The density of the model system as well as the packing efficiency is discussed as the reason for this observed increase.

In general, Green-Kubo methods are a reliable tool for describing the transport properties of elastomers and may serve as a prediction tool for new elastomer composite materials.

4 Experimental Methods

Only a few literature reviews with a direct focus of the measurement of thermal conductivities of polymers are available. One early work from Anderson [27] reviews numerous methods for measuring κ as well as comments on different kinds of polymers as thermoplastics, rubbers, and thermosets. A recent review for measurement techniques can be found in Henry [26].

4.1 Classical Equilibrium and Transient Methods

In general, a classical experimental set-up is easy to establish: a sample is placed in contact with a heating element or a heating sensor, and changes of the temperature are measured during heat flux. However, several artefacts may occur due to finite geometries and unknown interfacial thermal resistances [18].

One precondition for the measurement of thermal conductivities is: two bodies at different temperatures are brought in physical contact and the heat flux is measured. Anderson [27] differentiates between equilibrium methods (guarded hot plate) and transient methods (thermal conductivity probe, Powell comparator). One major difference between the steady-state and the transient approach is the time needed for the measurements, which is shorter for the transient one. Both approaches have advantages (hot plate: accuracy; transient: fast, size of the sample not decisive) as well as disadvantages (hot plate: slow; transient: requires a set of standards).

The most accurate method for measuring the thermal conductivities of homogeneous samples is the guarded hot plate method, which was also described by the American Society for Testing Materials, where a detailed set-up can be found [27, 108].

An early paper on the measurement of amorphous elastomers was published by Hennig et al. [88]. The authors developed a tool for measuring stretched and strained specimens. The main focus of their investigations was to describe anisotropy phenomena in amorphous materials like polystyrene PS, polymethylmethacrylate PMMA, polyvinylchloride PVC, and polycarbonate. The results have been related to the molecular structures of the corresponding solid materials showing that the thermal conductivity was greater in the direction of stretch and lower perpendicular to the stretch direction compared to the unstretched isotropic state. Further experiments have been done with respect to the glass transition temperature especially for polyvinylchloride (PVC). PVC was heated in the stretched state above the glass transition temperature. The anisotropy of the thermal conductivity did not change after this treatment, thus, it was concluded that also strained polymers possess an anisotropy in their thermal conductivity. For the measurement, a two-plate apparatus was used with known thermal resistivity of one plate thereby allowing an improved measurement. A relation of the anisotropy of the thermal conductivity to the thermal expansivity was given, see the following equation [88]

$$\frac{\kappa_0}{\kappa_{\text{II}\ (\perp)}} = 0.8\frac{\beta_{\text{II}(\perp)}}{\beta_0} + 0.2 \tag{13}$$

where κ_0 is the isotropic unstretched thermal conductivity, κ_{II} is the thermal conductivity for parallel treatment, $\kappa_{\perp}$ is the thermal conductivity for perpendicular treatment, $\beta_{\text{II}(\perp)}$ is the coefficient of linear thermal expansivity for perpendicular treatment, and β_0 is the linear thermal expansivity for the unstretched sample.

The hot wire method as a direct measurement method was applied by dos Santos et al. [14]. Advantages and disadvantages of the hot wire approach are presented. Advantages are easy experimental set-up and usability for a wide range of solids. Disadvantages are the sample size (mainly kg) and long measurement times for reaching a thermal equilibrium. Melting of polymers is also to be taken into account for crystalline or semicrystalline samples as for phase transitions, the thermodynamic parameters change significantly. Another problem is to maintain the surface contact between the sample and the plates. Details of all of these aspects are found in in dos Santos [109]. Besides the hot wire method, the laser flash technique (see also Sect. 4.2 in Chapter "2D Transient Thermal Analytical Solution of the Heat Build-Up in Cyclically Loaded Rubber Cylinder") was used by dos Santos et al. [110] showing numerous advantages compared to the hot wire set-up. Advantages of the laser flash technique are the sample sizes, e.g. very small samples are possible to measure, quick thermal equilibrium, ability to measure solid and molten samples, and useful also for semicrystalline samples.

The transient hot wire method observes the temperature change of a thin, conducting wire, which is placed in a single homogenous material of interest. The wire and sample are in thermal equilibrium first, and are then subjected to a temperature increase by sending an electrical current through the wire. It heats up and generates a time-dependent temperature field inside the wire and the material of interest. The temperatures are measured and represent in its evolution the thermal conductivity of the sample. The measurement data are then transferred to a model, and the thermal conductivity can be extracted from it.

One prerequisite for measuring thermal conductivities according to the hot wire approach is the perfect contact between the sample and the heating device as well as the thermometer. However, this contact often produces a resistance, which is caused by a thin layer of air when surfaces are not perfectly plane. Several improvements have been suggested and are summarized in [111]. Velliadou et al. [111] used an optimized set-up for measuring solid and liquid samples of hexadecan-1-ol and octadecan-1-ol and evaluated the data with a theoretical model. It is an absolute measure and deviations have been around 1%.

In a review paper, Assaeel et al. [112] present the historical evolution of the transient hot wire method. As the method is a standard technique, standard norms exist and describe the correct use of this experimental technique [113, 114]. Vasilev [103, 115] also used the hot wire method to investigate elastomers and was able to prove his theoretically determined values for several elastomers via the measurements.

Eiermann [116, 117] measured amorphous high polymers (natural rubber, Vulkollan – a polyurethane, polyisobutylene, atactic polypropylene, PVC, and polychlorotrifluoroethylene) in a wide temperature range from −190°C to 90°C. The detailed measurement procedure is based on the quasi-stationary approach of Gast et al. [118] and comprises five discs or plates (two heating discs, two sample discs and one reservoir) surrounded by an adiabatic shell. The heat flux is directed from the heating plates to the reservoir crossing the samples. Temperature differences between heating plates and reservoirs are measured. Details can be found in [116, 117]. For amorphous polymers, a break occurred in the thermal conductivity at the glass-transition temperature [116]. A model was set-up to explain the behaviour of κ in amorphous polymers. Each bond between adjacent atoms was interpreted as a thermal resistance. The overall thermal resistance of the macroscopic polymer is then the sum of all thermal resistances. The single thermal resistance depends on bonding forces (see Sect. 2.6 in Chapter "Effects of Temperature on Uniaxial and Multiaxial Fatigue of Natural Rubber Under Relaxing and Non-relaxing Loadings"), which is at least the reason why force fields are able to reproduce κ as a physical property. The bond forces vary depending on the temperature. Especially around glass transition temperature, the spring constant representing the bond strength is changed due to the changed space between the atoms. A higher κ, which is typically found for crystalline polymers, was therefore attributed to the higher density causing a higher spring constant. Further details are in [116].

Based on Hellwege [119] and Knappe [120], Hansen et al. [121] measured linear amorphous polymers and derived a theory for κ on a simple molecular structure. The molecular weight, the orientation of the molecules and the per cent crystallinity concept have been analysed for PE. The authors correlate their results also to the phonon theory of energy transport in dielectric crystals (see Sect. 2.3 in Chapter "Effects of Temperature on Uniaxial and Multiaxial Fatigue of Natural Rubber Under Relaxing and Non-relaxing Loadings").

Early measurements of rubbers, where the temperature-dependent stress behaviour was the focus, have been performed by Dauphine et al. [27, 122]. It was shown that the thermal conductivity of natural rubber remained constant with stretch, whereas modified rubbers showed a decrease of κ under all cure conditions [27]. Furthermore, the vulcanization and its influence on κ were investigated [27]. Under elongation, an increase of κ was found. The increase of κ was seen to be less for highly vulcanized samples, which was attributed to a more pronounced orientation of molecular chain segments in the elongation direction, and a decreasing number of scattering centres in the molecules [27]. Thermodynamic aspects of the κ behaviour of rubbers and its relation to thermodynamics have been addressed by Weir et al. [123] who described the thermal conductivity around the glass transition temperature with an interferometer, and DiMarzio [124] who made a theoretical approach to explain results from experiments. (The theoretical approach mentioned here based on Flory-Higgins and derived theories was not reviewed in detail in the previous chapters.)

Quite often, measurements of thermal conductivities on rubber samples are done on modified samples. Rubber materials which have been modified by fillers showed that depending on the filler types a general increase of κ was reached [27]. For rubbers modified with sulphur, the samples with higher sulphur contents have a maximal effect of κ increase [125].

Dawson et al. [126] dealt intensively with the thermal conductivity behaviour of amorphous polymers at temperatures above the glass transition temperature, which is not well described. Various references exist which show different κ for the same polymer. The authors describe a new measurement technique, the heat transfer coefficient (HTC) apparatus, which is preferably useful for measuring amorphous polymers above the glass transition temperature. Details can be found in [126].

Rubberlike materials were measured and theoretically evaluated by Rehner [44] in 1947 without using a certain designation of the method. In fact, sample discs have been brought in contact with a thermocouple, cured overnight (details see [44]) and then subjected to a sudden temperature difference. The Fourier equation was solved including the geometry of the samples and allowed to interpret the measurement data sufficiently.

4.2 Laser or Spectroscopy-Related Methods

New methods for the measurement of heat flow and its derived properties employ elaborated spectroscopy, laser or laser pulse, respectively.

The use of laser and its certain properties is part of the so-called flash method. The flash method is widely used for measuring thermal diffusivities of a thin sample material with small diameters. The sample material is submitted to a laser pulse on one of its faces. The thermal conductivity is obtained after measurement through a linear relation between κ and the thermal diffusivity [127]. Here, semicrystalline polymers have been investigated to describe the anisotropy in physical properties when polymers like PET, POM, PP, and PCTFE are drawn or extruded. In the oriented polymer, there may occur amorphous portions, which have an influence on the final thermal conductivity of the polymer. Choy et al. [128] also measured polymers with the flash method and evaluated the data via a modified Maxwell model.

One field of interest which also needs to be investigated in the nanoscale range is the semiconductor field and its combination with polymers and elastomers as for instance in soft robotics. Measurements done for this kind of solid materials may serve as an example to imitate measurements in the field of polymer science. As example here, the thermal conductivity of nanowires may be seen as a model. Measurements of thermal conductivities for individual nanowires were done to check out the dependence of κ on the wire diameter down to diameters of 22 nm [15]. Those kinds of measurements are challenging as κ is one order of magnitude smaller than in bulk, due to strong boundary scattering effects and phonon confinement effects. Moreover, measurements may be subject to surface oxide coatings, which opens new scattering pathways. Most measurements in this area have been naturally performed on Si nanowires. Aside from the experimental set-up, an intensive discussion of mechanistic aspects can be found here as for instance the discussion of Umklapp effects, the decrease of κ depending on the dimensionality, and strong boundary effects. At small diameters, the temperature dependence of κ was found which is consistent to the temperature dependence of the specific heat for 1D systems [15].

Venerus et al. [18] present a novel optical technique, which is sensitive and non-invasive and is applied to deformed elastomers and avoids issues associated with traditional methods that need contact between the sample and the measurement device. The authors developed two measurement techniques employing the forced Rayleigh scattering (FRS) and the infrared thermography (IRT). FRS as a time-dependent method creates an optical grating within the sample and detects its dynamics [18]. Two beams from a coherent laser form a grating at the intersection point. The sample depending on its thickness absorbs a small fraction of the energy of the electromagnetic wave. The absorbed energy delivers a sinusoidal temperature field with a certain amplitude. The temperature modulation leads to a modulation of the refractive index. A photodetector measures the intensity of the diffracted beam. Thermal diffusivity tensors can thus be measured in the direction both parallel and perpendicular to the direction of elongation [18]. The IRT as a steady-state method works with laser as well and passes through a sample where a small fraction of the energy is absorbed as well. An infrared camera on the opposite side measures the surface temperature field. The power of the laser needs to be adjusted to result in a reasonable signal-to-noise ratio. The result of the measurement is an isotherm which is evaluated based on a solution of the steady-state, three-dimensional temperature equation [18] and allows the measurement of thermal conductivities. Both techniques gave similar results for crosslinked PBD showing a heat capacity which is independent of deformation. Crosslinked PIP however, showed a difference between FRS and IRT, which was interpreted to be a deformation-dependent heat capacity. The authors also validated the stress-thermal rule, which predicts a linear relationship between thermal conductivities and stress tensors. The rule was shown to be valid for the three investigated crosslinked elastomers (polydimethylsiloxane PDMS, polybutadiene PBD, and polysioprene PIP) [18].

Acknowledgements [58] Funding of DFG (Deutsche Forschungsgemeinschaft) is acknowledged via the Research Training Group "Interactive Fiber-Rubber Composites" Project 380321452/GRK2430.

References

1. Zhang B, Mao P, Liang Y, He Y, Liu W, Liu Z (2019) Modulating thermal transport in polymers and interfaces: theories, simulations, and experiments. ES Energy Environ 5:37–55
2. Huang X, Jiang P, Tanaka T (2011) A review of dielectric polymer composites with high thermal conductivity. IEEE Electr Insul Mag **27**:8–16
3. Mark JE (2007) Physical properties of polymers handbook. Springer, New York
4. Koh YK (2012) Thermal conductivity and phonon transport. In: Bhushan B (ed) Encyclopedia of nanotechnology. Springer, Dordrecht, pp 2704–2711. https://doi.org/10.1007/978-90-481-9751-4_277
5. Chen H, Ginzburg VV, Yang J, Yang Y, Liu W, Huang Y, Du L, Chen B (2016) Mechanism of thermal conductivity reduction in few-layer graphene. Prog Polym Sci 59:41–85
6. Lv W, Winters RM, Angelis FD, Weinberg G, Henry A (2017) J Phys Chem A 121:5586–5596
7. Chen Q, Wang C, Li Y, Feng L, Huang S (2023) Performance development of polyurethane elastomer composites in different construction and curing environments. Construct Build Mater 365:130047
8. Kamble VG, Mersch J, Tahir M, Stöckelhuber KW, Das A, Wießner S (2022) Development of liquid diene rubber based highly deformable interactive fiber-elastomer composites. Materials 15:390
9. GRK 2430 website: https://tu-dresden.de/ing/forschung/graduiertenkollegs/grk2430
10. Cevallos JG, Bergles AE, Bar-Cohen A, Rodgers P, Gupta SK (2012) Polymer heat exchangers – history, opportunities, and challenges. Heat Transfer Eng 33:1075–1093
11. Zhu J, Hippalgaonkar K, Shen S, Wang K, Abate Y, Lee S, Wu J, Yin X, Majumdar A, Zhang X (2014) Temperature-gated thermal rectifier for active heat flow control. Nano Lett 14(8): 4867–4872
12. Chen J, Meng L (2022) Computational study of the thermal rectification properties of a graphene-based nanostructure. ACS Omega 7(32):28030–28040
13. Roberts NA, Walker DG (2011) A review of thermal rectification observations and models in solid materials. Int J Therm Sci 50(5):648–662
14. dos Santos WN, de Sousa JA, Gregorio R (2013) Thermal conductivity behaviour of polymers around glass transition and crystalline melting temperatures. Polymer Testing 32(5):987–994
15. Dresselhaus MS, Lin Y-L, Rabin O, Black MR, Kong J, Dresselhaus G (2010) Nanowires. In: Bhushan B (ed) Springer handbook of nanotechnology, vol 2010. Springer, pp 145–147
16. Feng T, He J, Rai A, Hun D, Liu J, Shrestha SS (2020) Size effects in the thermal conductivity of amorphous polymers. Phys Rev Appl 14:044023-1–044023-14
17. Zhang T, Wu X, Luo T (2014) Polymer nanofibers with outstanding thermal conductivity and thermal stability: fundamental linkage between molecular characteristics and macroscopic thermal properties. J Phys Chem C 118(36):21148–21159
18. Venerus D, Simavilla DN, Schieber JD (2019) Thermal transport in cross-linked elastomers subjected to elongational deformations. Rubber Chem Technol 92(4):639–652
19. Chen H, Ginzburg VV, Yang J, Yang Y, Liu W, Huang Y, Du L, Chen B (2016) Thermal conductivity of polymer-based composites: fundamentals and applications. Prog Polym Sci 59:41–85
20. Wingert MC, Zheng J, Kwon S, Chen R (2016) Thermal transport in amorphous materials: a review. Semicond Sci Technol 31:113003
21. Campaioli F, Cole JH (2021) Exciton transport in amorphous polymers and the role of morphology and thermalisation. New J Phys 23:113038
22. Spitalsky Z, Tasis D, Papagelis K, Galiotis C (2010) Carbon nanotube–polymer composites: chemistry, processing, mechanical and electrical properties. Prog Polym Sci 35:357–401
23. Breuer O, Sundararaj U (2004) Big returns from small fibers: a review of polymer/carbon nanotube composites. Polym Compos 25:630–645

24. Bagchi A, Nomura S (2006) On the effective thermal conductivity of carbon nanotube reinforced polymer composites. Compos Sci Technol 66:1703–1712
25. Mehra N, Mu L, Ji T, Yang X, Kong J, Gu J, Zhu J (2018) Thermal transport in polymeric materials and across composite interfaces. Appl Mater Today 12:92–130. https://doi.org/10.1016/j.apmt.2018.04.004
26. Henry A (2013) Thermal transport in polymers. Ann Rev Heat Transfer 17:485–520
27. Anderson DR (1966) Thermal conductivity of polymers. Chem Rev 66:677–690
28. Nellis GF, Klein SA (2021) Introduction to engineering heat transfer. Cambridge University Press
29. Hänel D (2004) Molekulare Gasdynamik. Einführung in die kinetische Theorie der Gase und Lattice-Boltzmann-Methoden. Springer, Berlin
30. Hofmann P (2013) Einführung in die Festkörperphysik. Wiley-VCH, Weinheim
31. Kaviany M (2014) Phonon energy storage, transport and transformation kinetics. In: Heat transfer physics. Cambridge University Press, pp 173–305
32. Dove MT (1993) Introduction to lattice dynamics. Cambridge University Press, Cambridge
33. Peierls RE (1996) Quantum theory of solids, international series of monographs on physics. Clarendon Press
34. Kittel C (2005) Introduction to solid state physics.8th edn. Wiley
35. Peters MH (2005) Molecular thermodynamic and transport phenomena. Complexity of scales in space and time. McGraw Hill
36. Bird RB, Stewart WE, Lightfood EN, Klingenberg DJ (2015) Introductory transport phenomena. Chapter 9 Thermal conductivity and the mechanisms of energy transport. Wiley, pp 253–278
37. Hauke G (2008) An introduction to fluid mechanics and transport phenomena. Springer
38. Wang X (2012) Experimental micro/nanoscale thermal transport. Wiley
39. Eiermann K (1961) Wärmeleitung in Kunststoffen in Abhängigkeit von Struktur. Temperatur und Vorgeschichte Kunststoffe 51(9):512–517
40. Kingery WD, McQuarrie MC (1954) Thermal conductivity: I, Concepts of measurement and factors affecting thermal conductivity of ceramic materials. J Am Ceram Soc 37:67–72
41. Yamamoto O (1971) Thermal conductivity of cross-linked polymers. Polym J 2:509–517
42. Wada Y, Itami A, Nishi T, Nagai S (1969) Grüneisen constant and thermal properties of crystalline and glassy polymers. J Polym Sci Part A-2 7:201–208
43. Tarasov VV (1950) Theory of the heat capacity of chain and layer structures. J Phys Chem (USSR) Zh Fiz Khim 24:111–128
44. Rehner J (1947) Heat conduction and molecular structure in rubberlike polymers. J Polym Sci 2:263–274
45. Beek WJ, Muttzall KMK, van Heuven JW (2000) Transport phenomena. Chapter III Heat transport.2nd edn. Wiley
46. Debye P (1912) Zur Theorie der spezifischen Wärmen. Ann Phys 344(14):789–839
47. Simon SH (2013) The Oxford solid state basics.1st edn. Oxford University Press, Oxford, p 82
48. Franklin A, Knox E (2018) Emergence without limits: The case of phonon. Stud Hist Philos Sci B Stud Hist Philos Mod Phys 64:68–78
49. Peierls R (1929) Zur kinetischen Theorie der Waermeleitung in Kristallen. Ann Phys 395: 1055–1101
50. Zeng L, Collins KC, Hu Y, Luckyanova MN, Maznev AA, Huberman S, Chiloyan V, Zhou J, Huang X, Nelson KA, Chen G (2015) Measuring phonon mean free path distributions by probing quasiballistic phonon transport in grating nanostructures. Sci Rep 5:17131. https://doi.org/10.1038/srep17131
51. Xu Y, Chung DDL, Mroz C (2001) Thermally conducting aluminum nitride polymer-matrix composites. Compos A: Appl Sci Manuf 32(12):1749–1757. https://doi.org/10.1016/S1359-835X(01)00023-9
52. Liu Y, He H (2021) Approach to phonon relaxation time and mean free path in nonlinear lattices. Chin Phys Lett 38. 044401-1-5

53. Eiermann K (1964) Modellmäßige Deutung der Wärmeleitfähigkeit von Hochpolymeren. Kolloid-ZuZPolymere 198:5–16. https://doi.org/10.1007/BF01499450
54. Kittel C (1949) Interpretation of the thermal conductivity of glasses. Phys Rev 75(6):972–974
55. Hertel P (2017) Quantum theory and statistical thermodynamics. Principles and worked examples. Springer
56. Choy CL (1977) Thermal conductivity of polymers. Polymer 18:984–1004
57. Balandin AA (2011) Thermal properties of graphene and nanostructured carbon materials. Nat Mater 10:569–581
58. Lindsay L, Broido DA (2012) Theory of thermal transport in multilayer hexagonal boron nitride and nanotubes. Phys Rev B Condens Matter 85. 035436-8
59. Lindsay L, Broido DA, Mingo N (2011) Flexural phonons and thermal transport in multilayer graphene and graphite. Phys Rev B Condens Matter 8. 235428-5
60. Singh D, Murthy JY, Fisher TS (2011) Mechanism of thermal conductivity reduction in few-layer graphene. J Appl Phys 110. 044317-8
61. Wang X, Kaviany M, Huang B (2017) Phonon coupling and transport in individual polyethylen chains: a comparison study with the bulk crystal. Nanoscale 9:18022–18031
62. Berman R (1973) Heat conductivity of non-metallic crystals. Contemp Phys 14(2):101–117. https://doi.org/10.1080/00107517308213728
63. Henry A, Chen G (2009) Anomalous heat conduction in polyethylene chains: theory and molecular dynamics simulations. Phys Rev B 79. 144305-10
64. Singh V, Bougher TL, Weathers A, Cai Y, Bi K, Pettes MT, McMenamin SA, Lv W, Resler DP, Gattuso TR, Altman DH, Sandhage KH, Shi L, Henry A, Cola BA (2014) High thermal conductivity of chain-oriented amorphous polythiophene. Nat Nanotechnol 9:384–390
65. Choy CL, Chen FC, Luk WH (1980) Thermal conductivity of oriented crystalline polymers. J Polym Sci Part B Polymer Phys 18:1187–1207
66. Klemens PG (1951) The thermal conductivity of dielectric solids at low temperatures (theoretical). Proc R Soc London Ser A: Math Phys Sci 208:108–133
67. Henry A, Chen G, Plimpton SJ, Thompson A (2010) 1D-to-3D transition of phonon heat conduction in polyethylene using molecular dynamics simulations. Phys Rev B 82:144308–144305
68. Wang X, Ho V, Segalman RA, Cahill DG (2013) Thermal conductivity of high-modulus polymer fibers. Macromolecules 46:4937–4943
69. Choy CL, Luk WH, Chen FC (1978) Thermal conductivity of highly oriented polyethylene. Polymer 19:155–162
70. Yu J, Sundqvist B, Tonpheng B, Andersson O (2014) Thermal conductivity of highly crystallized polyethylene. Polymer 55:195–200
71. Zhang S, Minus ML, Zhu L, Wong C-P, Kumar S (2008) Polymer transcrystallinity induced by carbon nanotubes. Polymer 49:1356–1364
72. Chen G (2005) Nanoscale energy transport and conversion: a parallel treatment of electrons, molecules, phonons, and photons. Oxford University Press
73. Han Z, Fina A (2011) Thermal conductivity of carbon nanotubes and their polymer nanocomposites: a review. Prog Polym Sci 36(7):914–944. https://doi.org/10.1016/j.progpolymsci.2010.11.004
74. Hansen D, Ho CC (1965) Thermal conductivity of high polymers. J Polym Sci Part A General Papers 3:659–670
75. Shao C, Yu X, Yang N, Yue Y, Bao H (2017) A review of thermal transport in low-dimensional materials under external perturbation: effect of strain, substrate, and clustering. Nanosc Microsc Therm 21:201–236
76. Zhang Z, Ouyang Y, Guo Y, Nakayama T, Nomura M, Volz S, Chen J (2020) Hydrodynamic phonon transport in bulk crystalline polymers. Phys Rev B 102. 195302-1-8
77. Slack GA (1979) The thermal conductivity of nonmetallic crystals. Solid State Phys 34:1–71
78. Cahill DG, Watson SK, Pohl RO (1992) Lower limit to the thermal conductivity of disordered crystals. Phys Rev B 46(10):6131–6140

79. Allen PB, Feldmann JL, Fabian J, Wooten F (1999) Diffusions, locons, and propagons: character of atomic vibrations in amorphous Si. Philos Mag B Phys Cond Matter Stat Mech Electron Opt Mag Props 79(11/12):1715–1731
80. Wingert MC, Zheng J, Kwon S, Chen R (2016) Thermal transport in amorphous materials: a review. Semicond Sci Technol 31(11):113003/1–113003/16
81. Kikugawa G, Desai TG, Keblinski P, Ohara T (2013) Effect of crosslink formation on heat conduction in amorphous polymers. J Appl Phys 114(3):034302–034306. https://doi.org/10.1063/1.4813505
82. Rashidi V, Coyle EJ, Sebeck K, Kieffer J, Pipe KP (2017) Thermal conductance in cross-linked polymers: effects of non-bonding interactions. J Phys Chem B 121:4600–4609
83. Kim GH, Lee D, Shanker A, Shao L, Kwon MS, Gidley D, Kim J, Pipe KP (2015) High thermal conductivity in amorphous polymer blends by engineered interchain interactions. Nat Mater 14(3):295–300. https://doi.org/10.1038/nmat4141
84. Zhang L, Ruesch M, Zhang X, Bai Z, Liu L (2015) Tuning thermal conductivity of crystalline polymer nanofibers by interchain hydrogen bonding. RSC Adv 5:87981–87986
85. Zhang T, Luo T (2012) Morphology-influenced thermal conductivity of polyethylene single chains and crystalline fibers. J Appl Phys 112(9):094304–094304. https://doi.org/10.1063/1.4759293
86. Agari Y, Ueda A, Omura Y, Nagai S (1997) Thermal diffusivity and conductivity of PMMA/PC blends. Polymer 38:801–807
87. Kausch-Blecken von Schmeling HH (1970) Über die Zusammenhänge von makroskopischer und molekularer Anisotropie in Hochpolymeren. Kolloid-Z uZPolymere 237:251–266
88. Hennig J, Knappe W (1964) Anisotropy of thermal conductivity in stretched amorphous linear polymers and in strained elastomers. J Polym Sci C Polym Symp 6:167–174
89. Venerus DC, Simavilla DN, Schieber JD (2019) Thermal transport in cross-linked elastomers subjected to elongational deformations. Rubber Chem Technol 92(4):639–652
90. Eiermann K (1964) Thermal conductivity of high polymers. J Polym Sci Part C Polym Symp 6:157–165
91. Überreiter K, Otto-Laupenmühlen E (1953) Spezifische warme, spezifisches Volumen, Temperatur-und Wärmeleitfähigkeit von Hochpolymeren, Teil II. Kettenlangenabhängigkeit bei fraktionierten Polystyrolen. Z Naturforsch Teil A 8:664–673
92. Breitkopf C (2024) Theoretical characterization of heat transport in polymers – an review. Feature paper. Thermo 4(1):31–47. https://doi.org/10.3390/thermo4010004
93. Drude P (1900) Zur Elektronentheorie der Metalle. Ann Phys 306(3):566–613
94. Di Ventra M (2008) Electrical transport in nanoscale systems. Cambridge University Press. ISBN-13 978-0-521-89634-4
95. Kubo R (1957) Statistical mechanical theory of irreversible processes I, general theory and simple applications to magnetic and conduction problems. J Physical Soc Japan 1:570–586
96. Green MS (1954) Markoff random processes and the statistical mechanics of time-dependent phenomena. II. Irreversible processes in fluids. J Chem Phys 22(3):398–413
97. Wikipedia. https://en.wikipedia.org/wiki/Green%E2%80%93Kubo_relations
98. Schelling PK, Phillpot SR, Keblinski P (2002) Comparison of atomic-level simulation methods for computing thermal conductivity. Phys Rev B 65. 144306-1-12
99. Evans DJ, Morriss G (2008) Statistical mechanics of nonequilibrium methods. Cambridge University Press. ISBN-13 978-0-521-85791-8
100. Hoover WG, Evans DJ, Hickman RB, Ladd AJC, Ashurst WT, Moran B (1980) Lennard-Jones triple-point bulk and shear viscosities. Green-Kubo theory, Hamiltonian mechanics, and nonequilibrium molecular dynamics. Phys Rev A 22:1690–1697
101. Zwanzig R (1964) Elementary derivation of time-correlation formulas for transport coefficients. J Chem Phys 40(9):2527–2533
102. Vasilev A, Lorenz T, Breitkopf C (2022) Prediction of thermal conductivities of rubbers by MD simulations-new insights. Polymers 14(10) 2046-1-10. https://doi.org/10.3390/polym14102046

103. Vasilev A, Lorenz T, Kamble VG, Wießner S, Breitkopf C (2021) Thermal conductivity of polybutadiene rubber from molecular dynamics simulations and measurements by the heat flow meter method. Materials 14: 7737-1-10. https://doi.org/10.3390/ma14247737
104. Vasilev A, Lorenz T, Breitkopf C (2020) Thermal conductivity of polyisoprene and polybutadiene from molecular dynamics simulations and transient measurements. Polymers 12. 1081-1-9
105. Alamfard T, Lorenz T, Breitkopf C (2023) Thermal conductivities of uniform and random sulfur crosslinking in polybutadiene by molecular dynamic simulation. Polymers 15:2058-1-15
106. Kanan A, Vasilev A, Breitkopf C, Kaliske M (2022) Thermo-electro-mechanical simulation of electro-active composites. Materials (Basel) 15(3):783. https://doi.org/10.3390/ma15030783
107. Alamfard T, Lorenz T, Breitkopf C (2024) Glass transition temperatures and thermal conductivities of polybutadiene crosslinked with randomly distributed different lengths of sulfur chains by molecular dynamics simulations. Polymers 16:1–22
108. ASTM (1965) ASTM Std C 177-63, Part 14, 15–26
109. dos Santos WN (2005) Thermal properties of melt polymers by the hot wire technique. Polym Test 24(7):932–941
110. dos Santos WN, Mummery P Wallwork, thermal diffusivity of polymers by the laser flash technique. Polym Test 24(5):628–634
111. Velliadou D, Antoniadis KD, Assimopoulou AN, Assimopoulou, Assael MJ, Wakeham WA (2022) Accurate measurements of the thermal conductivity of hexadecan-1-ol and octadecan-1-ol in the solid and liquid phases. Int J Thermophys 43. 124-1-18
112. Assael MJ, Antoniadis KD, Wakeham WA (2010) Historical evolution of the transient hot-wire technique. Int J Thermophys 31(6):1051–1072
113. DIN EN ISO 8894-1:2010-10 (2010) Feuerfeste Werkstoffe – Bestimmung der Wärmeleitfähigkeit – Teil 1: Heißdrahtverfahren (Kreuzverfahren und Widerstandsthermometer-Verfahren), Beuth-Verlag
114. ASTM D7896-19: Standard test method for thermal conductivity, thermal diffusivity, and volumetric heat capacity of engine coolants and related fluids by transient hot wire liquid thermal conductivity method. https://www.astm.org/d7896-19.html
115. Vasilev A (2022) Theoretical modeling and experimental characterization of heat transport in interactive fiber rubber composites. PhD. TU Dresden. Germany
116. Eiermann K (1964) Thermal conductivity of high polymers. J Polymer Sci Part C 6:157–165
117. Eiermann K, Hellwege KH, Knappe W (1961) Quasistationäre Messung der Wärmeleitfähigkeit von Kunststoffen im Temperaturbereich von −180 °C bis +90°C. Kolloid-Zeitschrift 174:134–142
118. Gast T, Hellwege KH, Kohlhepp E (1957) Wärmeleitfähigkeit von Polyvinylchlorid mit und ohne Weichmacher im Temperaturbereich von 70 bis 360°K. Kolloid-Zeitschrift 152:24–31
119. Hellwege KH (1959) Quasi-stationary measurement of the specific heat and heat conductivity of plastics. Zeitschrift für Angewandte Physik 11:285–290
120. Knappe W (1960) Bestimmung der thermischen Kenngrößen schlecht wärmeleitender Stoffe mit einer Zweiplattenapparatur ohne Schutzring. Z Angew Phys 12:508–514
121. Hansen D, Ho CC (1965) Thermal conductivity of high polymers. J Polym Sci A Gen Pap 3: 659–670
122. Dauphinee TM, Ivery DG, Smith HD (1950) The thermal conductivity of elastomers under stretch and at low temperatures. Can J Res 28A:596–615
123. Weir CE, Leser WH, Wood IA (1950) Crystallization and second-order transitions in silicone rubbers. J Res Natl Bur Stand. Research Paper RP2084 44:367–372
124. DiMarzio EA (1964) On the second-order transition of a rubber. J Res Natl Bur Stand A Phys Chem 68A(6):611–617
125. Zanemonets NA, Fogel VO (1959) Thermophysical characteristics and thermal effects of reaction of vulcanization of hard rubber compositions from styrene butadiene rubber. Kauchuk i Rezina 18(2):21–24

126. Dawson A, Ride M, Allen C (2017) The measurement of the thermal conductivity of amorphous polymers above glass transition temperature. NPL Rep MAT 7
127. Choy CL, Chen FC, Luk WH (1980) Thermal conductivity of oriented crystalline polymers. J Polym Sci Polym Phys Edn 18:1187–1207
128. Choy CL, Chen FC, Luk WH (1980) Thermal conductivity of oriented crystalline polymers. Aust J Polit Sci 18:1187–1207

Adv Polym Sci (2024) 294: 251–324
https://doi.org/10.1007/12_2024_175

Published online: 26 July 2024

A Review of Thermal Effects on Elastomer Durability

W. V. Mars and T. G. Ebbott

Contents

W. V. Mars (✉) and T. G. Ebbott
Endurica LLC, Findlay, OH, USA
e-mail: wvmars@endurica.com

Abstract The low thermal conductivity of elastomers means that frequently the material operates subject to a temperature gradient with high temperatures in the interior. Other times, elastomers are called to operate directly in higher temperature environments which can affect their properties or invoke aging. Thus, the temperature impact on material behavior is an important consideration in characterization and analysis. Temperature has many influences that can strongly impact structural response and durability, both in helpful and in detrimental ways. These include: elastic and dissipative properties, thermal expansion/contraction, strength and crack growth properties, strain crystallization, aging, and the resistance to thermal runaway. There are at least two parameters without a temperature sensitivity: intrinsic strength and crack precursor size. This review considers the characterization and modeling of the impact of temperature on durability, and includes examples computed with the Endurica fatigue solvers.

Keyword Fatigue life prediction · Materials testing · Simulation · Temperature dependence

1 Overview

Rubber's low thermal conductivity and its tendency to dissipate mechanical work input as heat mean that large temperature gradients are common in applications. This fact, combined with the strong dependence of many of rubber's properties on temperature, means that successful characterization and simulation must take these circumstances into account. Therefore, a complete scientific consideration must consider the elastic, dissipative, thermal transport, strength and crack growth, strain crystallization, aging, and the thermal reversion limit.

Temperature impacts rubber's mechanical response in two main ways: (1) there is an immediate, reversible change due to temperature and (2) there may be a long term, irreversible change associated with the chemistry of the rubber compound (aging). One of the most well-known changes in immediate response is observed in the viscoelastic properties. The viscoelastic properties of rubber exhibit a time- (or frequency-) temperature dependence that originates in the rate of thermal motions of long-chain molecules. For example, a decrease in dynamic stiffness that is caused by an increase in temperature (and therefore a larger number of molecular collisions) may be equivalently achieved by lower frequency or longer time (again leading to a larger, similar number of molecular collisions). Aging is driven by irreversible chemical reactions that will depend on time and temperature, as well as on the details of the reactants. For example, aging under aerobic (oxidative) conditions can differ significantly from aging under anaerobic (without oxygen) conditions.

Colder temperatures also impact the response of elastomers, especially when the colder temperature induces response close to the glass transition. In this work, we will focus on the impact of elevated temperatures on rubber response, and will limit discussion on cold temperature effects, such as operation near or below the glass transition temperature.

2 Elastic Behavior

Perhaps the most unique and valuable quality of rubber is its high elasticity. Rubber's elasticity permits applications where large deformations must be repeatedly accommodated without failure, and where stored elastic potential energy must be efficiently and repeatedly returned to restore the initial state that preceded a loading event.

2.1 *Hyperelasticity*

Because of rubber's elasticity, mechanical energy can be stored in and recovered from the material. On loading, the work done by a reaction force (or stress) moving through a corresponding displacement (or strain) is stored in the material as elastic potential. On unloading, work is recovered from the material at the expense of the elastic potential energy. To the extent that the elastic response governs, the stress state is a function of the strain state and the temperature.

2.1.1 Deformation and Strain Energy Density

All strain states can be characterized [1] in terms of 3 principal stretch ratios λ_1, λ_2, and λ_3. Equivalently, they can also be characterized in terms of the invariants of the Cauchy–Green deformation tensor I_1, I_2, and I_3, and the volume ratio J, as follows:

$$I_1 = \lambda_1^2 + \lambda_2^2 + \lambda_3^2 \tag{1}$$

$$I_2 = \lambda_1^2\lambda_2^2 + \lambda_2^2\lambda_3^2 + \lambda_3^2\lambda_1^2 \tag{2}$$

$$I_3 = J^2 = \lambda_1^2\lambda_2^2\lambda_3^2 \tag{3}$$

In the context of nonlinear elasticity, the stored strain energy density W is a function of the strain state [2] and of the temperature θ, so that we may write.

$$W = W(I_1, I_2, I_3, \theta) = W(\lambda_1, \lambda_2, \lambda_3, \theta) \tag{4}$$

Many forms for W are known and practiced, and excellent reviews of the available options are already published [3]. Because the energy function is multivariate, it is generally necessary to specify stress–strain behavior for several modes of deformation (simple, planar, and biaxial tension, for example) [4, 5]. Specification of a single mode of deformation is generally risky, as it under-constrains the energy function, too easily resulting in highly inaccurate extrapolations.

2.1.2 Temperature Dependence: Entropy Elasticity and Energy Elasticity

Rubber's elasticity derives fundamentally from the fact that long-chain molecules of a given molecular weight are much more likely to be found in states with shorter end-to-end distances than in states with long end-to-end distances [6, 7]. Thus, the random thermal motions that accompany temperatures above the glass transition tend to favor retraction over extension, resulting in the recovery of original dimensions when load is removed. Equivalently, rubber's deformed state is a lower entropy state than its higher entropy undeformed state, so that when entropy increases in the unavoidable progression of time, rubber tends to return to its undeformed state.

Under certain simplifying assumptions, these arguments lead to the neo-Hookean stored energy function

$$W = W(I_1, \theta) = \frac{1}{2} nk\theta(I_1 - 3) \tag{5}$$

$$G = nk\theta \tag{6}$$

n = Number of elastomer chains per unit volume
k = Boltzmann constant
θ = Absolute temperature
G = Shear modulus

Equation (5) predicts that rubber becomes stiffer as temperature increases, which to a large degree accurately predicts the behavior of gum rubbers with respect to temperature [8]. For materials filled with a sufficient proportion of carbon black filler, however, an increase in temperature is often associated with a decrease in stiffness. This behavior is understood to arise from a reduction of elastic energy storage capacity in the bonds mediating the material's load carrying capacity, and possibly associated with viscous effects [9]. The effect seems to be most strongly manifest in association with fillers, suggesting that filler–matrix interface activity governs the phenomenon (reflecting the observation that neither the filler itself nor the matrix rubber itself exhibits a decrease of stiffness with temperature increase). Figure 1 illustrates typical behavior for both cases.

Li et al. [10] have proposed a modified version of the Arruda—Boyce eight-chain hyperelastic law (combined with a modified Killian term) that is capable of representing both the entropy and energy contributions to the strain energy density function of temperature variations [11, 12]. Table 1 gives model values for a filled rubber. The unfilled rubber is included, but since it only exhibits entropy elasticity, the energy elasticity terms are not needed. It is interesting to note that a decrease of stress at fixed strain with increasing temperature correlates with the trend of viscous behavior, as pointed out by Le Cam et al. [9].

$$W = G\left[\lambda_m^2\left(\beta\sqrt{\frac{I_1}{3\lambda_m^2}} + \ln\frac{\beta}{\sinh\beta}\right) - \frac{a}{b+1}\left(\frac{I_1-3}{2}\right)^{b+1}\right] \tag{7}$$

$$G = nk\theta = G_0\left[1 + C_1\left(\frac{\Delta\theta}{\theta_0}\right) + C_1\left(\frac{\Delta\theta}{\theta_0}\right)^2\right] \tag{8}$$

$$\lambda_m = K_1\Delta\theta + \lambda_{m0} \tag{9}$$

$$a = K_2\Delta\theta + a_0 \tag{10}$$

$$b = K_3\Delta\theta + b_0 \tag{11}$$

2.2 *Thermal Expansion/Contraction*

Self-stresses induced by thermal expansion or contraction can strongly influence part durability. For example, when a rubber part is released from the mold, it will likely be in a stress-free state, but differences in expansion coefficient between embedded or bonded metal or polymer cord interfaces can lead to self-stresses upon cooling to room temperature. Compressive self-stresses can be beneficial. Tensile self-stresses can be harmful or beneficial, depending on strain crystallization.

In the unstressed state, the simplest model of reversible behavior with respect to a temperature change $\Delta\theta$ is often written in terms of a linear expansion coefficient α [21, 22]. Over a large range of temperature, a nonlinear relationship may be appropriate due to transitions in the expansion coefficient, see Fig. 2 which shows thermal expansion results on two rubbers in the unstressed state [16].

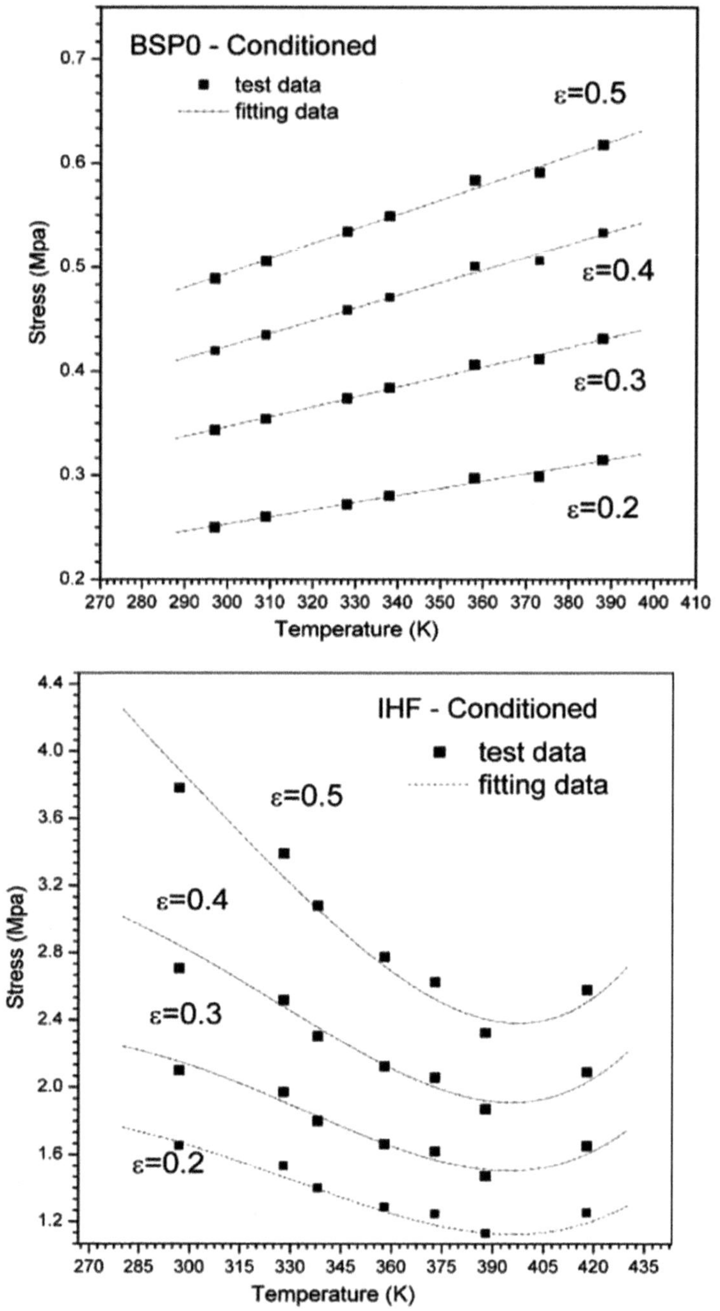

Fig. 1 Simple tension stress–temperature curves at 4 strain levels (ε) for unfilled (top) and highly filled (bottom) natural rubber [10]. The trend with respect to temperature is opposite due to domination of entropy elasticity for unfilled rubber (top) and of energy elasticity for filled rubber (bottom)

Table 1 Material parameters for temperature-dependent modified Arruda–Boyce model [10]

Property	Gum NR (BSP0)	Filled NR (IHF)	Units
G	1.0	35.329	MPa
λ_m	–	1.81	Dimensionless
a_0	–	1.268	Dimensionless
b_0	–	0.0254	Dimensionless
C_1	–	−3.3031	Dimensionless
C_2	–	4.0972	Dimensionless
K_1	–	0.0017	1/°C
K_2	–	−0.0016	1/°C
K_3	–	-1.4728×10^{-5}	1/°C

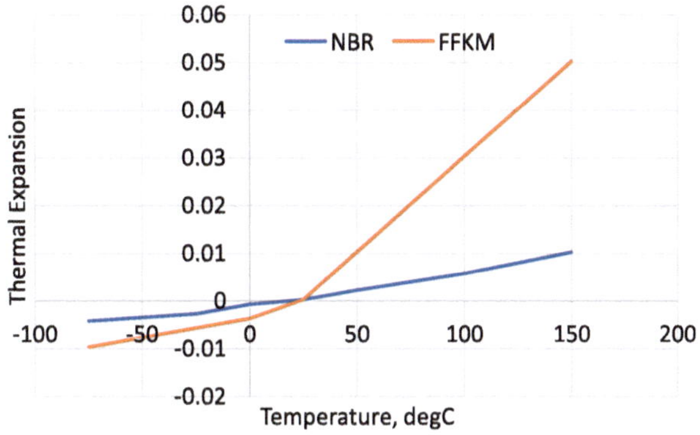

Fig. 2 Thermal expansion vs temperature for Nitrile (NBR) and Kalrez (FFKM), as measured via thermal mechanical analysis (TMA) on unstressed material [16]

$$\varepsilon_\theta = \alpha \Delta\theta \tag{12}$$

Above a small strain, the entropic nature of rubber's elasticity is manifest as a negative coefficient – i.e. the rubber contracts upon heating. This effect is the basis for several heat engine and heat pump designs [23–25]. The presence of carbon black filler in the rubber has a large effect on the strain dependence of the expansion coefficient [21], see Fig. 3. Carbon black itself is thought to have a much lower expansion coefficient than typical polymers, and therefore higher loadings of carbon black filler are observed to reduce the expansion coefficient of rubber [22].

Linear straining ε_θ (i.e., strain along one direction due to temperature) can be related to volumetric thermal straining ($J_\theta - 1$) using

$$J_\theta = (1 + \varepsilon_\theta)^3 \tag{13}$$

The elastic component J_e and the thermal component J_θ of the volume ratio are related through a multiplicative split, i.e.

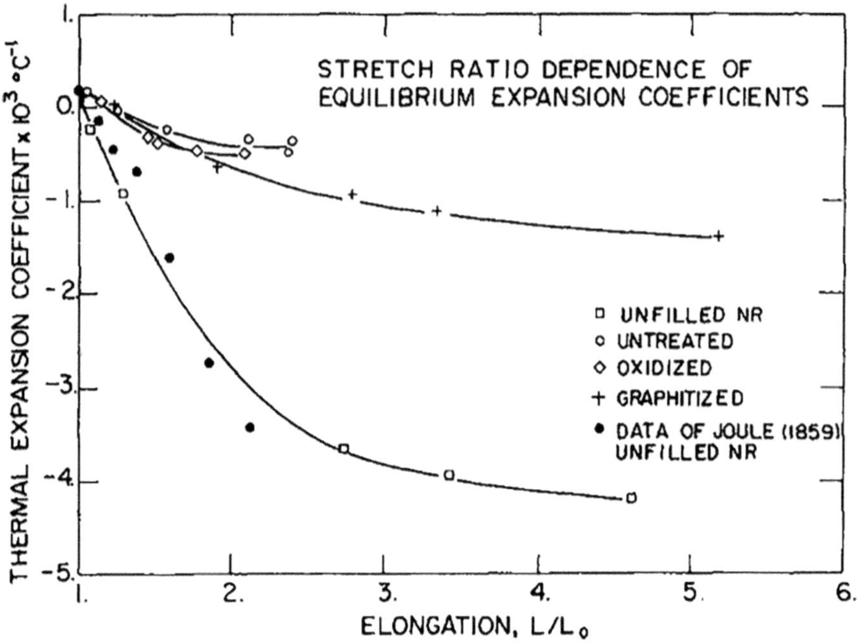

Fig. 3 Coefficient of thermal expansion vs elongation for unfilled and carbon black filled (50 phr) natural rubber [21]

$$J = J_e J_\theta \tag{14}$$

It should also be mentioned that irreversible thermal straining associated with shape memory effects can sometimes be a further consideration. For example, upon curing, some polymer cords may shrink irreversibly. The phenomenon can be used to advantageous effect to induce beneficial compressive stresses at the critical belt edge region in tires. Nylon overwrap in tires is a common example [26, 27].

2.3 *Mullins Effect*

Rubber's steady-state cyclic stress–strain response is generally softer than its initial primary response by an amount that depends on the most extreme loading encountered during prior history [13, 14]. The effect can be modeled via a pseudo elastic model, first proposed by Ogden and Roxburgh [15], in which the instantaneous energy function W corresponding to steady-state cyclic loading is assumed to derive from the energy function $\tilde{W}$ of the material's initial (or primary) response.

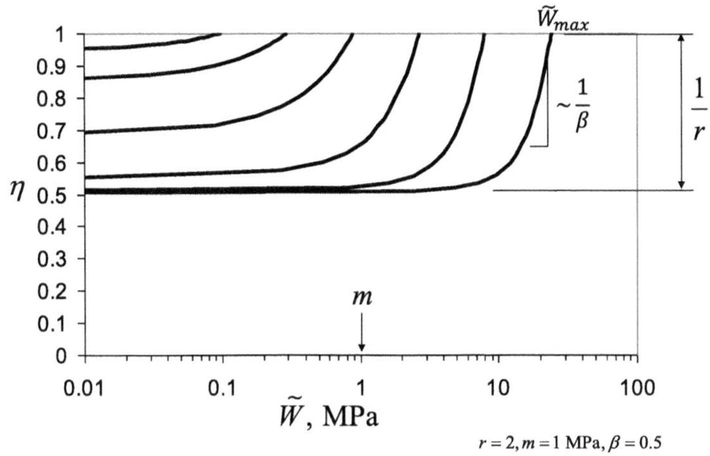

Fig. 4 Physical interpretation of Mullins model parameters

$$W = \eta\tilde{W} + \phi(\eta) \tag{15}$$

The difference $\eta\tilde{W} - W = -\phi$ is the energy loss associated with one-time softening of the material. η is a function that takes values in the range $0 < \eta < 1$ and specifies the ratio of the cyclic stabilized stress to the original unsoftened stress.

$$\eta = 1 - \frac{1}{r} erf\left(\frac{\tilde{W}_{\max} - \tilde{W}}{m + \beta\tilde{W}_{\max}}\right) \tag{16}$$

Equation (16) defines a family of curves, with each corresponding to a distinct value of $\tilde{W}_{\max}$, the prior all time maximum strain energy density on the primary curve. A typical family of curves is shown in Fig. 4. The dimensionless parameter r governs the degree to which it is possible to soften the material, i.e. the overall strength of the Mullins effect. Mullins effect is minimized when $r \to \infty$, and maximized when $r \to 1$. The parameter m specifies a strain energy density level that marks a transition point. When $m > \beta\tilde{W}_{\max}$, the cyclic stress–strain curve is only partially softened relative to the potential softening defined by r, and increases in $\tilde{W}_{\max}$ result in further stiffness decreases. In contrast, when $m < \beta\tilde{W}_{\max}$, softening of the cyclic stress–strain curve is limited by the asymptote defined by r, with further increases of $\tilde{W}_{\max}$ resulting in no further stiffness decrease. The parameter β controls the slope $\partial\eta/\partial\tilde{W}$ as $\tilde{W} \to \tilde{W}_{\max}$, according to equation (17).

$$\frac{\partial\eta}{\partial\tilde{W}} = \frac{2}{r\sqrt{\pi}\beta\tilde{W}_{\max}} \tag{17}$$

There is evidence that as temperature increases, the Mullins effect is reduced. Elhard et al. [16] reported the Ogden–Roxburgh parameters at two temperatures for a

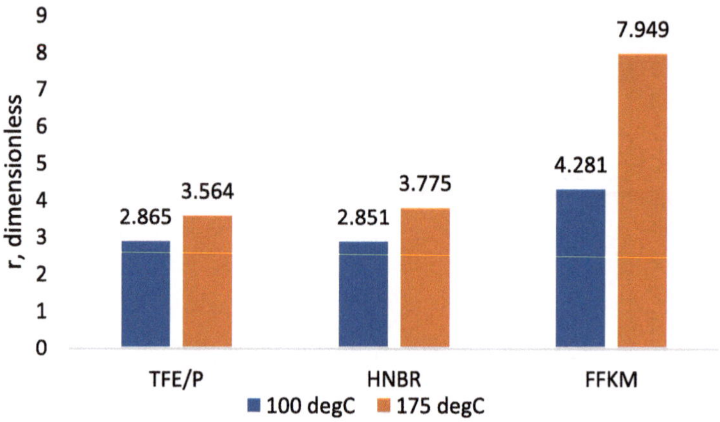

Fig. 5 Variation of the r Ogden–Roxburgh parameter with temperature for 3 rubber compounds [16]. A higher value of r indicates a reduced Mullins effect when temperature is increased

Table 2 Dependence of Ogden–Roxburgh parameters on temperature for 3 rubber compounds [16]

		100 C			175 C		
Material	Durometer	r	m, MPa	β	r	m, MPa	β
TFE/P[a]	80	2.87	0.240	0.1	3.56	0.146	0.1
HNBR	90	2.85	0.360	0.1	3.78	0.363	0.1
FFKM[b]	90	4.28	0.187	0.1	7.95	0.090	0.1

[a] Tetrafluoroethylene and propylene copolymer, AGC Chemicals trade name Aflas
[b] Perfluoroelastomer, DuPont trade name Kalrez

number of rubber compounds used in sealing applications. Figure 5 shows typical results from Table 2. The parameter r was always increased at higher temperature, indicating a reduced Mullins effect.

The effect of temperature on the Mullins effect can also be seen in the series of cyclic stress–strain loops reported by Huang et al. [17], as shown in Fig. 6. Clement et al. [18] interpreted the effect in terms of changes to the entropic component of the elasticity, see Fig. 7.

Although the Mullins effect is often idealized as irreversible, the original stress–strain behavior may sometimes be recovered at least partially over long time scales following a sufficient rest period at elevated temperature, as shown by Mullins [19] and by Houwink [20].

2.4 *Mode of Control and Elasticity*

Applied forces and displacements vary in proportions that depend on an application's mode of control and material stiffness. Thus, when temperature changes give

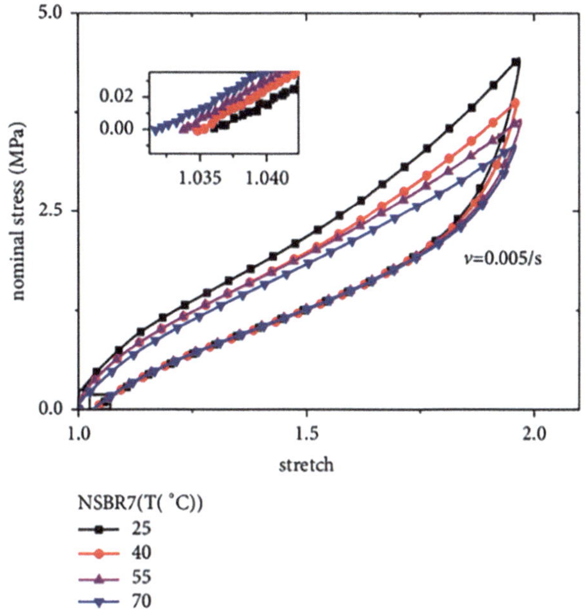

Fig. 6 Effect of temperature on silica-filled polydimethylsiloxane (PDMS) [17]

rise to stiffness changes, equilibration requires that stresses σ and strains ε adapt depending on their mode of control. Mode of control can be defined by considering the relationships between stress, strain, and stiffness during a process [28]. For a linear elastic material with Young's modulus E, whose behavior is defined by

$$\sigma = E\varepsilon \tag{18}$$

And whose strain energy density W is given by

$$W = \frac{1}{2}E\varepsilon = \frac{1}{2}\frac{\sigma}{E} \tag{19}$$

we may define the mode of control X in terms of a variable p representing "that which is held constant." This definition leads to

$$p = \varepsilon E^{\frac{1-X}{2}} = \sigma E^{-\frac{1+X}{2}} \tag{20}$$

Several familiar modes of control can be recovered by specializing to particular choices of X.

$$\text{For } X = -1 \text{ (stress control)}, p = \varepsilon E = \sigma \tag{21}$$

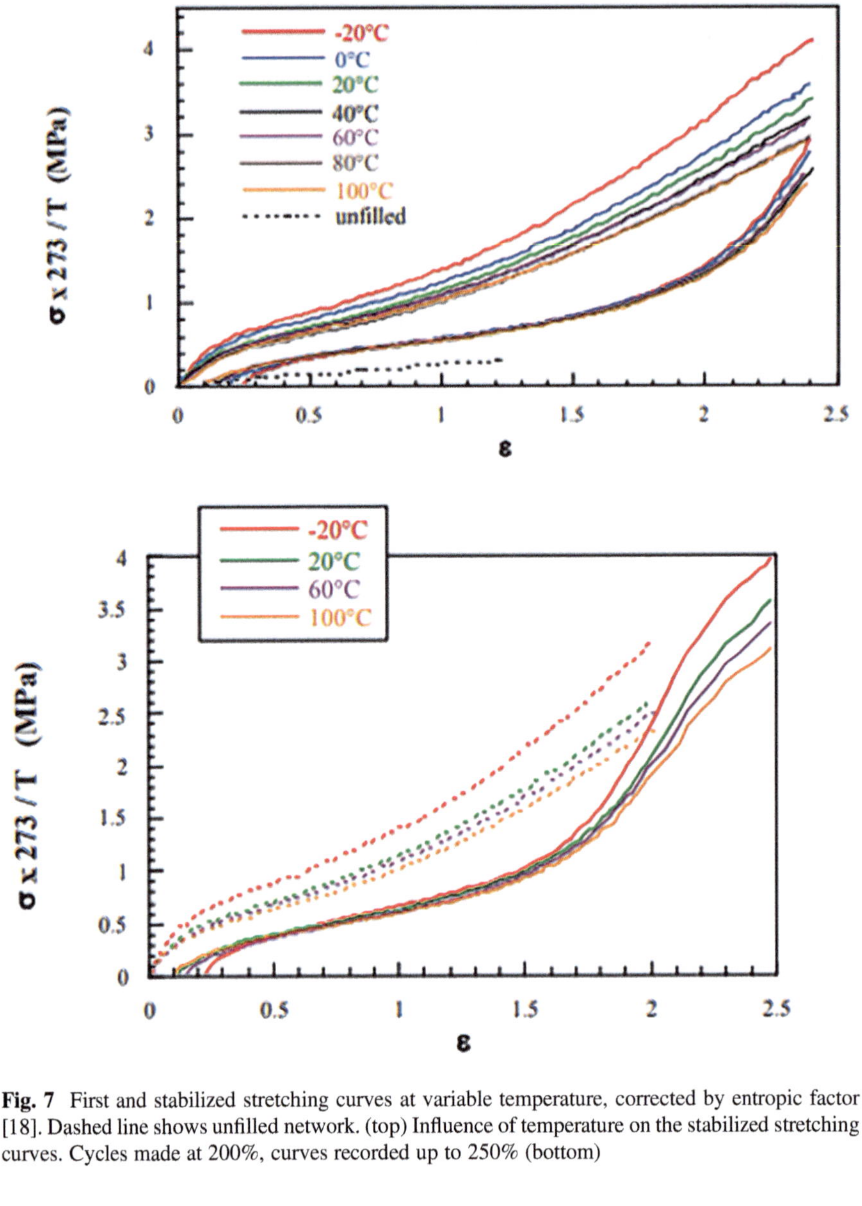

Fig. 7 First and stabilized stretching curves at variable temperature, corrected by entropic factor [18]. Dashed line shows unfilled network. (top) Influence of temperature on the stabilized stretching curves. Cycles made at 200%, curves recorded up to 250% (bottom)

$$\text{For } X = 1 \text{ (strain control)}, p = \varepsilon = \sigma E^{-1} \tag{22}$$

$$\text{For } X = 0 \text{ (energy control)}, p = \varepsilon E^{1/2} = \sigma E^{-1/2} = \sqrt{2W} \tag{23}$$

A continuous spectrum of intermediate control modes can also be recognized by choosing X in the range $-1 < X < 1$.

When variation of the temperature θ causes a change in rubber's stiffness E, the following relationships indicate the corresponding changes to strain, stress, and strain energy density.

$$\frac{\varepsilon(\theta)}{\varepsilon(\theta_0)} = \left(\frac{E(\theta)}{E(\theta_0)}\right)^{\frac{X-1}{2}} \tag{24}$$

$$\frac{\sigma(\theta)}{\sigma(\theta_0)} = \left(\frac{E(\theta)}{E(\theta_0)}\right)^{\frac{X+1}{2}} \tag{25}$$

$$\frac{W(\theta)}{W(\theta_0)} = \left(\frac{E(\theta)}{E(\theta_0)}\right)^{X} \tag{26}$$

The last equation can be solved to give a convenient definition for the mode of control X in terms of stored strain energy density $W(\theta_0)$ and $W(\theta)$, respectively, before and after a variation of stiffness from $E(\theta_0)$ to $E(\theta)$.

$$X = \frac{\ln(W(\theta)/W(\theta_0))}{\ln(E(\theta)/E(\theta_0))} \tag{27}$$

In general, the mode of control varies from point to point in a part, and it can be computed via a finite element analysis of the change in stored energy density in each element due to a variation of stiffness. Whether a change of stiffness (due to temperature, due to changing the compound, or due to any other reason) is beneficial or detrimental to self-heating or durability depends on the mode of control.

3 Dissipative Properties

Although rubber's defining characteristic is its relatively high elasticity, it also possesses a dissipative character that can be useful in applications requiring damping, or inconvenient in applications where energy conservation is desired. Energy dissipated appears as heat that leads to temperature rise. The resulting temperature gradients in turn cause material property variations that then must be considered in material selection and product design.

3.1 Self-heating

When material with dissipative behavior undergoes periodic loading at frequency ω, the work input on the loading stroke W_L is greater than the work recovered W_U on the unloading stroke, as shown in Fig. 8. On a volumetric basis, we compute the hysteresis H by integrating the stress history with respect to strain, as follows:

$$H = W_L - W_U = \oint \sigma d\varepsilon \tag{28}$$

For the case of purely elastic behavior without crystallization, we have $W_L = W_U$, giving $H = 0$. The volumetric heat rate $\dot{q}$ is related to the hysteresis as

$$\dot{q} = H\omega \tag{29}$$

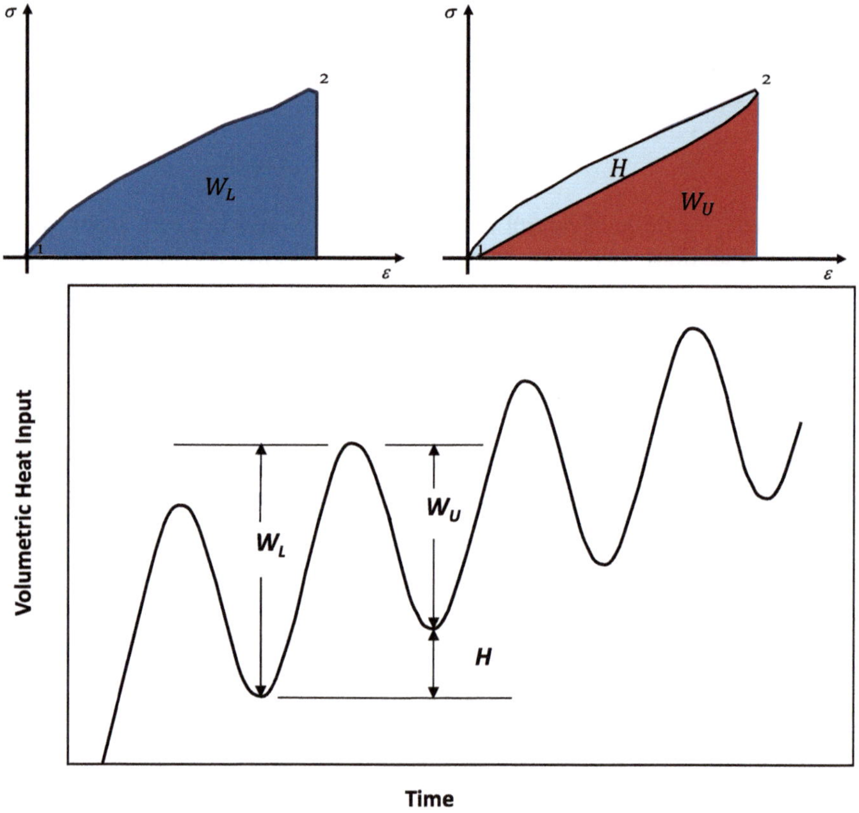

Fig. 8 The work done on the loading stroke W_L is given by the blue area under the stress–strain curve (top left). The work recovered on the unloading stroke W_U is given by the red area under the stress–strain curve, and the heat energy remaining in the material as hysteresis H is given by the light blue area (top right). The heat energy input during periodic straining increases by amount H per cycle (bottom)

In general, an accurate description of the viscoelastic response of filled rubber compounds at moderate strains requires a nonlinear viscoelastic theory. Two examples are the Polymer Rheological Framework (PRF) [29] and the Comprehensive Constitutive Model [30]. The latter is written by integrating the stress–strain law, which is itself a hereditary integral, giving the history of energy input $\Psi(t)$, also known as the free energy, as:

$$\Psi(t) = f_D \int_0^t du \int_0^t ds \mu(t^* - s^*, t^* - u^*) \frac{d(h\underline{X})}{ds} : \frac{d(h\underline{X})}{du} + \Psi_\infty \tag{30}$$

where f_D is a dimensional factor representing the Mullins effect, $\mu(t^* - s^*, t^* - u^*)$ is a memory function in terms of reduced time that accounts for rubber's temperature–frequency dependence, h is a function of strain history that represents Payne effect, and X is a generalized strain measure. Implementation and use of these and similar approaches can be complex.

Over the years, several approximations have been popular, especially for the analysis of rolling resistance and temperature rise in tires [31–34]. For linear viscoelasticity, the stress history $\sigma_{ij}(t)$ can be calculated by integrating the shear relaxation modulus G and the bulk relaxation modulus K.

$$\sigma_{ij}(t) = \int_{-\infty}^{t} G(t-\tau) \frac{d\epsilon_{ij}}{d\tau} d\tau + \int_{-\infty}^{t} K(t-\tau) \frac{d\epsilon_{ii}}{d\tau} d\tau \tag{31}$$

Assuming that volumetric straining is purely elastic (i.e., $d\epsilon_{ii} = 0$), the energy loss per cycle H can be estimated from the linear viscoelastic loss modulus G'' by [33]

$$H = \oint \left(\underline{\sigma} : \frac{d}{dt} \underline{\varepsilon} \right) dt \cong 2\pi G''(\epsilon_{eq}, a_T \omega) \underline{\varepsilon}_0 : \underline{\varepsilon}_0 \tag{32}$$

here the ϵ_{ij} are the small strain measures, and the ϵ_0 are the amplitudes of a sinusoidal cycle. A quick check on Eq. 32 shows that it reproduces the more familiar uniaxial results and that it is invariant for coordinate transformations. For non-sinusoidal periodic deformations, a Fourier series representation has been used [33].

In general, G'', as calculated from a typical Dynamic Mechanical Analysis (DMA) test, will depend on temperature and frequency, which may still be treated as linearly viscoelastic. Temperature and frequency dependency is accounted for by the reduced frequency $a_T\omega$, where a_T is a nondimensional shift factor computed from the Williams–Landel–Ferry (WLF) equation [35, 36]. The loss modulus G'' might also depend on strain amplitude (Payne [37], or Fletcher–Gent [38] softening) and on mean strain [32].

3.2 Strain Amplitude Dependence

The nonlinear strain amplitude dependence of G'' is frequently characterized via a DMA test, producing the plot shown in Fig. 9. For filled rubbers, a model proposed originally by Kraus [39] and modified later by Ulmer [40] is frequently quite accurate.

$$G''(\varepsilon_0) = G''_\infty + \frac{(G''_{max} - G''_\infty)(\varepsilon_0/\varepsilon_{0c})^m + \Delta G''_U}{(\varepsilon_0/\varepsilon_{0c})^{2m} + 1} \tag{32}$$

The strain amplitude dependence of rubber's dynamic shear storage and loss moduli (i.e., Payne or Fletcher–Gent effect) should not be confused with the worst-ever-strain dependence (i.e., Mullins effect). Although both result in softening of the elastic response, the dynamic response is reversible on short time scales, whereas the dependence on worst-ever strain implies irreversible softening, as shown in Fig. 10.

3.3 Temperature and Rate Dependence of Self-heating

The viscoelastic character of rubber depends on temperature and rate or time. At temperatures well above the glass transition (i.e., the rubbery plateau), dissipative behavior tends to decrease as temperature increases. This trend has sometimes [32, 34] been described via an exponential dependence on temperature θ and strain rate $\dot{\varepsilon}$

$$\frac{G''(\theta)}{G''(\theta_0)} = e^{r(\theta - \theta_0) + Z(\dot{\varepsilon} - \dot{\varepsilon}_0)} \tag{33}$$

Typical values for the material parameters r and Z for tire compounds are given in Table 3.

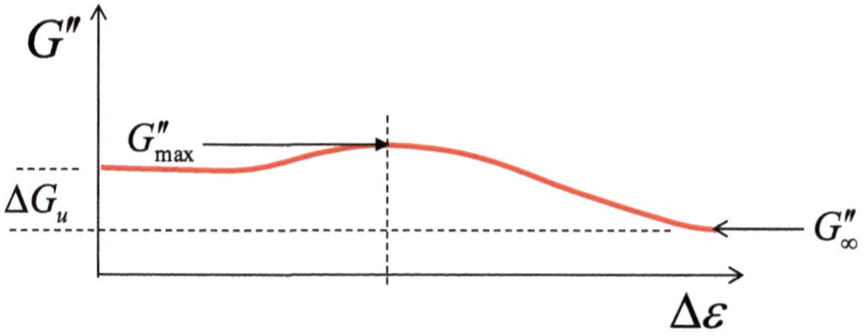

Fig. 9 Physical meaning of Kraus model parameters

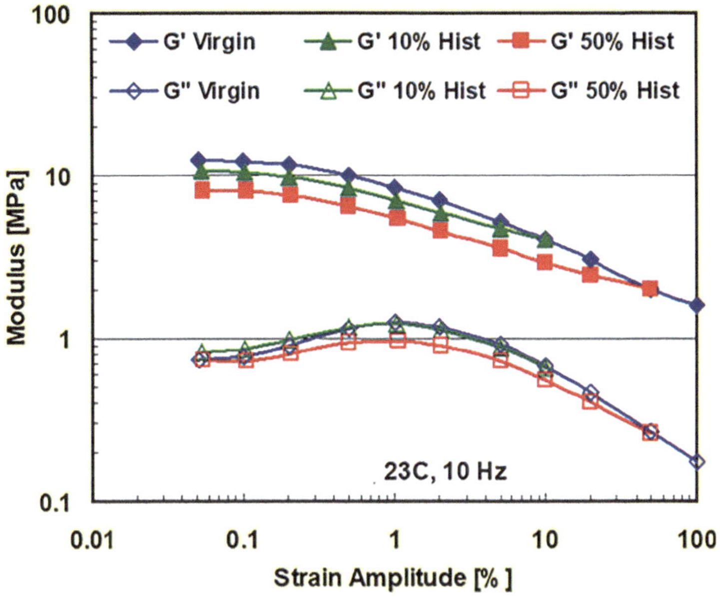

Fig. 10 Payne effect plot showing variation of storage and loss modulus as a function of strain amplitude and strain history (Mullins effect)

Table 3 Temperature and rate sensitivities for several tire rubbers [34]

Rubber	Temperature sensitivity r, 1/°C	Rate sensitivity Z, s
Tread	−0.015	0.032
Subtread	−0.011	0.026
Innerliner	−0.012	0.028
Sidewall	−0.0077	0.021
Bead filler	−0.0076	0.011

For large $\theta - \theta_0$ or $\dot{\varepsilon} - \dot{\varepsilon}_0$, this model may extrapolate poorly over the full range of possible behavior. However, it has been demonstrated to be sufficient for simulating heat buildup under conditions typical to tires.

More generally, a time–temperature, or rate–temperature equivalence exists for many polymers, i.e. raising the frequency produces the same property change as lowering the temperature. The rate–temperature equivalence arises from the fact that the mobility and the rate φ of polymer segmental jumps decreases as the temperature θ decreases relative to the glass transition temperature θ_g. Williams, Landel, and Ferry [35] described this in the following form:

$$\ln\left(\frac{\varphi(\theta)}{\varphi(\theta_g)}\right) = \frac{A(\theta - \theta_g)}{B + \theta - \theta_g} \tag{34}$$

As the temperature decreases and approaches the glass transition temperature, the rate of segmental jumps $\varphi(\theta)$ is so slow that the polymer is unable to uncoil to accommodate a call for deformation at nominal rates, and the material behaves more like a glass than a rubbery elastomer.

This logic implies an equivalence [41] of viscoelastic properties Q such that:

$$Q(\omega, \theta) = Q(a_T \omega, \theta_0) \tag{35}$$

Q can be any of the linear viscoelastic properties or functions, and the shift factor a_{T} is defined by:

$$\log(a_T) = \frac{-c_1(\theta - \theta_0)}{c_2 + (\theta - \theta_0)} \tag{36}$$

With these relationships, the material property, for example the loss modulus G'', as a function of frequency ω can be predicted at any temperature θ by shifting the known function at temperature θ_0 by an amount a_{T} along the frequency axis. This is shown in Fig. 11. Since the plot is plotted in terms of Log (ω), Log(a_{T}) provides the number of decades in frequency to shift the master curve to represent properties at temperature, θ.

Laboratory equipment to measure G' and G'' are typically limited in frequency range, but have a large temperature range, such that the glass transition behavior can be characterized by testing at low temperatures. Using a series of tests with the available frequency range, say 0.1 Hz to 1,000 Hz, and conducted at multiple temperatures, and shifting the results using Eq. 36 allows one to construct the so-called master curve as shown in Fig. 11. Once the master curve is known, the viscoelastic response can be predicted over a wide range of temperatures and frequencies.

There are some limitations to this approach. It works well for an unfilled, single polymer rubber compound. Polymer blends, filled polymers, and the usual combination of materials in an industrial rubber compound may reduce accuracy, so experimental verification is recommended when possible.

3.4 Mode of Control and Viscoelasticity

For products that undergo cyclic or periodic deformations, the concept of "Deformation Index" introduced by Futamura [42–44] is useful for elastomer durability analysis, since the mode of deformation control (e.g., stress, strain, energy) for portions or a part in service can be determined. This mode of control can impact choices for compounds for certain regions of a product, since a stiffer compound will generally perform better under stress control, while a more compliant compound will perform better under strain control. With the implementation of the Deformation

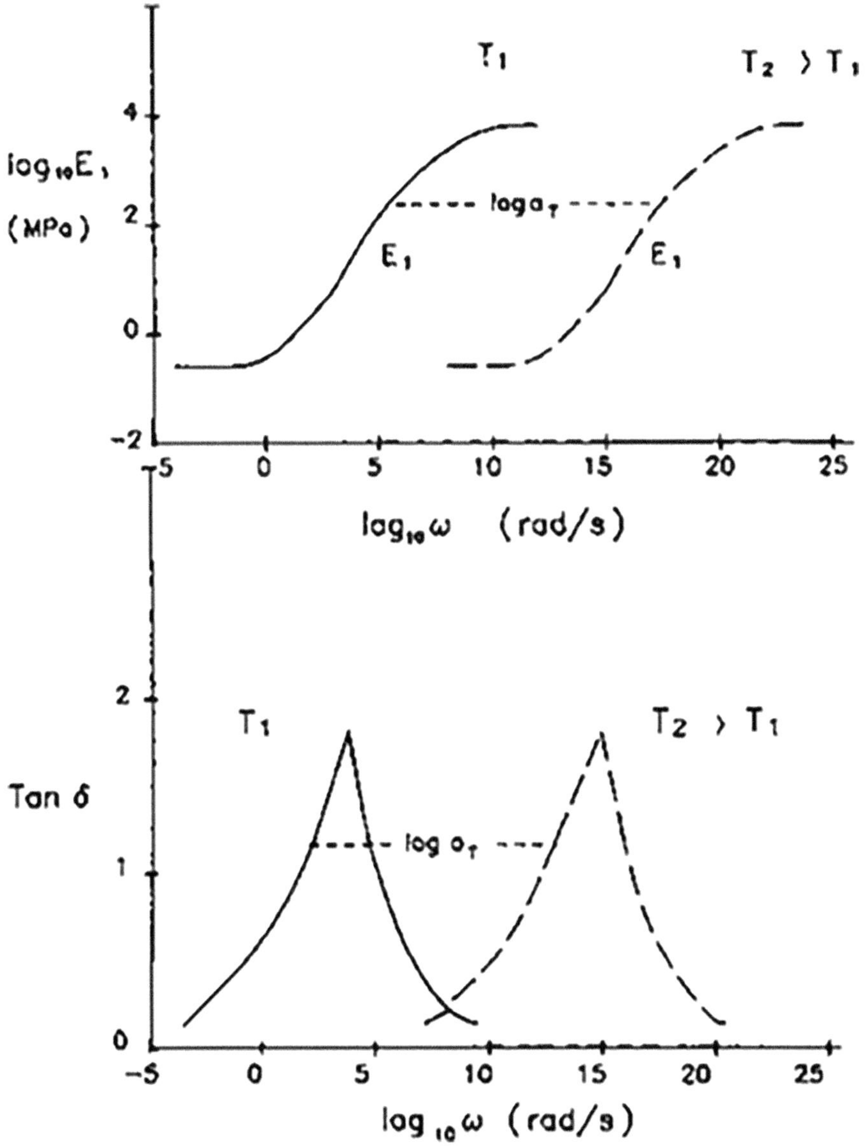

Fig. 11 Example Williams–Landel–Ferry (WLF) shifts [41]

Index calculation with the finite element analysis [43], the degree of stress–strain or energy control can be evaluated and displayed as a distribution over the volume of the part.

The work by Futamura and co-workers showed that the heat rate depends on the linear viscoelastic moduli and on the deformation index n as follows:

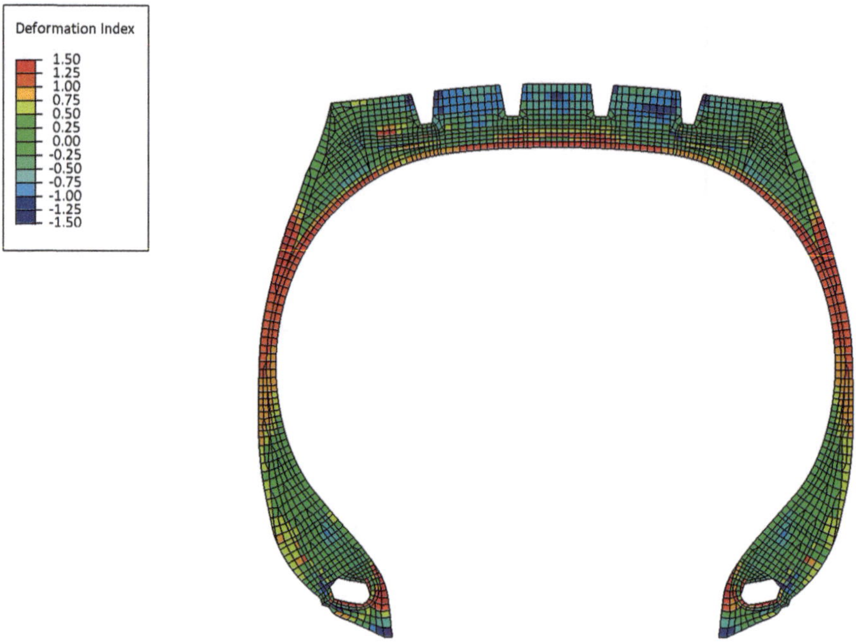

Fig. 12 Contours showing deformation index X on truck tire cross section, derived from Endurica-computed heat rates

$$\frac{\dot{q}(\theta)}{\dot{q}(\theta_0)} = \frac{G''(\theta)}{G''(\theta_0)} \left(\frac{G^*(\theta_0)}{G^*(\theta)} \right)^n \tag{37}$$

The above equation can be derived from the mode of control definition Eq. 27 with the identification

$$n = 1 - X \tag{38}$$

To solve Eq. 37 for n, two finite element analyses are run, a control and a variation. The elastic property G^* of the variation is perturbed, while the loss modulus, G'' is held constant. A scheme with separate but coupled elastic and energy dissipation analyses such as that in Ebbott [33] makes these computations straightforward. Figure 12 shows a plot of the Deformation Index for a typical truck tire. As expected, the sidewalls are mostly in strain control mode and the tread is leaning toward stress control mode. Since the ribs in this tire model are solid, there is an amount of deformation control due to the curvature change from circular to flat. If the tire was more lugged, the tread would be further into stress control mode for a constant applied vertical load. The belt edge region operates close to an energy control mode.

4 Strength and Crack Growth Properties

There are limits to rubber's longevity, and these limits are a primary consideration when selecting a material or designing a part or system. The potential for the growth of cracks is of central importance. The characterization and analysis of cracks is greatly simplified when framed in terms of the tearing energy [45] T (sometimes also called Griffith's energy release rate [46]).

$$T = -\frac{d(U - V)}{dA} \tag{40}$$

U is the potential energy stored in the structure, V is the external work done on the structure, and A is the crack surface area.

Equation 40 may be computed with the finite element analysis [47–51], or specialized to closed form [52, 53] for common situations like edge cracked planar tension (Eq. 41) or edge cracked simple tension (Eq. 42).

$$T = Wh \tag{41}$$

$$T = 2kWc \tag{42}$$

$$k = \frac{\pi}{\sqrt{\lambda}} \tag{43}$$

W is the strain energy density, h is the gauge height of the planar tension test specimen, a is the length of the crack, and λ is the extension ratio.

4.1 *Intrinsic Strength* T_0 *(Fatigue Limit)*

Crack growth in rubber requires at least the breakage of all polymer chains that cross the plane of the crack [54]. In addition, dissipation associated with the crack tip process zone also draws power, increasing the tearing energy required for crack growth [55]. Figure 13 illustrates the process by which crack growth occurs.

The minimum tearing energy input requirement T_0 for producing crack growth (i.e., the *intrinsic strength*) is equal to the product of the per-chain bond breaking energy of the polymer and the number of chains broken per unit of crack surface area. Neither of these parameters is sensitive to temperature within the range of typical application (except for the case where thermally activated aging changes cross-link density/average chain length between cross-links). We therefore do not see dependence of the intrinsic strength on temperature.

Indeed, whenever temperature dependence of the tearing energy T at constant crack growth rate has been observed in experiments, it is interpreted as evidence of process zone dissipation [56–59]. Accordingly, it is customary to write

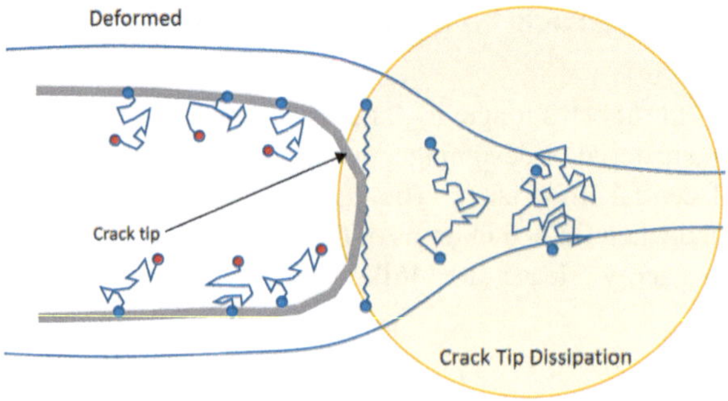

Fig. 13 Crack growth requires energy sufficient to break chains crossing the crack plane and to power the dissipation requirements of the crack tip process zone. The chain breaking energy does not depend on temperature. Crack tip dissipation is viscoelastic in nature and thus depends on temperature and rate

$$T(\theta, \dot{\varepsilon})\Big|_{\frac{dc}{dN}} = T_0 + T_z(\theta, \dot{\varepsilon}) \tag{44}$$

where T_0 is the intrinsic strength and T_z is the part of the tearing energy contributed by process zone dissipation at constant crack growth rate *dc*/*dN*.

Experimental determination of the intrinsic strength requires the minimization, or ideally, the reduction to zero of T_z. This can be accomplished through various means, including direct fatigue cycling at low strain [60, 61] (hence the nomenclature *fatigue limit*), slow tearing of solvent-swollen specimens at elevated temperature [56–59], and cutting with a sharp blade [62, 63].

4.2 Critical Tear Strength T_c

The critical tear strength T_c quantifies the maximum tearing energy that can be carried by the rubber without spontaneous crack growth and rupture. Because tearing behavior is dominated by the dissipative T_z term of Eq. 44, the critical tearing energy exhibits strong dependence on rate and temperature [64, 65], consistent with the implications of the WLF relationship of Eq. 36. Generally, an increase of temperature leads to a decrease of critical tear strength. The example of tear strength dependence on rate and temperature for unfilled natural rubber [66] is shown in Fig. 14.

Just as the temperature effects on dissipation may be approximated with the exponential of Eq. 33 over a certain range, a similar rule may be applied to the critical tear strength T_c.

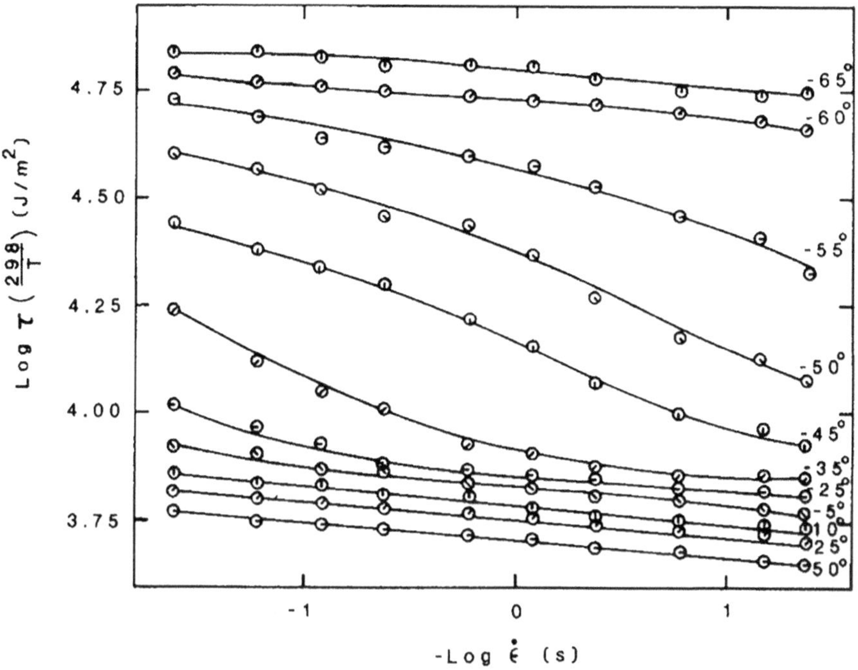

Fig. 14 Critical tear strength dependence on strain rate and temperature for unfilled natural rubber [66]

$$\frac{T_c(\theta)}{T_c(\theta_0)} = e^{-\frac{C}{F}(\theta-\theta_0)} \tag{45}$$

C is the temperature sensitivity coefficient of the fatigue crack growth rate law and F is the power-law slope to be discussed in the next section.

The reduction of critical tear strength with temperature for several elastomers [16] is shown in Fig. 15 and in Table 4.

4.3 Crack Growth Rate

For tearing energies intermediate between the tear strength upper limit T_c and the intrinsic strength T_0, the rate of crack growth r varies as a power-law function with characteristic index F, as shown in Fig. 16. This relationship was first observed by Thomas [67], and it is therefore called the Thomas law.

$$r = \frac{dc}{dN} = r_c\left(\frac{T}{T_c}\right)^F \tag{46}$$

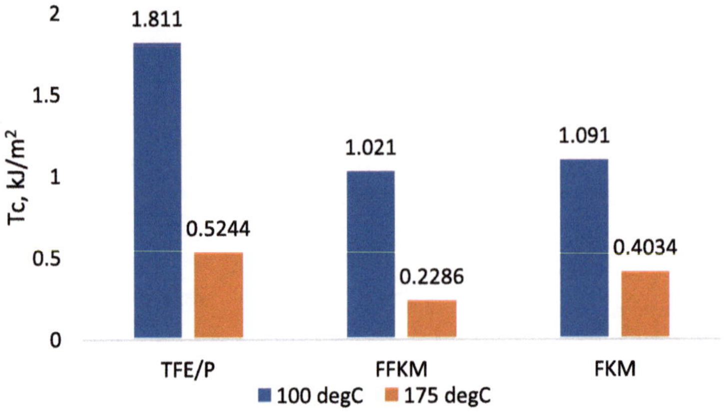

Fig. 15 Typical effect of temperature on tear strength Tc [16]

Table 4 Dependence of tear strength Tc on temperature for three rubber compounds [16]

Material	Durometer	100°C	175°C
		Critical tearing energy, kJ/m^2	Critical tearing energy, kJ/m^2
TFE/P[a]	80	1.811	0.524
FFKM[b]	90	1.021	0.229
FKM[c]	90	1.091	0.403

[a] Tetrafluoroethylene and propylene copolymer, AGC Chemicals trade name Aflas
[b] Perfluoroelastomer, DuPont trade name Kalrez
[c] Fluoroelastomer, Chemours trade name Viton

Fig. 16 Parameters of the Lake–Lindley fatigue crack growth rate law

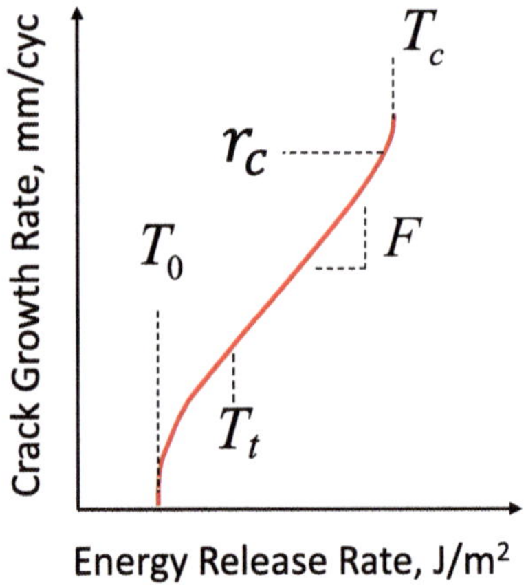

When Eq. 46 is augmented by specifying the intrinsic strength T_0, it is called the Lake–Lindley law [68, 69], and crack growth rates for below a transition energy T_t are calculated by means of

$$r = \frac{dc}{dN} = A(T - T_0) \tag{47}$$

The transition energy and the constant A can both be obtained by requiring continuity of the crack growth rate r and its derivative dr/dT at the transition energy. This leads to

$$T_t = \frac{T_0}{1 - \frac{1}{F}} \tag{48}$$

and

$$A = \frac{r_c T_t^F}{T_c^F (T_t - T_o)} \tag{49}$$

Combining Eqs. (45) or (47) with (46), it can be shown that the temperature effect on the crack growth rate law can be written as

$$\frac{r(\theta)}{r(\theta_0)} = e^{C(\theta - \theta_0)} \tag{50}$$

That the crack growth rate approximately follows Eq. 50 can be seen in the experiments by Young [70, 71] in Fig. 17. His results for natural rubber roughly indicate $C = \ln\left(\frac{r(\theta)}{r(\theta_0)}\right)/(\theta - \theta_0) = \ln(50)/(75°C) = 0.052$ 1/°C, and his results for

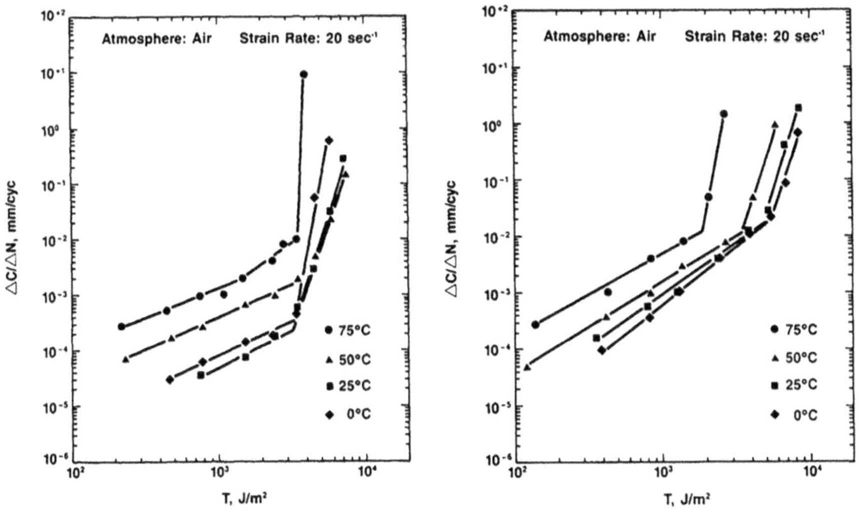

Fig. 17 Dependence of fatigue crack growth rate curves on temperature [70, 71] for natural rubber (left) and butadiene rubber (right)

butadiene rubber roughly indicate $C = \ln(5)/(75\ ^\circ\mathrm{C}) = 0.021\ 1/^\circ\mathrm{C}$. The positive sign on C indicates that increasing temperature results in increasing crack growth rate.

The effect of temperature on the crack growth behavior of rubber is the subject of a number of recent studies [72–77].

4.4 Crack Precursor Size

Rupture and fatigue failure are the culmination of a crack growth process that begins with crack precursors present naturally in rubber [78, 79]. Although crack precursors initially possess a wide range of shapes and a distribution of sizes [80, 81], they may for purposes of characterization and analysis be (1) idealized as penny-shaped cracks having a single characteristic diameter (their size), (2) summarized by the maximum size occurring within a given volume, and (3) statistically described via the Weibull distribution [82–84]. The precursor size refers to size in the unstressed reference configuration, and it can be assumed that neither instantaneous temperature nor aging has any effect on the crack precursor size.

4.5 Wöhler Curves

Rubber's fatigue behavior is sometimes reported in the form of a Wöhler curve. A Wöhler curve plots the fatigue life as a function of strain or stress. Typically, Wöhler curve results are derived from tension experiments on apparently "uncracked" dumbbell specimens. Measurements in the literature of Wöhler curve temperature dependence are relatively rare, perhaps owing to the poor economics of the testing method relative to alternatives (i.e., the need to test a large number of individual specimens at many strain levels), and to the fact that the temperature dependence of a Wöhler curve confounds at least three distinct effects: (1) temperature dependence of the fatigue crack growth rate curve, (2) self-heating of the testing specimens used in fatigue testing, and (3) temperature dependence of stress–strain behavior. Fortunately, Wöhler curves can easily be computed from crack growth experiments by integrating the crack growth rate law [78] via

$$N = \int_{c_0}^{c_f} \frac{1}{r} dc \tag{51}$$

Lake and Lindley reported the experimental fatigue life versus temperature results plotted in Fig. 18 for natural rubber and for styrene butadiene rubber. Their results were simulated with the Endurica CL fatigue solver. The material properties for these simulations are summarized in Table 5.

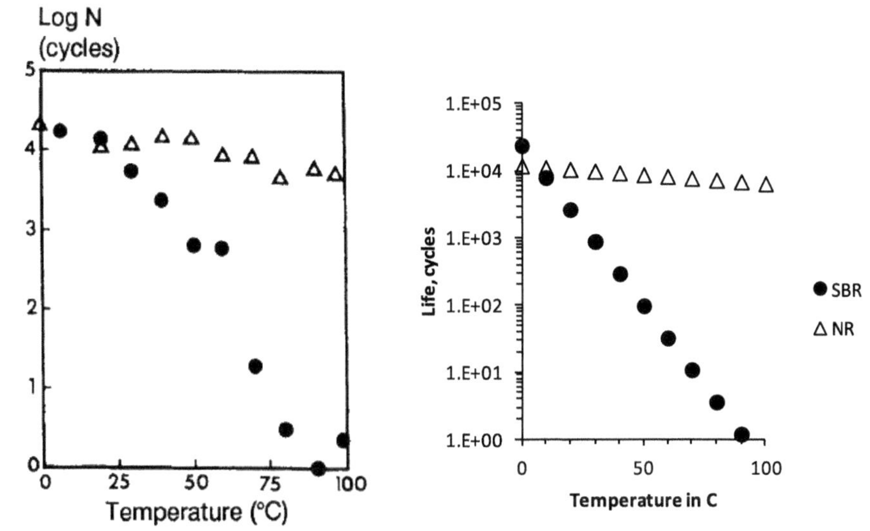

Fig. 18 Effect of temperature on crack nucleation life of filled NR and filled SBR [78]. Solid circles = SBR at 175% peak strain. Open triangles = NR at 250% peak strain. Minimum strain = zero in all cases. Experimental results (left). Simulated results (right)

Table 5 Material properties for simulating Lake and Lindley's fatigue vs. temperature experiments [78] with Endurica CL

	Natural rubber	Styrene butadiene rubber
Shear modulus G, MPa	0.26	0.26
Precursor crack size c_0, mm	0.025	0.055
End of life crack size c_f, mm	1.0	1.0
Reference temperature θ_0, °C	20	20
Temperature sensitivity coefficient, 1/°C	0.0061	0.110
Tear strength T_c, kJ/m^2	10.0	2.5
Intrinsic strength T_0, kJ/m^2	0.04	0.06
Crack growth rate r_c at T_c, mm/cyc	5×10^{-3}	50×10^{-3}
Thomas law slope F	2	4
Strain crystallization	Yes	No

5 Strain-Induced Crystallization

Some rubbers, natural rubber being the most well-known example, exhibit a phenomenon in which stretching can induce spontaneous crystallization [85]. The effect has profound impacts on fatigue behavior due to greatly increased resistance of the crystallized material to crack growth when operating under nonrelaxing conditions [86].

5.1 Experimental Characterization

The crystallization effect has a strong dependence on temperature, with higher temperatures tending to suppress crystallization and its associated benefit [87]. The phenomenon may equivalently be described as a strain-dependent melting temperature of the rubber's crystalline phase [88, 89] as shown by the experimental results of Gent and Zhang in Fig. 19.

The strain crystallization phenomenon and its dependence on rate and temperature has been directly studied via wide angle X-ray scattering (WAXS) methods [90–93]. Figure 20 shows typical WAXS results. Bright spots in the image represent locations to which X-rays were preferentially scattered due to the crystalline structure. The size and brightness of these spots directly indicate the degree of crystallinity. Such measurements have led to advances in characterization and simulation of strain crystallization effects [89, 94].

5.2 Models for Nonrelaxing Fatigue Crack Growth Rate Behavior

The effect of strain crystallization on fatigue behavior is most apparent in nonrelaxing experiments where the tearing energy $T(t)$ is at all times greater than zero, with positive cycle minimum T_{min} and maximum T_{max}. To a first approximation, the effect is manifest as a dependence of the crack growth rate law exponent

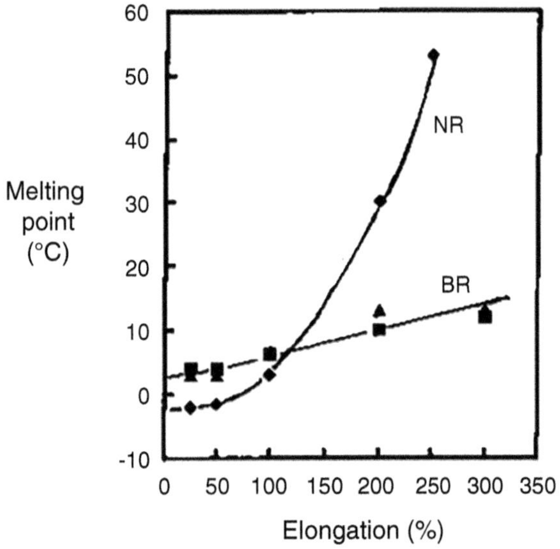

Fig. 19 Crystalline phase melting temperature vs. elongation for NR and BR [88]

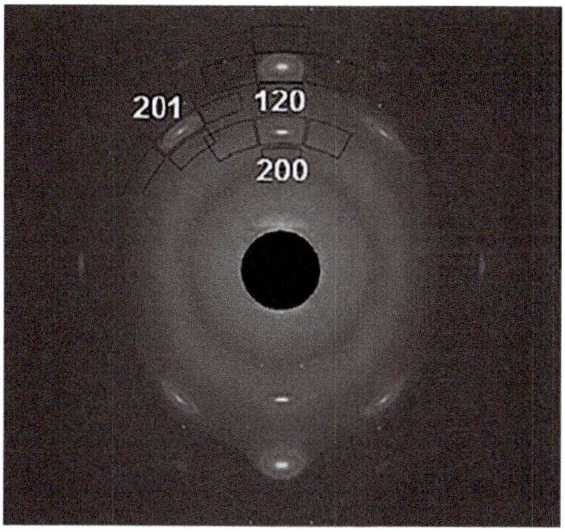

Fig. 20 Wide angle X-ray scattering (WAXS) of natural rubber. Bright spots in the image indicate strain-induced crystallization. Crystallographic indices are also noted [90]

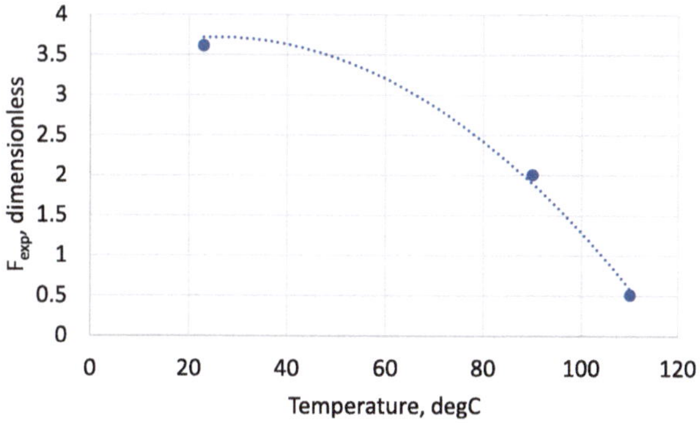

Fig. 21 Strain crystallization strength F_{exp}, as a function of temperature θ, as inferred for natural rubber from results of Ruellan et al. [98, 99]

F on the quotient $R = T_{min}/T_{max}$. A simple form of the Mars–Fatemi model [95–97] can capture the primary effect with a single material parameter F_{exp}.

$$F(R) = F_0 e^{F_{exp}(\theta)R} \tag{52}$$

F_0 is the power-law slope of the fully relaxing crack growth rate law. The nondimensional parameter F_{exp} specifies the strength of the crystallization effect. It can be defined as a function of temperature θ, as in Fig. 21. These results were derived from experiments reported by Ruellan et al. [98, 99].

The effect can also be described more generally (in cases where the crack growth rate law deviates from the power law) by means of a shift function $x(R)$. The shift x specifies the fraction by which the nonrelaxing crack growth rate curve at tearing energy $T(R)$ is shifted relative to the left and right bounds, respectively, the equivalent fully relaxing tearing energy T_{eq} and the rupture tearing energy T_c.

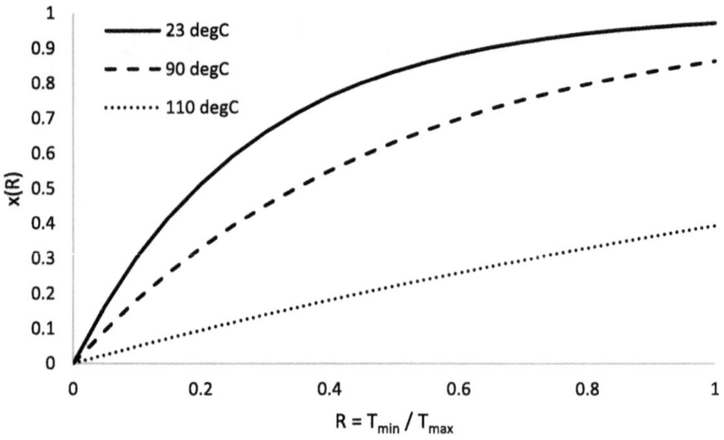

Fig. 22 Strain crystallization strength $x(R)$, for three temperatures, as inferred from results of Ruellan et al. [98, 99]

$$x(R) = \frac{\log(T) - \log(T_{\mathrm{eq}})}{\log(T_c) - \log(T_{\mathrm{eq}})} \tag{53}$$

Recognizing that the shift function can also parameterize the Thomas law, it is possible to derive the following mapping between the Mars–Fatemi model and the shift factor x for the Ruellan data set, as shown in Fig. 22.

$$x(R, \theta) = 1 - \frac{F_0}{F(\mathrm{R}, \theta)} \tag{54}$$

5.3 Haigh Diagram

Rubber's Haigh diagram plots contours of constant fatigue life under cyclic uniaxial loading as a function of mean strain (abscissa) and of strain amplitude (ordinate) [100]. The large experimental effort required to produce such a map limits its full adoption to routine practice, but the Haigh diagram can provide an intuitive display of the fatigue life dependence on mean strain, strain amplitude, and temperature. The results from Ruellan et al. [98, 99], shown in Fig. 23, are perhaps the most complete Haigh diagram available to date for natural rubber. Their results clearly show the reduction of the crystallization effect as temperature increases. Note that at lower temperatures, in the region where $R > 0$, increases in mean strain result in an increase in fatigue life. But at higher temperatures, increasing mean strain is always associated with decreasing life because strain crystallization is suppressed.

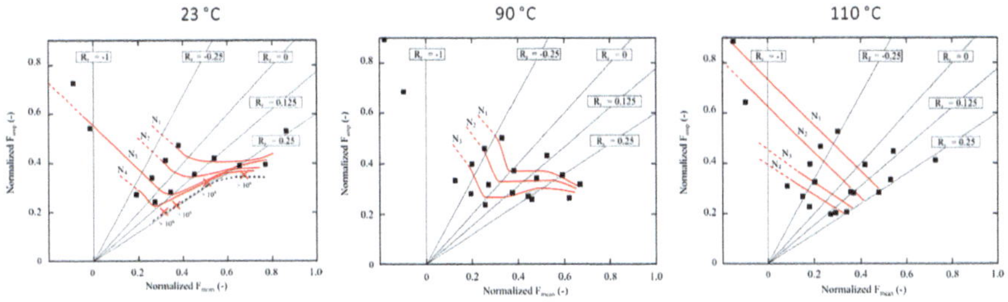

Fig. 23 Haigh diagram dependence on temperature, as observed by Ruellan et al. [98, 99]

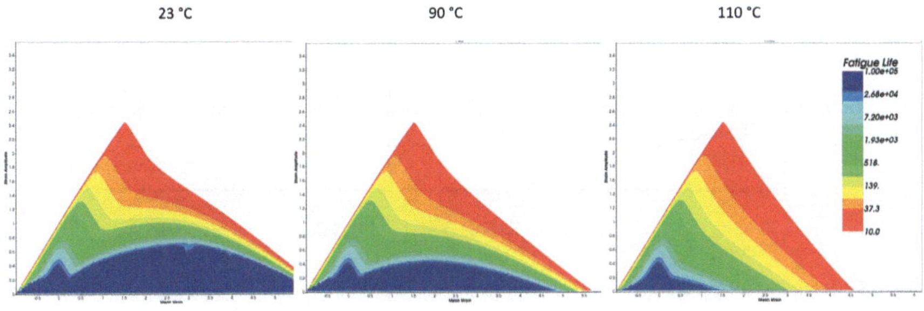

Fig. 24 Endurica-computed Haigh diagram dependence on temperature

Ruellan's experimental observations can be replicated via the strain crystallization models defined by Eqs. 52 and 54, when combined with critical plane analysis [101] and evaluation of Eq. 51. The simulation results were computed with the Endurica CL fatigue solver and are shown in Fig. 24.

6 Aging

Rubber's material properties arise largely from underlying chemical structures that, while stable at short and medium time scales and cooler temperatures, may evolve significantly at longer time scales and hotter temperatures [6, 7]. The presence of oxygen, for example, can impact aging chemistry for many elastomers [102, 103]. Accordingly, aging is often categorized as either aerobic (including oxygen) or anaerobic (without oxygen) [104].

Oxidative aging has at least two mechanisms. One creates more cross-links and tends to make the elastomer harder and more brittle. The second cleaves the C=C double bond resulting in chain scission and creating lower molecular weight material with reduced desirable properties.

Anaerobic aging in sulfur-cured systems is due to the evolution of the type and density of cross-links. The mechanisms of anaerobic aging are largely an extension of the cure process [41, 105], in which the post-cure behavior observed in a moving

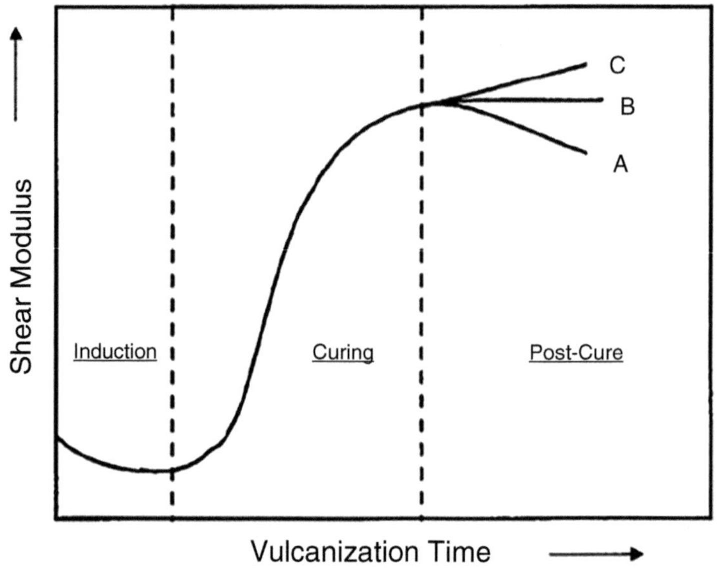

Fig. 25 Typical sulfur cure curves with different post-cure behavior [41, 105]: (**a**) Reversion and loss of stiffness, (**b**) stable, and (**c**) additional cross-linking, or "marching modulus"

die rheometer (MDR) may show reversion, stability, or a "marching modulus," as shown in Fig. 25.

For purposes of characterization, simulation, materials selection, and product design, it is necessary to quantify the time rate of material property evolution (i.e., the kinetics), as a function of thermal history, and to predict the likely pathways for material property evolution.

6.1 Arrhenius Law for Chemical Kinetics

The rate of a chemical reaction depends on temperature in accordance with the rate of molecular thermal motion [106, 107], and on the kinetic energy threshold for the conversion to a product of the reactants. Although in rubber many molecular species and reactions may occur simultaneously, characterization and analysis can often be simplified by assuming a single effective chemical process with a reaction rate κ and an activation energy E_a. The Arrhenius acceleration factor a_T is the ratio of the reaction rate κ at absolute temperature θ to the reaction rate κ_0 at absolute temperature θ_0 is given by

Table 6 Activation energy for various polymers

Polymer	Activation energy, kJ mol^{-1} K^{-1}	Reference
33% ACN NBR [a]	76–81	[109]
BIIR [b]	53–61	[110]
Sulfur cure	54–65	
Efficient cure	41–59	
Peroxide cure		
NR [c]	98–102	[111]
In air	90	[112]
In seawater	63	[112]
Tire compound	55	[116]
EPDM [d]	92–120	[113, 114]
CR [e]	96	[114]
SBR [f]	80	[115]
Tire compound	81–89	[116]
BR [g]	57	[116]
Tire compound		

[a] Acrylic/nitrile copolymer
[b] Butyl rubber
[c] Natural rubber
[d] Ethylene propylene diene terpolymer rubber
[e] Chloroprene rubber
[f] Styrene butadiene rubber
[g] Butadiene rubber

$$\frac{\kappa}{\kappa_0} = a_T = e^{\frac{E_a}{\mathcal{R}}\left(\frac{1}{\theta} - \frac{1}{\theta_0}\right)} \tag{55}$$

$\mathcal{R}$ is the universal gas constant, which has the value $\mathcal{R} = 8.314\,\mathrm{J\,K^{-1}\,mol^{-1}}$. Care should be taken when applying Eq. 55, especially when there is simultaneous diffusion that might change the chemistry and therefore the activation energy [108]. Table 6 gives typical values of E_a for several common elastomers.

6.2 Aging Master Curves

For a constant temperature exposure, the evolution with time τ of any material property that is determined by chemical/molecular structure may be represented with a master curve indexed to a single reference temperature. For the purpose of fatigue life prediction, the key parameters are modulus E, intrinsic strength T_0, and critical tear strength T_c.

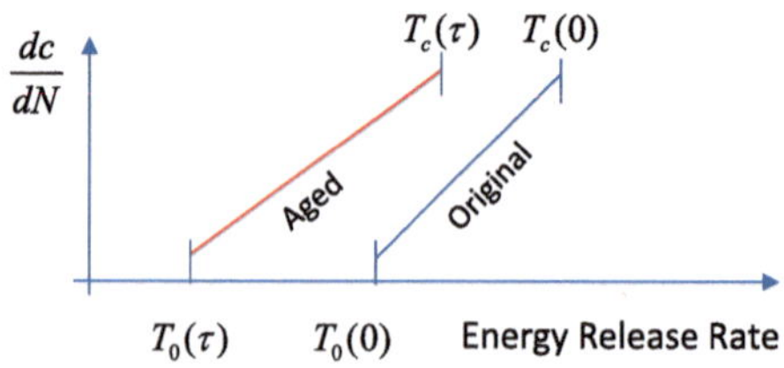

Fig. 26 Typical evolution of fatigue crack growth rate curve with aging

$$E(\theta,\tau) = E(a_T(\theta)\tau) \tag{56}$$

$$T_0(\theta,\tau) = T_0(a_T(\theta)\tau) \tag{57}$$

$$T_c(\theta,\tau) = T_c(a_T(\theta)\tau) \tag{58}$$

Knowing the evolution with exposure time and temperature of the upper and lower bounds of the crack growth rate curve, respectively, the tear strength T_c and the fatigue threshold T_0, the evolution of the power-law slope F can be estimated as

$$F(\tau) = F_0 \frac{\log(T_c(0)) - \log(T_0(0))}{\log(T_c(\tau)) - \log(T_0(\tau))} \tag{59}$$

In this way, an exposure time-dependent crack growth rate curve is defined in Eq. 60, and shown in Fig. 26.

$$\frac{dc}{dN} = r_c \left(\frac{T_{\max}}{T_c(\tau)}\right)^{F(\tau)} \tag{60}$$

Using Eqs. 56–58, together with the Arrhenius shift defined in Eq. 55, prediction of the aged material property at a temperature other than the reference temperature can be accomplished. The left-hand side plot in Fig. 27 shows shore hardness data collected at 70°C and 100°C [117]. A common master curve $E(a_T(\theta)\tau)$ is followed by both, but the curve for 100°C progresses much faster, having $a_T(100\ °\ \mathrm{C})/a_T(70\ °\ \mathrm{C})$ of approximately 150 h/25 h = 6 (determined by inspection of the left-hand plot). The right-hand side of the plot shows the evolution of the fatigue crack growth rate curve with exposure at 100°C in air.

Aging can have a dramatic effect on strength properties. Consider the results in Fig. 28 below. These results are from specimens extracted from a commercial truck tire at the belt edge [117]. This region is thought to remain largely anaerobic over the life of the laboratory test. The critical tearing energy decreases significantly with in situ anaerobic aging. Figure 29 shows the corresponding measurements of the evolution of the cross-link types. The increase in mono-sulfidic cross-links with reduced toughness is thought to be the driver of the reduced critical energy release rate.

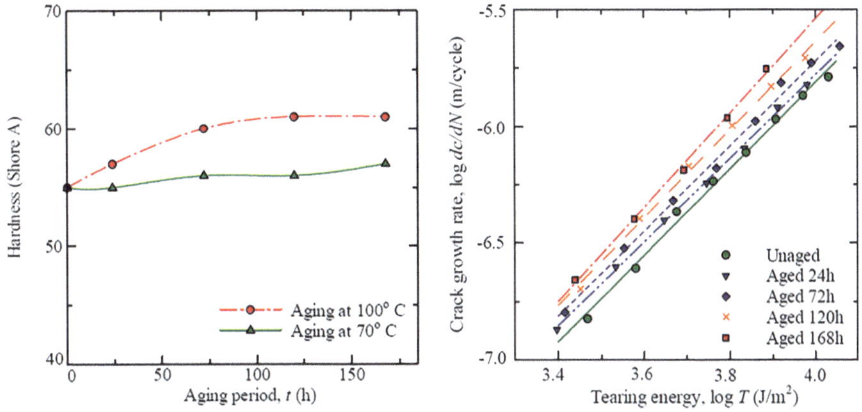

Fig. 27 Evolution of hardness for two exposure temperatures (left), and fatigue crack growth rate curve with aging at 100°C in air for natural rubber [117]

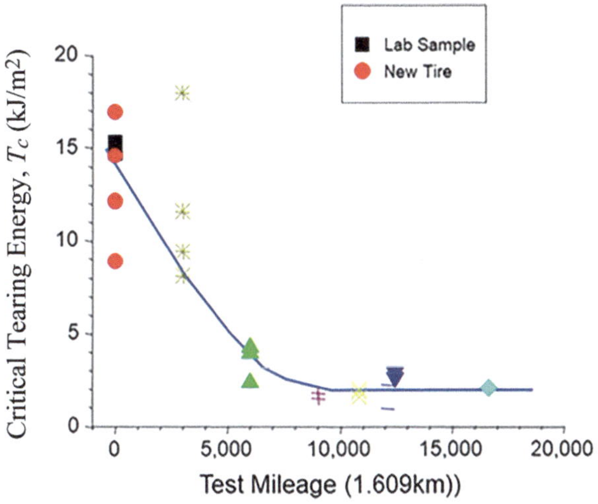

Fig. 28 Evolution of critical tearing energy as a function of test mileage in the belt edge region of a truck tire, assumed to age anaerobically [117]

6.3 *Accrual of Aging Under Variable Temperature History*

When the temperature $\theta(t)$ varies in time, the total accrued age τ on the constant temperature aging master curves in Eqs. 56–58 may be estimated by integrating the relative reaction rate in Eq. 55, as follows:

$$\tau = \int_0^t \frac{\kappa(t)}{\kappa_0} dt = \int_0^t e^{\frac{E_a}{\mathcal{R}}\left(\frac{1}{\theta(t)} - \frac{1}{\theta_0}\right)} dt \tag{61}$$

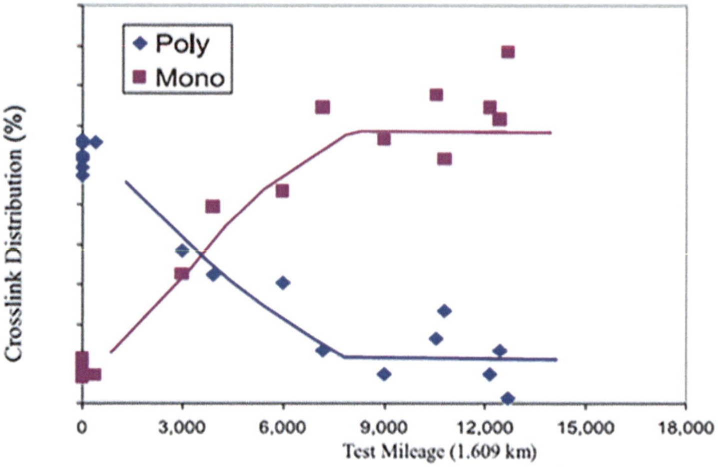

Fig. 29 Evolution of sulfur cross-link types as a function of test mileage in the belt edge region of a truck tire (measurements by E. Terrill) [117]

Evolving material properties may thus be evaluated from their corresponding master curves, and integration of the crack growth rate law can then be made, yielding an account of the coupled aging-fatigue problem with simultaneously evolving crack length and material properties.

$$N_f = \int_{c_0}^{c_f} \frac{1}{r(T(c,\tau),\tau)} dc \tag{62}$$

The Endurica CL fatigue solver [119] can solve Eq. 62 for the case where it is not necessary to update the stress–strain fields of the structural simulation. For example, this approach could be taken if the stiffness and stress–strain fields do not change much with aging during the life of the part.

An incremental solution scheme [120, 121] is required, however, in order to allow for updating of the stress–strain fields, and to solve the more general problem.

$$\Delta c = \int_{N_i}^{N_{i+1}} r(T(c,\tau),\tau) dN \tag{63}$$

As shown in Fig. 30, the Endurica DT solver [122] offers a cosimulation workflow in which both the finite element and fatigue analyses are updated from one increment to the next in order to track the growth of cracks and the accrual of exposure time until failure. Each increment of the analysis consists of the following steps: (1) compute stress σ, strain ε, and temperature θ fields via a finite element

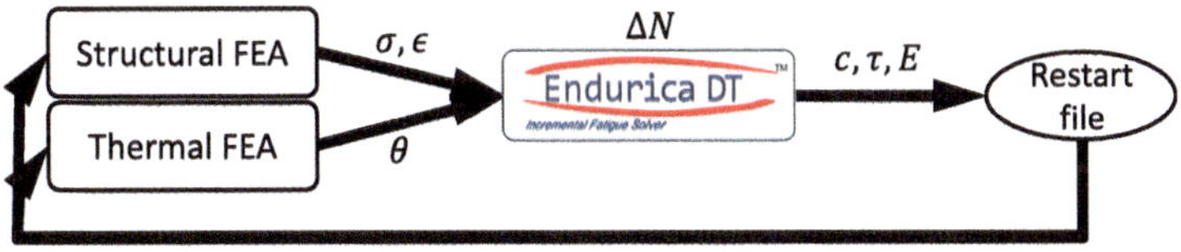

Fig. 30 Incremental solution scheme for updating structural and thermal simulations with evolving stiffness and updating the fatigue simulation with updated crack growth properties

Table 7 Diffusion coefficients for O_2 gas at 25°C for various polymers [125]

Polymer	O_2 diffusion coefficient @ 25°C, $cm^2\ s^{-1}$
NR	1.5×10^{-6}
NBR	$0.3–0.9 \times 10^{-6}$
BIIR	0.08×10^{-6}

solver using material properties evaluated at the current time, (2) compute for the duration of the increment the new crack length c, the new equivalent reference exposure time τ, and the new modulus E, intrinsic strength T_0 and critical tear strength T_c, and (3) repeat until a failure condition is reached (i.e., crack too long, properties too degraded, etc.).

6.4 *Diffusion*

The evolution of rubber's chemical structures and the associated material properties depend on the availability of reactive chemical species [105]. Common species include (but are not limited to) cross-linking agents such as sulfur or peroxide, cure accelerators, antidegradents, as well as oxygen. The availability of chemical species for reaction depends on their concentration, which may vary depending on environmental conditions and on solubility of the species in rubber. The spatial and time variation of the concentration φ of each species can be investigated via Fick's diffusion equation [123, 124], which has the same form as the heat equation.

$$\frac{\partial \varphi}{\partial t} = D\nabla^2 \varphi \tag{64}$$

The parameter D is the diffusion coefficient which, owing to it being a direct reflection of the rate of molecular motions, varies in linear proportion to absolute temperature θ [125]. That is, for reference diffusion coefficient D_0 at reference temperature θ_0

$$D = D_0 \frac{\theta}{\theta_0} \tag{65}$$

Diffusion coefficients for oxygen in several polymers are given in Table 7.

6.5 *Oxidation*

The pathways along which rubber's stiffness and strength properties evolve depend strongly on the reaction and incorporation of oxygen with the material's molecular structure. Ahagon et al. [126] studied the effects of aging temperature under both aerobic and anaerobic conditions, and he identified three types of aging, as shown in Fig. 16. Type I is driven by oxidative mechanisms, generally leading to increased stiffness and cross-link density but reduced strength. Type II is driven by anaerobic cross-link evolution, leading to a reduction of stiffness and cross-link density, along with reduced strength. Type III shows intermediate behavior, reflecting the fact that the availability of oxygen at higher temperatures may be limited if the oxygen is entirely consumed early in the exposure period (Fig. 31).

Oxidation of rubber parts is often maximum near the surface and reduced or minimal on the interior of the rubber part, especially for aging at high temperature. This occurs because, as oxygen is consumed by reaction with the rubber near the exposed surface, the concentration of oxygen diffusing to the interior is reduced. This is known as Diffusion Limited Oxidation or DLO [127]. Consider the plots in Fig. 32, which show on the left the density of C=C double bonds as a function of position within a 4.8 mm thick sheet of polychloroprene rubber, and on the right the modulus as a function of position. As oxygen reacts with rubber near the surface, it consumes C=C double bonds in the main polymer chain and they therefore decrease in density as a result of the reaction [128]. As a consequence, additional cross-links are formed, resulting in increased modulus. This occurs only near the surface and the interior of the test sample remains unchanged. The temperature-dependent reaction rate and diffusion coefficient can be taken into account in simulations [129–131].

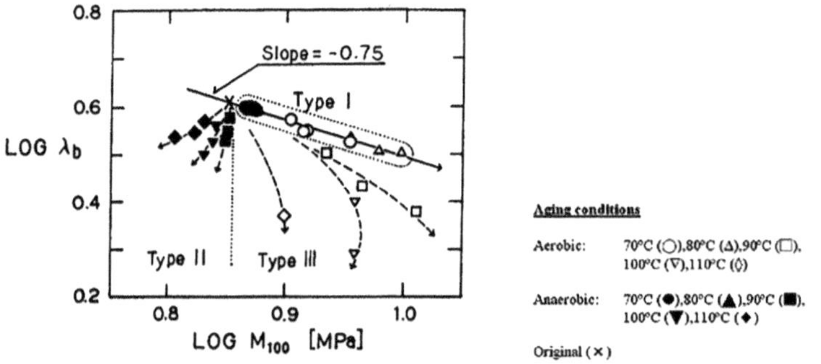

Fig. 31 Evolution of stretch at break and 100% modulus with aging under aerobic and anaerobic conditions for natural rubber [126]

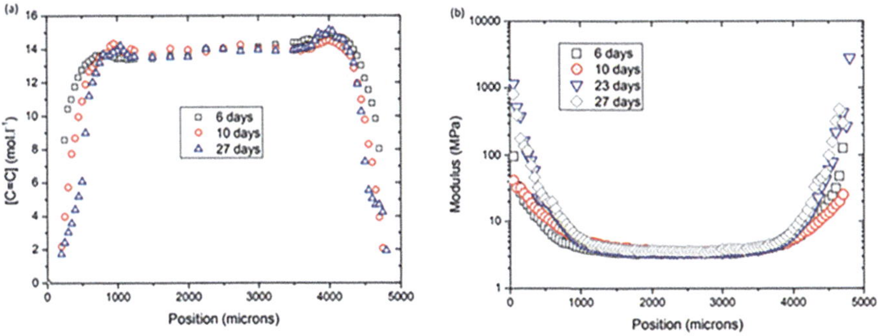

Fig. 32 Profile of double bonds (**a**) and modulus (**b**) through the thickness of a 4.8 mm thick sample during oxidation at 120°C in air [128]

Table 8 Some thermal properties for rubber at room temperature [133]

Polymer	Filler	K	C_p	α
Units		W/(m °C)	J/(kg °C)	mm^2/s
NR	Gum	0.15	1744	0.084
NR	40 phr N347	0.28	1,581	0.153
NR	40 phr silica	0.24	1,544	0.134
SBR	CB filled, 60 ShA	0.30	1,501	0.152
NBR	CB filled, 77 ShA	0.30	1,519	0.167

7 Thermal Transport Properties

Computing the temperature distribution in a body requires knowledge of the thermal transport properties of the material [132]: conductivity K, specific heat C_p, and diffusivity α. Table 8 provides typical thermal transport properties for a few selected compounds at room temperature. The thermal diffusivity can be related to conductivity and specific heat through

$$\alpha = \frac{K}{\rho C_p} \tag{66}$$

ρ is the mass density. Each of these parameters may depend on temperature.

7.1 Thermal Conductivity

The thermal conductivity K specifies the relationship between the heat flux rate $\dot{q}''$ per unit area via conduction and the spatial gradient of temperature $d\theta/dx$. It is the key parameter governing the steady-state conditions that will prevail once thermal transient effects have equilibrated.

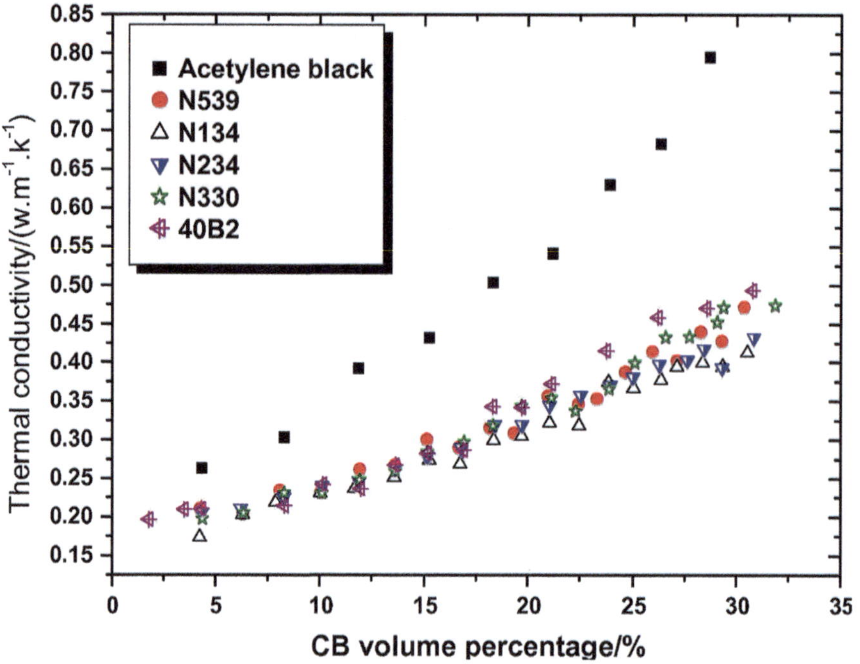

Fig. 33 Thermal conductivity dependence on filler loading in filled natural rubber for several grades of carbon black [136]

$$\dot{q}'' = K \frac{d\theta}{dx} \tag{67}$$

There are several methods for measuring thermal conductivity. Kerschbaumer et al. recently compared the steady-state heat flow meter method, with the transient plane source, transient line source, and laser flash analysis methods [133].

Saxena et al. [134] studied temperature dependence of SBR conductivity at several loadings of waste NR particles, ranging from 0 to 40 phr. Below the glass transition temperature, conductivity increases with temperature. But above the glass transition temperature, conductivity decreases with temperature.

Badawy [135] studied the effect of sulfur loading on carbon black filled SBR from room temperature to 150°C. At low loadings of sulfur (up to 2 phr), the conductivity showed a roughly 20% decrease with respect to increasing temperature. At 5 phr of sulfur, the conductivity showed a roughly 20% increase with increasing temperature.

The addition of carbon black fillers to a rubber compound tends to increase thermal conductivity [136], see Fig. 33.

The conductivity depends on temperature, especially for filled compounds, decreasing with increasing temperature [132], as shown in Fig. 34. This trend supports the hypothesis that the increased volume of the polymer matrix at higher

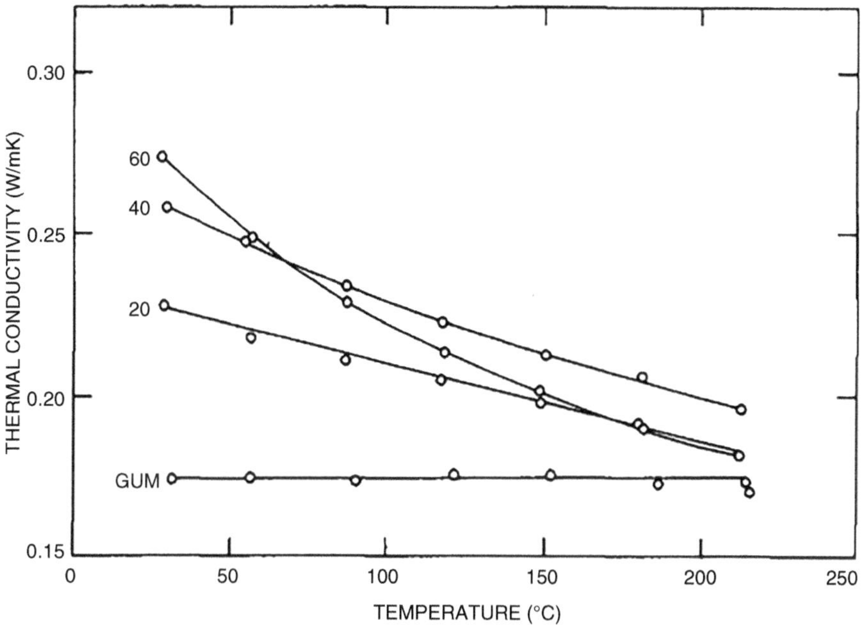

Fig. 34 Thermal conductivity of natural rubber compounds filled with N110 carbon black at different loading levels [132]

temperatures tends to move rigid filler particles apart, thereby reducing conduction paths through the filler network.

7.2 *Specific Heat*

The specific heat c_p specifies the relationship between the time rate of change of stored heat energy per unit volume $\dot{q}$ and the rate of change of temperature $d\theta/dt$. ρ is the mass density. Specific heat is a key parameter governing the initial and transient behavior of thermal systems.

$$\dot{q} = \rho c_p \frac{d\theta}{dt} \tag{68}$$

In rubber, the specific heat is largely determined by polymer type and filler content. When the specific heats of constituents are known, the rule of mixtures can be applied. Zaharescu et al. [137] demonstrated this principle for various blends of unfilled NR and EPDM, which by themselves show opposite sensitivities to temperature, as seen in Fig. 35.

Fig. 35 Molar specific heat temperature dependence [137] of unfilled natural rubber (o) and of unfilled EPDM (□)

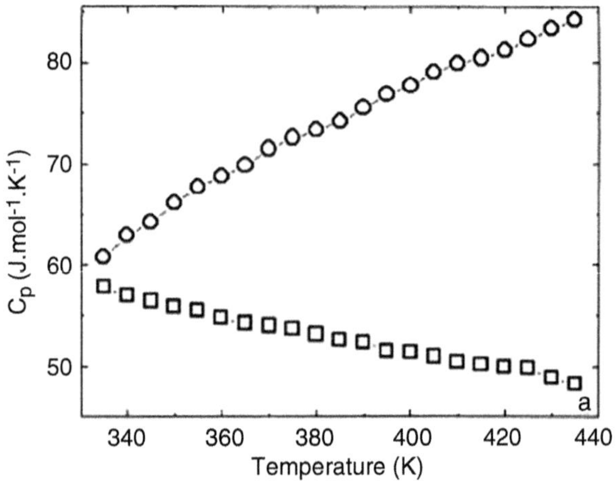

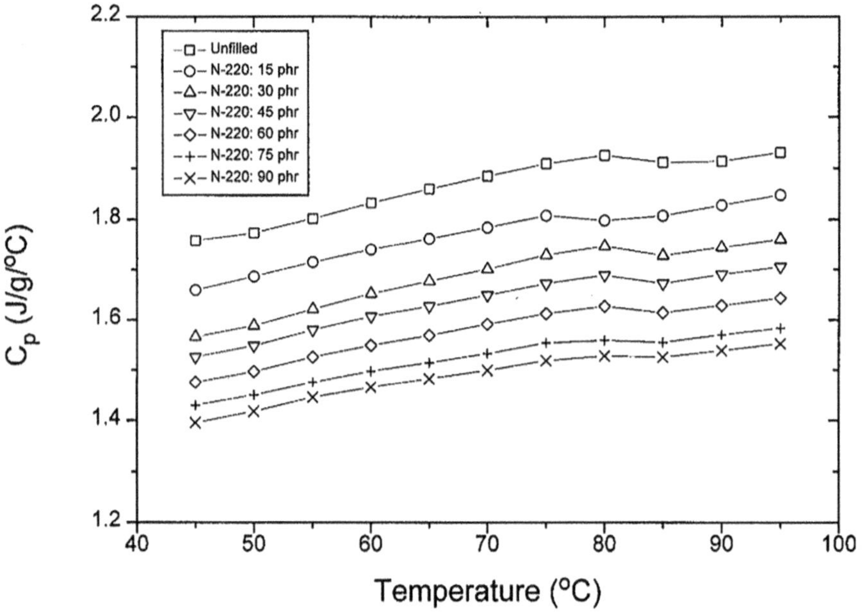

Fig. 36 Specific heat of N220 carbon black filled natural rubber [138]

Carbon black has a lower specific heat than rubber, so accordingly, filled rubber compounds tend to have a lower specific heat than filled compounds, as seen in the results of Nah et al. [138] in Fig. 36.

8 Thermal Analysis

If the rate of heat input to a rubber part is sufficiently large relative to the rate of heat removal, temperature will rise rapidly, and very rapid failure can occur. Fortunately, the causes and conditions of thermal runaway are well-established, so that unexpected failures can be avoided when sufficient care is taken to respect governing principles.

For problems involving nonuniform (in space or time) temperature fields, some version of the heat equation must be solved for the temperature distribution. The Eulerian form of the heat equation can be written as

$$\rho c_p \dot{\theta} - \nabla(k \nabla \theta) = \dot{q} \tag{69}$$

where $\dot{q}$ is the volumetric heating rate defined previously in Eq. 29.

So long as thermal runaway does not occur, then as the part heats, a steady-state temperature will be reached at which the input heat energy is in equilibrium with the heat energy carried away from the part by conduction, convection, and radiation. The steady-state solution of Eq. 69 can be obtained by taking $\dot{\theta} = 0$.

8.1 Steady-State Analysis

Many rubber products operate over long periods of time at steady-state temperature conditions. Steady-state response requires a balance between the heat generated by energy dissipation in the rubber and heat transferred away from the rubber article. The rate of heat transfer from the boundaries of a rubber article depends on the temperature difference between the surface and the ambient, and on the coefficients for convection and radiation. Conduction to other materials, for example rubber in contact with a steel rim or road surface, can also transfer heat. In addition to the dissipative behavior of the rubber itself, analysis of a steady-state temperature distribution requires knowledge of the geometry, thermal material properties, and boundary conditions (conduction, convection, radiation) of the rubber article.

Calculating the steady-state temperature field is a prerequisite for the analysis of durability in many applications, given the many strong temperature dependencies: stiffness, tear strength, crack growth rate law, strain crystallization, aging, etc. One of the early approaches was developed by Ebbott et al. [33] for axisymmetric geometry. This is a semi-coupled thermomechanical workflow that uses an approximate material representation for nonlinear viscoelasticity. The workflow is shown in Fig. 37. An equivalent nonlinear elastic representation is used in the structural model and the equivalent nonlinear viscoelastic model was described in Eq. 31. An iterative scheme is necessary due to temperature dependence of the elastic and dissipative material models. An example tire temperature distribution computed with this scheme is shown in Fig. 38. The effect of temperature on both the immediate

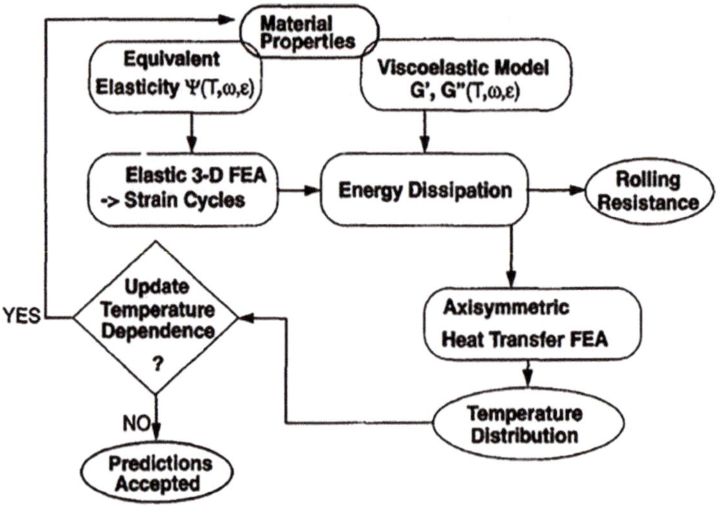

Fig. 37 Workflow for steady-state temperature prediction in a tire with axisymmetric tire geometry [33]

Fig. 38 Example temperature distribution for a commercial truck tire running against a cool road wheel [33]

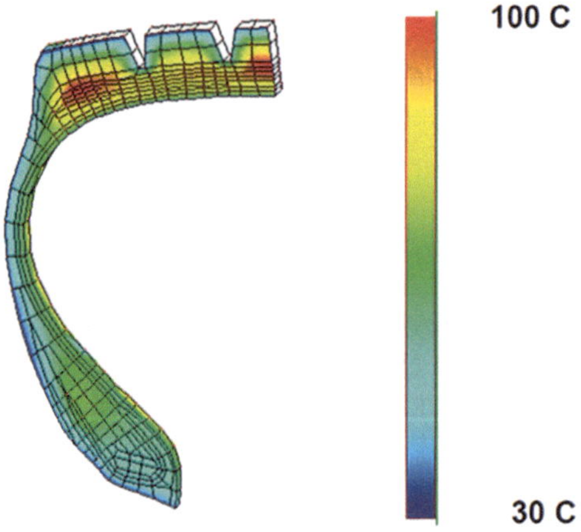

properties as well as the impact on aging should be considered when computing durability from the obtained steady-state temperature field. For purposes of computing heat rate, a more direct alternative to Eq. 32 would be a full nonlinear, time-dependent constitutive equation solved with the FEA solution. The computational cost would likely be higher, as would the model calibration requirements. The Parallel Rheological Framework model in ABAQUS [29, 31], or the Comprehensive Constitutive Equation [30] could serve for this type of workflow.

8.2 Transient Thermal Analysis

Temperature history often plays an important part in rubber part failures through its influence on both irreversible and reversible material property changes. The thermal history applied during tire vulcanization, for example, leads to spatial variations of cure state that might require consideration [139]. Similarly, thermal history experienced due to time-varying boundary conditions [140, 141] or to self-heating [142, 143] may lead to aging and material property spatial and time variation. In extreme cases, thermal runaway, reversion, and blowout may occur. An important parameter in transient thermal analyses is the thermal time constant, which governs how long it takes to reach steady-state conditions. The simulation workflow for transient problems is similar to the steady-state workflow shown in Fig. 37, as both are based on Eq. 69, with the difference that material property updates must be made incrementally at using a cosimulation scheme similar to that shown in Fig. 30.

8.3 Blowout Temperature θ_b

A full analysis of thermal runaway must consider the combined transient effects of viscoelastic heat production, thermal transport, material property evolution (due to both instantaneous and long-term thermal and cyclic effects), and mode of control. At high temperatures, it may also be necessary to consider latent heat due to phase change, and exothermic heat production due to chemical reaction and mode of control.

Since rubber is a relatively poor conductor of heat, and since the energy dissipation associated with cyclic loading can generate significant heat, the internal portions of thick rubber articles can become very hot. High temperatures can lead to reversion of the cure chemistry and reductions in stiffness and strength. Rubber compounds may also contain low molecular weight hydrocarbons like oils and waxes that can volatilize when the temperature becomes sufficiently high. If volatilized gases cannot escape the rubber via diffusion, then they may produce pressurized voids in the rubber. When pressure within a void exceeds about 5/6 of Young's modulus, the void may grow very rapidly and cause blowout [145, 146].

A common test for the resistance of rubber compounds to degradation from self-heating uses the Goodrich Flexometer [147] and is described in ASTM D623-99 [148]. The test is sometimes referred to as the Goodrich Blow-Out Test. It uses a cylindrical specimen with a diameter of 17.8 mm and a height of 25 mm. A static compressive load is applied followed by an oscillating displacement to produce, for example, a 25% cyclic strain amplitude. A record is made of temperature rise in the specimen and the experiment ends when the sample produces a visible blowout or a maximum time is exceeded.

The work by Gent and Hindi [149] and others [150, 151] showed that the blowout is due to volatilization and gas pressure as opposed to crack growth. Accordingly,

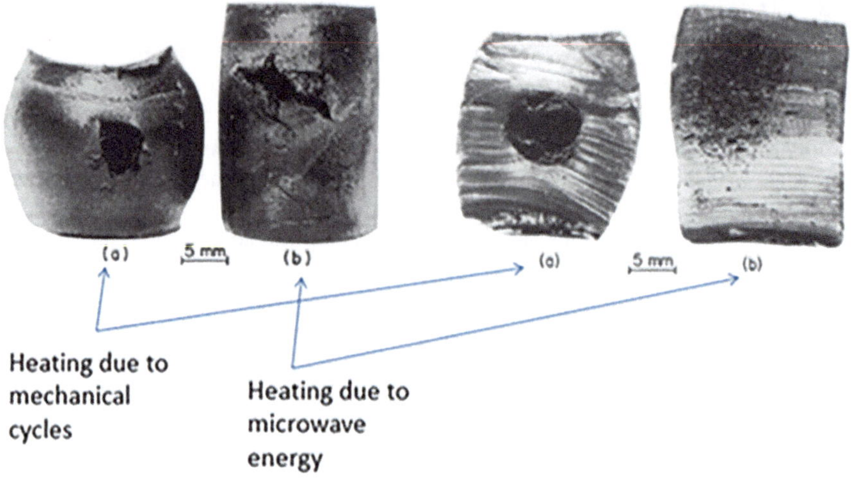

Fig. 39 Runaway temperature rise occurring due to rapid compression cycles (case a) induces a similar failure mode as microwave heating (case b) [149]

Table 9 Thermal blowout temperatures for several filled rubbers [149]

Polymer	Thermal blowout temperature θ_b, °C
NR	205
SBR	210
BR	195
IIR	180

the thermal runaway condition can be characterized by a critical temperature θ_b associated with thermal decomposition of the material. They did this by comparing blowout temperatures and test results of specimens tested in the Goodrich Flexometer and those placed in a microwave to induce heating and temperature rise. Images of tested specimen failure modes are shown in Fig. 39. Table 9 shows blowout temperatures for several filled polymers.

A recently developed alternative experiment, based on a rotating cylinder under bending (the Heat Build-up Analyser HBUA, produced by Coesfeld GmbH & Co. KG, Germany) [144], addresses two significant shortcomings of the Goodrich Flexometer experiment that (1) there is no way to observe the internal temperature, and (2) the strain and temperature fields throughout the specimen are complex and not adequately controlled (friction at the clamp may allow varying amounts of slip during compression, contributing an unknown amount of heating, and permitting an unknown amount of radial growth). The Coesfeld Heat Build-up Analyser will be further discussed in the upcoming applications section.

8.4 Initial Heat Rate and Blowout Time

When considering the possibility of thermal runaway and blowout as a failure mode, a simple criterion to check is the ratio of the thermally stored energy q_b associated with the blowout temperature θ_b to the initial adiabatic heat rate $\dot{q}$. This ratio gives the time it would take to reach the blowout temperature under the assumption that the initial heat rate is maintained and no heat leaves through conduction, convection, or radiation.

$$t_b = \frac{q_b}{\dot{q}} = \frac{\rho c(\theta_b - \theta_0)}{\dot{q}} \tag{70}$$

Although a full thermal analysis may sometimes be preferable, the above criterion has several advantages. (1) It can be evaluated locally at each point in a part (and can therefore be easily visualized as a contour plot). (2) It requires no knowledge of boundary conditions or part geometry. (3) It is likely to be conservative. Because this criterion neglects the mitigating effects of heat transport via conduction, convection, and radiation, as well as the effects of reduced heat production at higher temperature, it inevitably will predict thermal runaway times that are shorter than actual.

It is informative to compare the blowout time t_b to a given system's thermal time constant τ. When $t_b \ll \tau$, there is a substantial risk of thermal blowout. When $t_b \gg \tau$, additional analysis may show that conduction, convection, or radiation will remove sufficient heat from the system that blowout can be avoided.

9 Applications

Using the characterizations outlined above, most applications involving a temperature dependence can be predicted with sufficient accuracy to profitably inform development programs aimed at materials selection, design optimization, weight reduction, sustainability, failure mitigation, etc. It is rare that all characterizations would be strictly necessary in a single application. Rather, most commonly, engineers will judiciously select the needed characterizations, taking into account the unique circumstances of each application. For simulation, finite element solvers and the Endurica fatigue solvers [119, 122] offer temperature-dependent material models for all of the major behaviors discussed in this review. This section reviews several studies that have made use of the principles outlined above.

9.1 Transient Temperature Rise in Tires

Tire thermomechanical behavior and methods for steady-state and transient temperature prediction are well-established [33]. Due to the mission requirements for different types of tires, the transient thermal response can be very different. For a passenger car tire, one of the mission requirements is to operate continuously at rated load and highway speeds for long periods of time. Thus, the transient temperature rise as the tires "warm up" needs to reach a steady-state asymptote with maximum temperatures well below those that would promote damage and rapid aging. The rate of energy dissipation goes down as the tire temperature rises as indicated by Eqs. 33 or 35. This aspect, together with the rate that heat is taken from the system, largely due to convection, results in a stable steady-state response. An example of transient temperature rise in a passenger tire is shown in Fig. 40.

Tires for a large commercial aircraft have a mission profile that differs significantly from that of a passenger car tire. For example, the usage is only for short distances and short periods of time. Therefore, long-time steady-state operation is not required. Typical requirements include low weight, small size, heavy loads, and high speeds. These requirements result in a tire with many ply and belt layers, relatively thick cross section, and high relative deflection. Under continuous taxi operation at rated conditions, an aircraft tire design can fail due to the effects of high temperatures well before ever reaching steady-state operation. Figure 41 shows this transient temperature behavior for a high-speed taxi test [141]. The rate of energy dissipation in the tire exceeds the rate of heat taken away by convection, conduction, and radiation, and steady-state conditions are never achieved.

9.2 *Bending Rotating Tube Test*

Heat buildup and thermal runaway behavior can be efficiently characterized via an experiment in which a cylindrical specimen is bent and rotated about the cylinder's axis [144]. Typical geometry and boundary conditions are shown in Fig. 42. Bending produces a strain distribution with maximum tension and compression on the outside diameter, and zero strain along the cylinder's neutral axis, as shown in

Fig. 40 Transient passenger car tire temperature predictions [33]

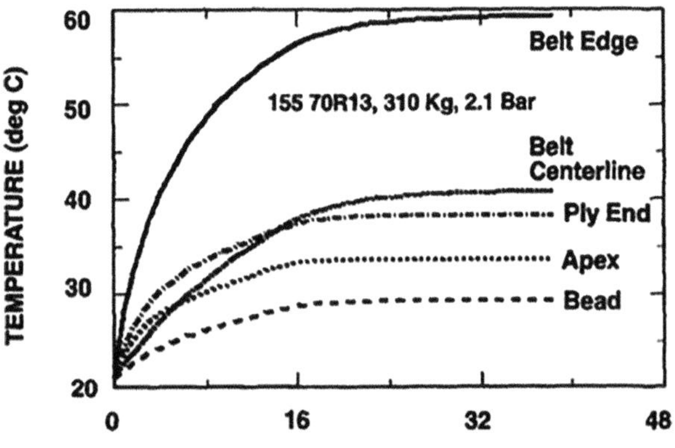

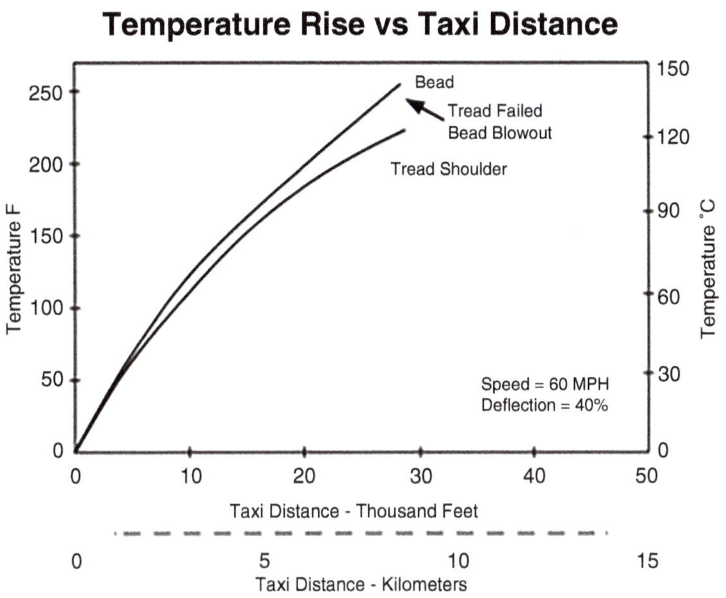

Fig. 41 Temperature rise vs. taxi distance for a commercial aircraft tire [141]

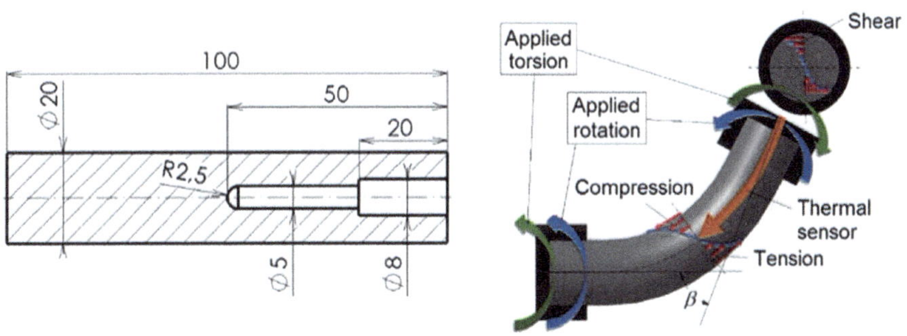

Fig. 42 Cylindrical specimen geometry and loading [144]

Fig. 43. Although the spatial strain distribution resulting from bending is not homogeneous, it is nevertheless well understood and relatively simple to calculate, especially under steady conditions when the cylinder rotates at constant velocity.

The specimen offers the advantage that a hole centered on the axis of the cylinder has only a tiny influence on the strain distribution, given that the hole reduces the moment of inertia by only $\frac{\Delta I}{I} = 1 - (d/D)^4 = 0.4\%$. Such a hole allows access to a temperature probe so that the internal temperature of the cylinder can be measured, while minimizing influence on the measured result.

Given that the strain amplitude increases with radius in a rotating, bent cylinder, and that volumetric heating varies approximately as the square of the strain

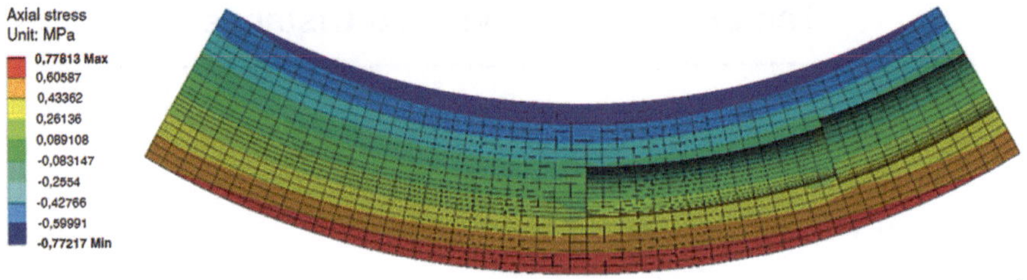

Fig. 43 Stress distribution in cylindrical specimen under bending [152]

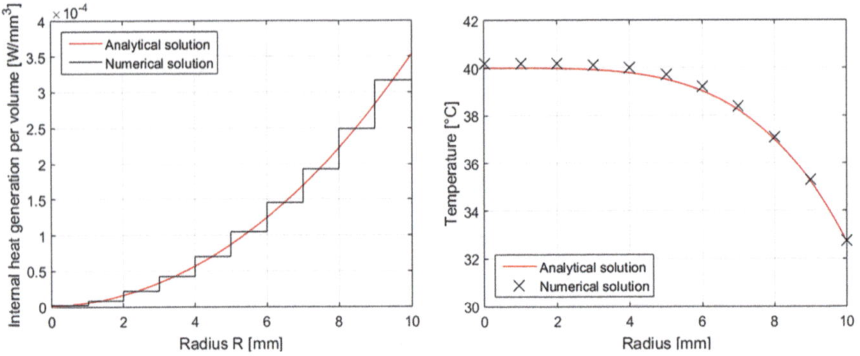

Fig. 44 Estimated heat generation and temperature distribution in rotating bending specimen [153]

amplitude, the distribution of volumetric heating can be estimated and reconciled against total power input to the specimen. Stocek, Peter, and Kratina [153] have demonstrated analytical and numerical solutions for the heat rate and steady-state temperature distributions in this specimen, as illustrated in Fig. 44. The highest heat rate occurs where the strain amplitude is highest, i.e. on the outside diameter. In contrast, the highest temperature occurs in the center of the specimen. The temperature distribution is rather constant in the center of the specimen, minimizing the potential for errors associated with placement of the temperature probe, and suggesting an adiabatic boundary condition on the inside diameter of the hole. The temperature on the outside diameter is reduced due to the effects of convective and radiative cooling.

The transient heat rate and thermal history of the specimen may also be studied [154, 155]. Typical observations are shown in Fig. 45. The initial rate of temperature rise is greatest on the outside diameter, where heat rate is maximum. However, surface temperature on the outside diameter tends to reach a steady-state value lower than the interior temperature, due to the removal of heat via convection and radiation.

If the specimen is driven under sufficiently severe conditions (i.e., large bending and large rotational frequency), then the thermal runaway/blowout temperature θ_b can be observed. Indeed, it is difficult to imagine an experiment that is better

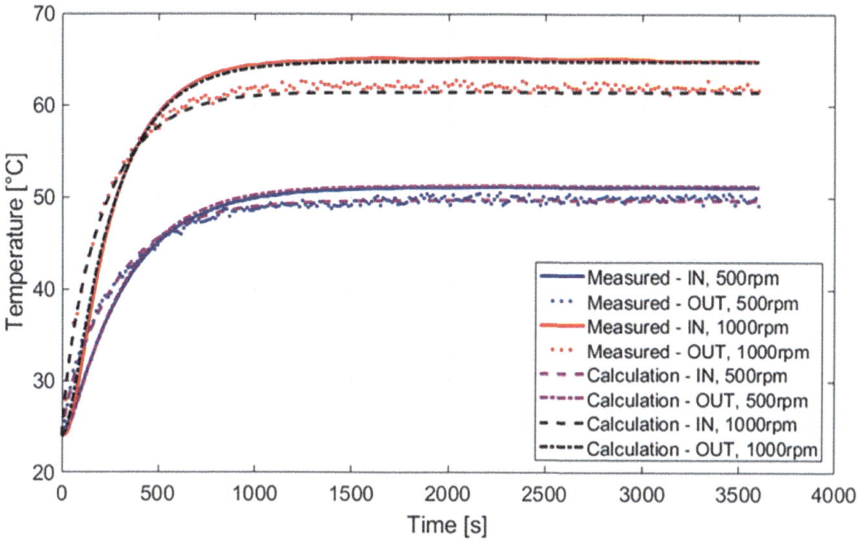

Fig. 45 Typical evolution of inside and outside temperatures [154]

optimized for this particular observation, given that the measurement of interior temperature otherwise usually involves significant strain and temperature field perturbations caused by the temperature probe.

9.3 Heat Buildup and Failure in Tank Track Pads

The rubber track of a military tank is an extreme application where high loads are applied at high frequency, resulting in coupled heat and fatigue related failure modes. Mars et al. [156] simulated self-heating and temperature-dependent fatigue performance using the Endurica fatigue solver and compared simulation results to component testing over a matrix of load and frequency. The material properties for the analysis are given in Table 10. The structural solution simulated a steel tank track wheel rolling over the track pad.

The resulting heat buildup and thermal solution are shown in Fig. 46. The external boundary was assumed to have a convection coefficient of 35 W m^{-2} K^{-1}. The self-heating rate law for given strain ε and frequency ω was given by

$$\dot{q} = H_0 \omega \left(\frac{\varepsilon}{\varepsilon_0} \right)^{\gamma} e^{r(\theta - \theta_0)} \quad (71)$$

With H_0 as the reference hysteresis, ε_0 the reference strain, γ the power-law slope (2 for linear viscoelasticity), and r the temperature sensitivity coefficient. Strain rate

Table 10 Material parameters for tank track pad analysis [156]

Property	Value	Units
Ogden Hyperelasticity	2.28, 0.054	MPa
μ_1, μ_2, μ_3	−1.008, 7.863	Dimensionless
$\alpha_1, \alpha_2, \alpha_3$		
Mullins	6.64	Dimensionless
r	0.56	MPa
m	0.029	Dimensionless
β		
Thermal transport	0.139	$W\ m^{-1}\ K^{-1}$
Conductivity k	3,002	$J\ kg^{-1}\ K^{-1}$
Specific heat C_p		
Lake–Lindley	8.342	$kJ\ m^{-2}$
Crack growth rate law	0.117	$kJ\ m^{-2}$
T_c	0.167	$kJ\ m^{-2}$
T_0	3.35	Dimensionless
T_t	0.58	mm/cyc
F	0.0506	1/°C
r_c	15.7×10^{-3}	mm
C	82	°C
c_0		
θ_0		
Self-heating law	0.0095	$mJ\ mm^{-3}\ cyc^{-1}$
H_0	0.10	Dimensionless
ε_0	1.94	Dimensionless
γ	−0.0074	1/°C
r	82	°C
θ_0		

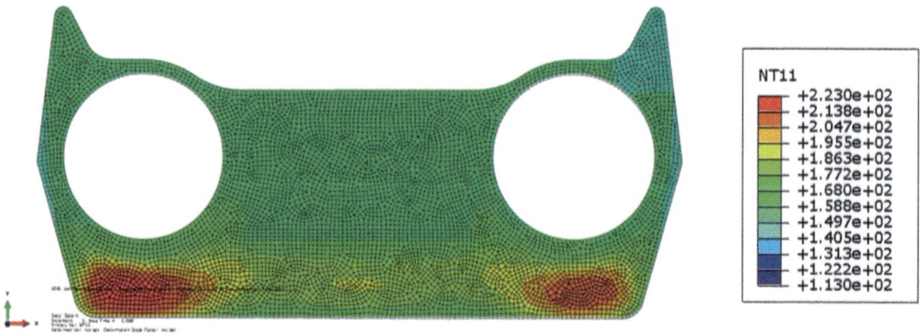

Fig. 46 Predicted temperature distribution (6,700 lbf = 29,800 N, 8 Hz) [156]. (blue = 45°C, green = 84°C, red = 122°C)

dependence was assumed negligible. Core temperatures were validated experimentally to within a few degrees across all operating conditions.

Two failure modes were observed in experiments: (1) an initiation of cracking at the bonded interface of the track shoe tube as shown in Fig. 47 and (2) a thermal

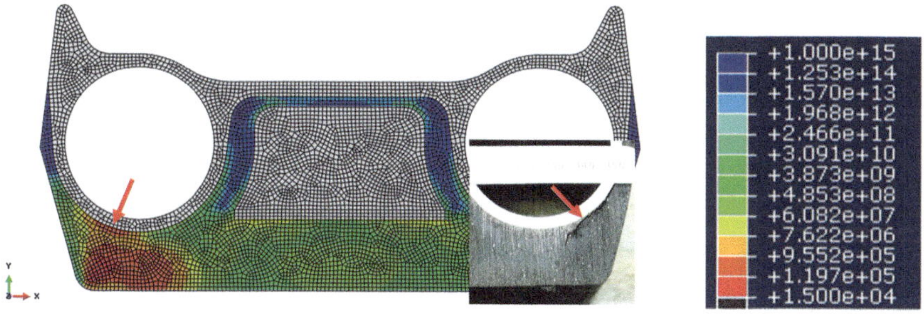

Fig. 47 Predicted fatigue life contours overlaid with observed failure mode. (6,700 lbf = 29,800, 8 Hz, observed minimum life =15,568 cycles) [156]

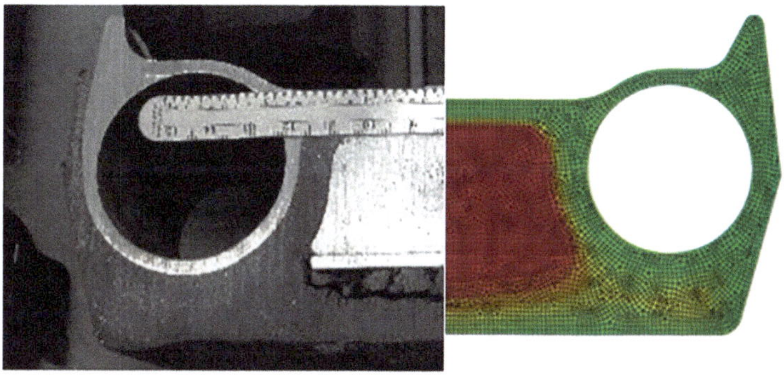

Fig. 48 Observed thermal runaway failure mode and predicted temperature distribution. (blue = 45°C, green = 84°C, red = 122°C) [156]

runaway occurring at the center of the shoe, as shown in Fig. 48. Both modes were successfully simulated.

9.4 *Heat Build-Up in a Solid Cross Section Tire*

He et al. [157] built a test rig for solid cross section tire specimens and compared rolling resistance and heat buildup results at steady state with Endurica CL simulations. The material was a silica filled S-SBR. Dissipative behavior was specified using the Kraus model defined in Eq. 32 and the rate–temperature law of Eq. 33, with the parameters given in Table 11.

The strain amplitude, temperature, and frequency dependence of the loss modulus is plotted in Fig. 49. The temperature dependence is strongly influenced by strain amplitude. For example, the temperature influence is much greater at 10^{-2} (1%) strain amplitude than it is at 10^{0} or 100% strain amplitude. The temperature and rate effects were quite adequately captured, as shown in Fig. 50.

Table 11 Material parameters for solid cross section tire self-heating analysis [156]

Property	Value	Units
Ogden hyperelasticity	0.547, 9.6 × 10^{-5}, 2.85 × 10^{-5}	MPa
μ_1, μ_2, μ_3	1.72, 19.5, −21.3	Dimensionless
α_1, α_2, α_3		
Kraus model	0.0573	MPa
G''_∞	0.305	MPa
G''_{max}	0.317	MPa
$\Delta G''_U$	0.0303	Dimensionless
ε_{0c}	0.557	Dimensionless
m	−0.0183	1/°C
r	0.022	s
Z	25	°C
θ_0		

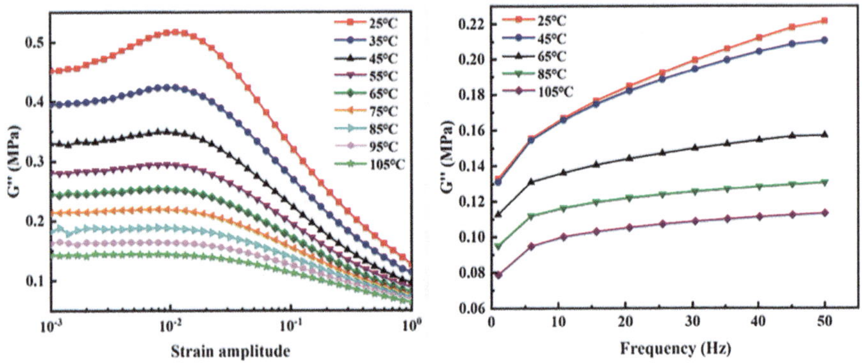

Fig. 49 Linear viscoelastic loss modulus vs. strain amplitude and temperature (left), and loss modulus vs. frequency (right), as determined via Dynamic Mechanical Analysis (DMA) measurements, used for fitting Kraus model parameters [157]

Steady-state surface temperature distribution of the solid tire recorded by an infrared camera under 30 kg of loading and at 600 rev/min is shown in Fig. 51, along with corresponding simulation results. Temperatures were correctly predicted to an accuracy of better than 5%, and rolling resistance to an accuracy of better than 10%.

9.5 *Incremental Residual Life Analysis for Tire Speed Rating and Impact Simulations*

Multistep tests such as those used in the tire industry to qualify durability and high speed performance [158, 159] require careful accounting using incremental analysis procedures such as those described in Fig. 30 and Eq. 63, and implemented in the Endurica DT [122] fatigue solver. In addition, because high tire speeds are likely to

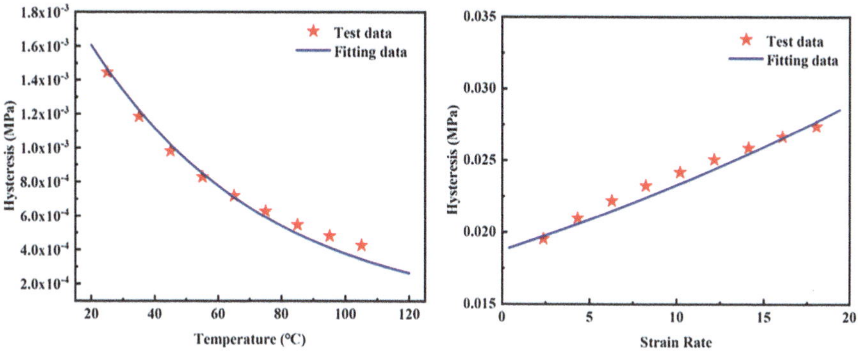

Fig. 50 Fit of Eq. 33 to temperature and strain rate measurements for constant strain amplitude [157]

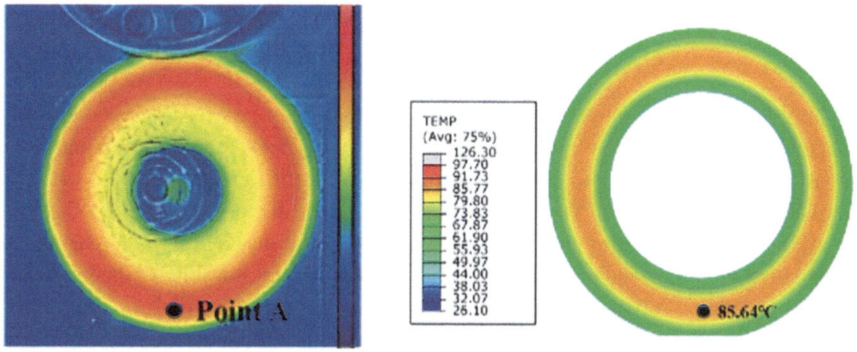

Fig. 51 Observed (left) and simulated (right) temperature distribution in solid cross section rubber wheel [157]

result in large temperature gradient effects, the strong temperature dependence of material properties must be considered.

Mars et al. [160] simulated a P195/75R17 passenger tire being run first on the FMVSS 139 speed qualification test [158], followed by the ECE 30 speed rating test [159]. The tire was modeled both with and without a nylon overwrap layer, and on both a 1.7 m test wheel and a flat road surface. The loads, speeds, and step duration of the operating schedule are given in Table 12. The inflation pressure was 220 kPa.

The structural finite element analysis was made using the steady-state rolling solution procedure in Abaqus, for each step of the test protocol in Table 12. The structural solutions show a standing wave that develops for speeds greater than 200 kph. The standing wave is especially destructive to the tire for several reasons: (1) it introduces multiple flex cycles per tire revolution, (2) it therefore causes a much higher heating rate, and (3) the high temperatures more rapidly degrade the tire.

Table 12 Multistep durability testing protocol combining the FMVSS 139 high-speed requirement and the UN ECE 30 speed rating test [158, 159]

Step/speed rating	% Rated load	Speed, kph	Step duration, min
1	85	80	120
2	85	140	30
3	85	150	30
4	85	160	30
5/Q	100	160	10
6/R	100	170	10
7/S	100	180	10
8/T	100	190	10
9/U	100	200	10
10/H	100	210	10
11/V	100	240	10
12/W	85	270	10
13/Y	85	300	10

In order to capture these effects, the crack growth rate laws for the belt, sidewall, and tread compounds were implemented for the Endurica DT fatigue solver based on measurements of tear strength T_c at 5 temperatures in the range from 0°C to 100°C, derived from the work of Young and Danik [71]. The Kraus law, i.e., Eq. 32, was used to describe the strain amplitude dependence of the loss modulus. Kraus model parameters were derived from published DMA data on carbon black filled solution styrene butadiene rubber (SSBR) [161]. As an alternative to fitting Eq. (45), experimentally determined values of T_c were implemented in the material definition as a tabular function.

Steady-state thermal solutions were then solved using Endurica to calculate heat rates, and Abaqus to solve the temperature distribution. Peak steady-state temperatures for each of the belt, sidewall, and tread are plotted in Fig. 52 versus test step for 3 cases: (1) tire operating on a 1.7 m road wheel (i.e., drum), tire on flat road—(2) with and (3) without a nylon overwrap layer. The nylon overwrap layer is a tire feature known to improve high speed rating. The strong heating effect induced by the standing waves is apparent beginning at step 10.

In order to accrue damage across all steps, taking into account each step's structural and thermal solutions, the incremental workflow shown in Fig. 53 was used. The figure shows the workflow for only the first two test steps. Subsequent steps continue the procedure until all steps have accumulated. The result of this workflow is the residual life of the tire plotted as function of time/step, as shown in Fig. 54. Residual life was defined as the number of tire revolutions, operating under the step 1 conditions, that it would take to produce failure starting from the end of the history up until that point in time. Once the residual life drops to zero, the tire has failed, and the test is over. Interestingly, the computation shows that the early steps of the testing procedure given in Table 12 do little to reduce the residual life until quite late in the test. It shows that a tire running on a test wheel ends up receiving a

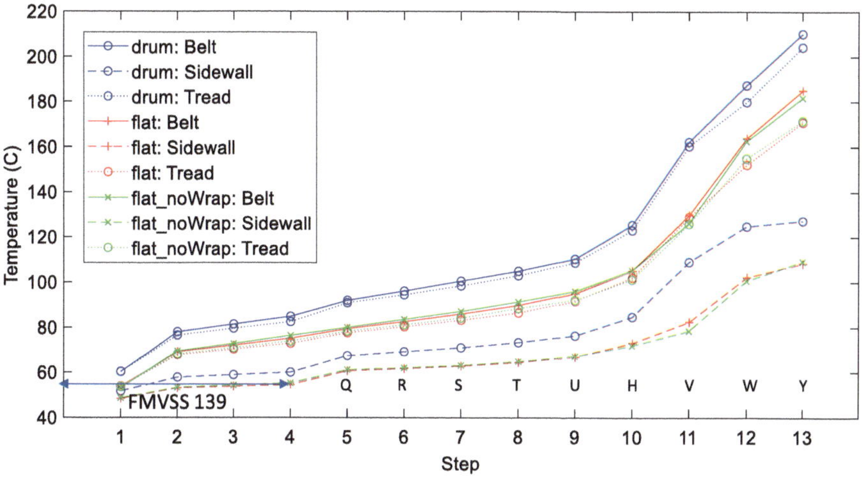

Fig. 52 Evolution of temperature vs. test step [160]

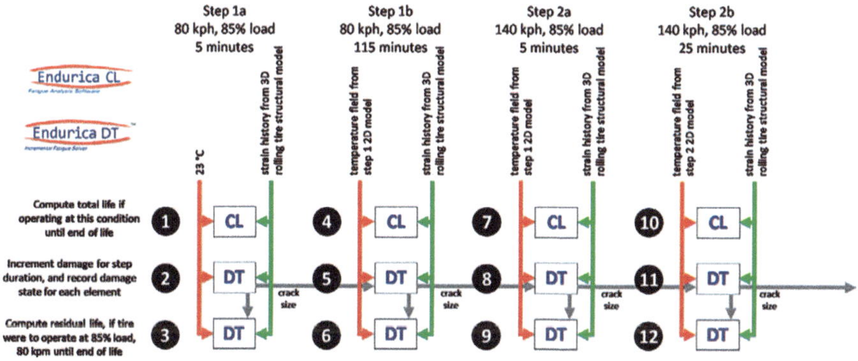

Fig. 53 High-speed durability analysis workflow for first two test steps [160]

speed rating that is one step lower than it might have received if the testing procedure had allowed testing on a flat surface. Finally, the analysis showed that the expected benefit of nylon overwrap for the subject tire was real, if modest.

The incremental workflow that was used above has also been used to simulate the effect on residual life of tire impacts [162] with a road hazard at varying speeds, impact angles, and worn tread depth states. The crack growth rate law used for the belt material during the high-rate impact was implemented via modification of the tear strength parameter during the impact event, based on experimental characterization of tear strength vs. sample strain rate, as shown in Fig. 55. Selected simulations at moment of impact are shown in Fig. 56. The study found that over the ranges considered, the impact angle had the largest effect on residual life, followed by the effect of impact speed, and finally by the effect of the worn tread depth of the tire.

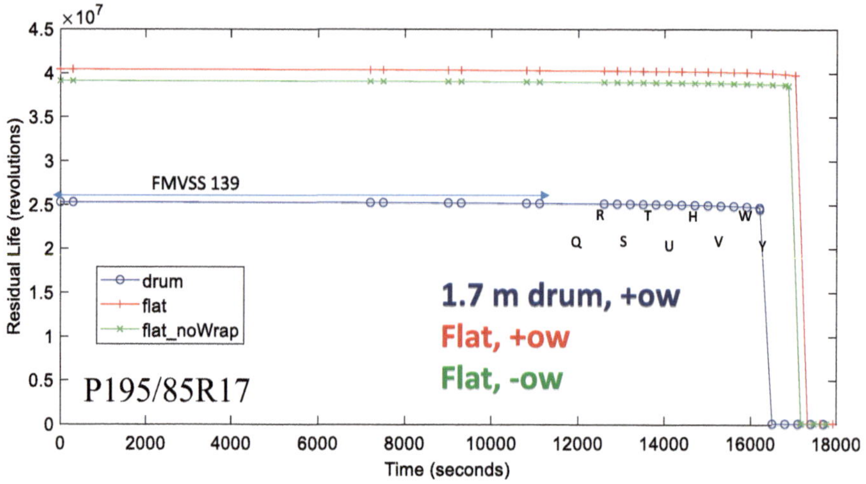

Fig. 54 Evolution of residual life vs. time/test step [160]

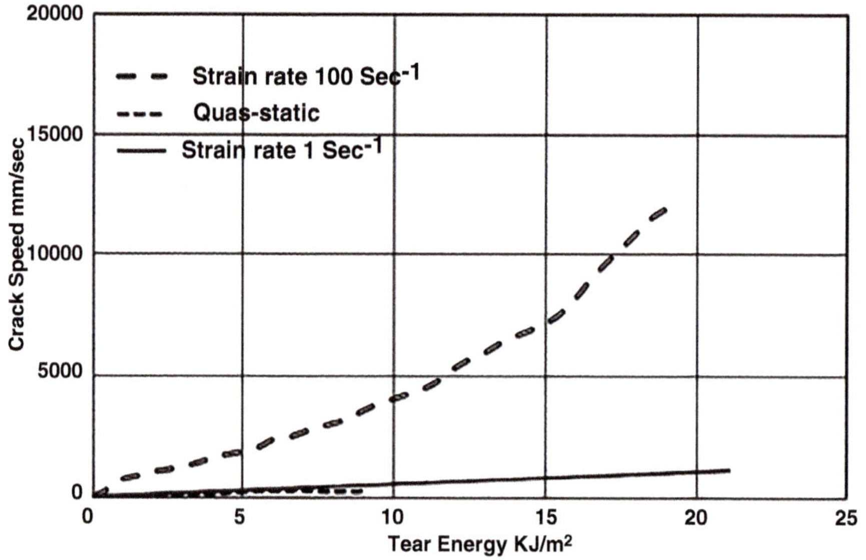

Fig. 55 Belt compound crack speed vs. critical tear strength was measured in quasistatic and in high-speed experiments [162]

9.6 Simulation of Simultaneous Aging and Fatigue

Passenger car and truck tires run at low strains and must typically endure $>10^8$ cycles over a span of several years. If a tire is re-treaded, the endurance requirements may

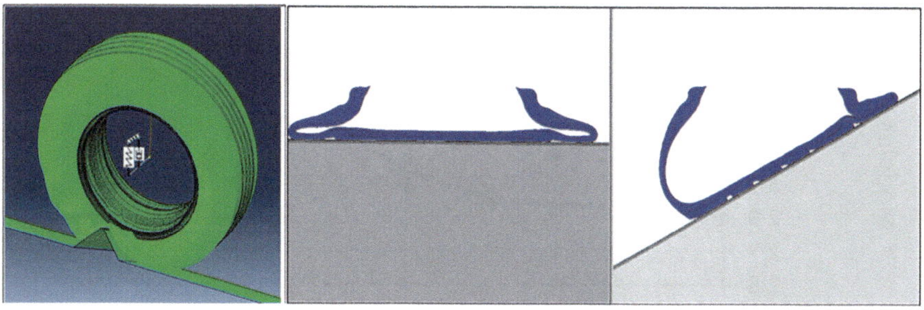

Fig. 56 Tire impact simulations were made to investigate post-impact residual life [162]

be even more demanding [163]. Fail safe design requires that the tire tread should wear down before the tire carcass loses structural integrity. In this context, both the time-dependent effects of aging and the cycle-dependent effects of fatigue must be considered, as well as the rate of treadwear. It may be the case that a new tire operates entirely below the mechanical fatigue threshold for a time and therefore initially exhibits no crack growth. The evolution of stiffness and of intrinsic strength due to thermochemistry following prolonged heat exposure may nevertheless result eventually in crack growth once the intrinsic strength degrades sufficiently relative to the operating condition.

Experimental measurements of the aerobic and anaerobic aging of a model tire belt compound were made at 80°C for the aerobic case, and at 120°C for the anaerobic case [164]. The compound was an N-326 carbon black (55 phr) and silica (10 phr) filled natural rubber, vulcanized at 150°C for 30 min.

The modulus E, intrinsic strength T_0, and tear strength T_c master curves are plotted in Fig. 57. The trends after 10 h are consistent with those reported by Ahagon [115] that aerobic aging causes embrittlement while anaerobic aging causes softening. During the first 10 h for the anaerobic case, an increase of stiffness and strength was observed. This was attributed to the specimen having started in a partially undercured state, such that the initial exposure served to complete the cure. The resulting material model definitions for simulation, based on Eqs. 55–58, for the aerobic and anaerobic cases are shown in Table 13.

The exposure time-dependent crack growth rate curves derived via Eq. 60 from these measurements are plotted in Fig. 58. In both the aerobic and anaerobic cases, the tear strength varied more with exposure than did the intrinsic strength, leading to a dependence of the power-law slope on aging, and a shifting of the crack growth rate curve depending on aged exposure time.

When the crack growth is accumulated simultaneously with evolution of the crack growth rate curve, the resulting fatigue life comes to depend on both strain and cycle rate. Figure 59 shows that for higher cycle rates, in this case >0.2 Hz, the strain-life curves do not depend on aging. This reflects the fact that faster cycle rates do not permit sufficient time for material aging to occur. However, as the cycle rate is reduced below 0.2 Hz, aging can be seen to exert growing influence with slower

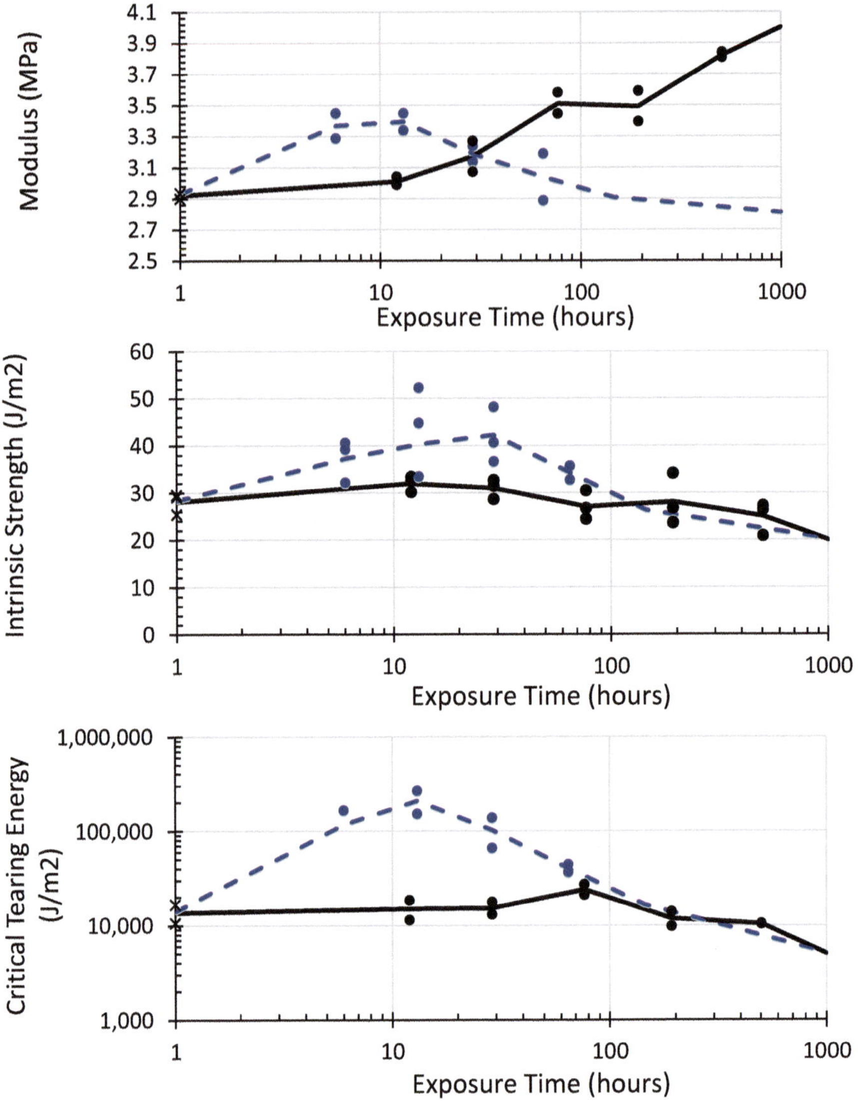

Fig. 57 Evolution of modulus E (top), intrinsic strength T_0 (middle), and critical tear strength T_c (bottom) with aging exposure time for aerobic (black solid line) and anaerobic (blue dashed line) conditions [164]. Experimental measurements provided by E. Terrill at ARDL

cycle rate. At the slowest rate, which is the equivalent of 1 cycle every 80 min, a significant change of the strain-life curve occurs. For the aerobic case, life is significantly reduced, particularly at low strain. For the anaerobic case, the tradeoffs between decrease of modulus and decrease of strength are more complex, yielding in fact a modest improvement of fatigue life and an increase of the strain-based fatigue

Table 13 Material properties for specifying aging simulation

Property	Aerobic case				Anaerobic case			
θ_0, °C	80				120			
E_a, J/mol	50×10^3				50×10^3			
Master curves	Time, hours @ 80°C	T_0, J/m2	T_c, J/m^2	E, MPa	Time, hours @ 120°C	T_0, J/m^2	T_c, J/m^2	E, MPa
	0	28	13,650	2.92	0	28	13,650	2.92
	12	32	14,950	3.01	6	37	113,050	3.36
	28.8	31	15,300	3.17	13	40	202,500	3.385
	76.8	27	23,950	3.51	28.8	42	99,050	3.18
	192	28	11,810	3.49	64.8	34	39,000	3.03
	504	25	10,350	3.82	144	26	16,000	2.90
	1,000	20	5,000	4.00	1,000	20	5,000	2.80

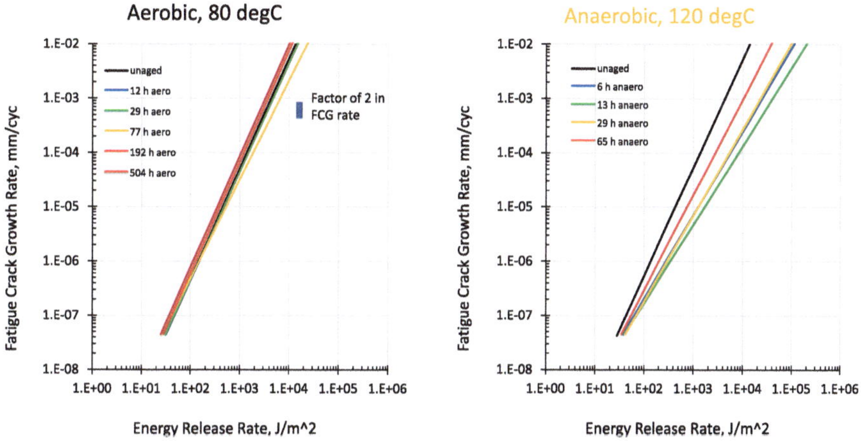

Fig. 58 Estimated aerobic (left) and anaerobic (right) evolution of fatigue crack growth rate curve with aging exposure time

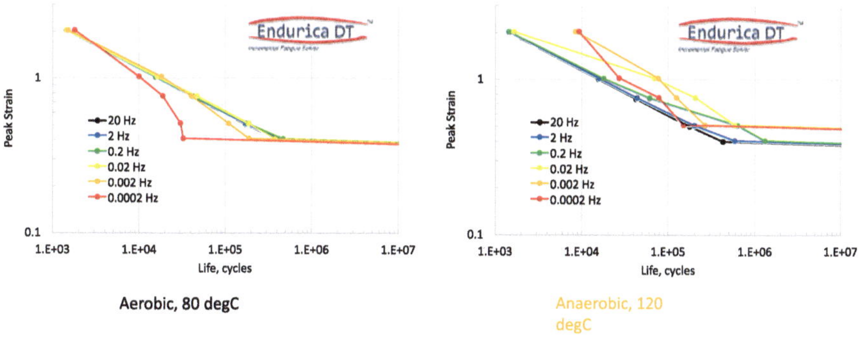

Fig. 59 Endurica-computed strain-life fatigue curves at varying cycle rate for aerobic (left) and anaerobic (right) cases

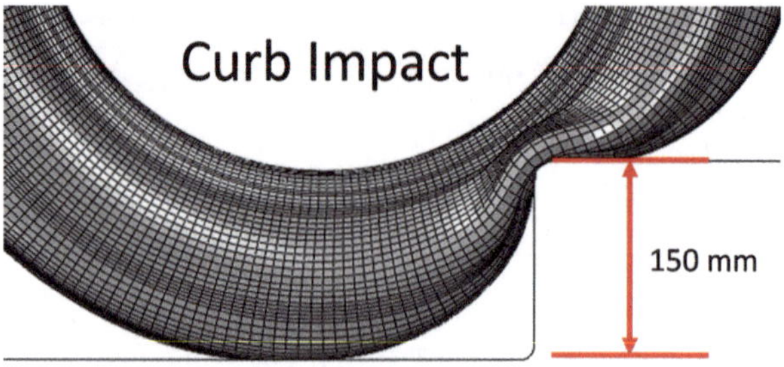

Fig. 60 Deformed tire at moment of curb impact at 135 kph [166]

threshold. The example illustrates the complexity of the interactions between aging and fatigue life, and the necessity to consider fully the details of how material properties evolve as cracks grow. It also illustrates one reason why Wöhler curves are not nearly an optimal fatigue life prediction strategy for rubber: when aging and fatigue combine (as they often will), the convolution of crack growth and aging effects leads to Wöhler curves that pertain only to very specific conditions. To derive a more general predictive capacity to handle combined aging and fatigue from Wöhler curves would require a deconvolution procedure that has not been developed and seems unlikely to be more cost effective than the crack growth procedures utilized here.

As a final example, the aging material models were implemented to a P195/75R17 passenger tire model to simulate the combined effects of oven pre-aging followed by 30 min of break-in/warm-up on a test wheel, followed by an impact, and finally followed by runout to tire end of life [165, 166]. The break-in and runout portions of the test were simulated using Abaqus/standard's steady-state rolling feature. The impact portion was simulated in Abaqus/explicit. The impact was run in straight-ahead free rolling against a 150 mm tall road hazard, shown in Fig. 60, at 220 kPa inflation, 85% of T&RA load, and a speed of 135 kph. The pre-age was run at 3 levels: 0, 12, and 120 h. A fourth model was also run with all aging effects turned off. The pre-age temperature was run at 3 levels: 35, 60, and 100°C. The aerobic aging model was used for tire compounds exposed to air. The anaerobic aging model was used for tire compounds on the interior of the tire.

The tire life contours were computed on the full 3D tire. For purposes of simplifying comparison of results, the shortest life along each streamline was collected on a tire cross section, so that a 2D plot could be constructed to indicate critical tire locations. These results are shown in Fig. 61. The most critical life from each case was then determined and summarized in the bar chart of Fig. 62.

Regarding failure mode, aging tended to emphasize a sidewall failure mode, which is unsurprising given the severity of the impact. Tire runout life following aging and impact depended on both the duration of the oven exposure and on the oven temperature. The effect of temperature was much greater than the effect of exposure time, within the ranges studied. In the simulation, aging was accrued both during the oven aging period, as well as during steady operation of the tire during break-in and run-out. Aging during the impact was neglected. The number of tire revolutions in the break-in period was 29,822. The number of tire revolutions

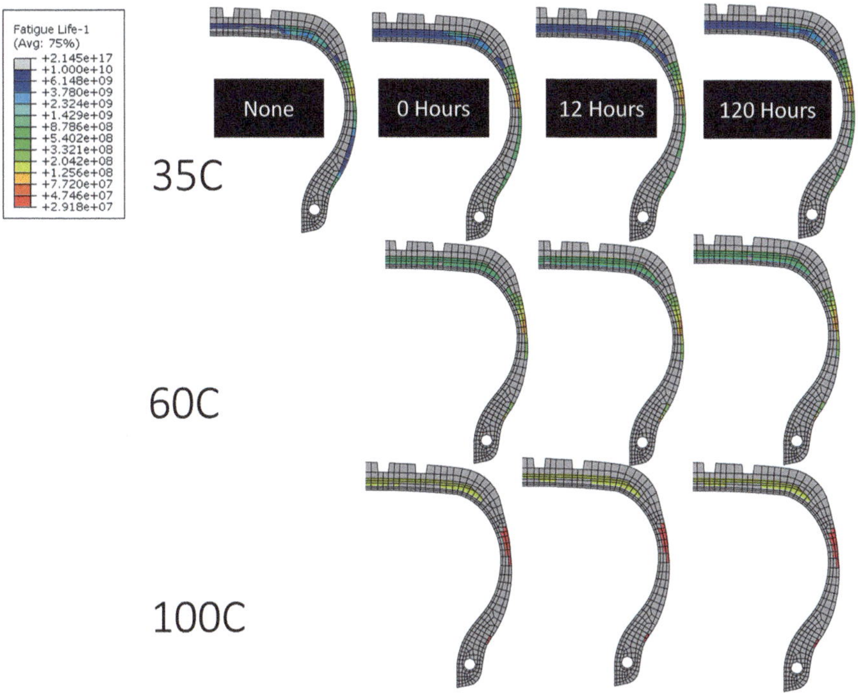

Fig. 61 Runout fatigue life following aging and impact plotted on tire cross section [166]. Leftmost column gives results for the case where aging effects are neglected. Right 3 columns give results for 0, 12, and 120 h of pre-aging at oven temperatures of 35, 60, and 100°C

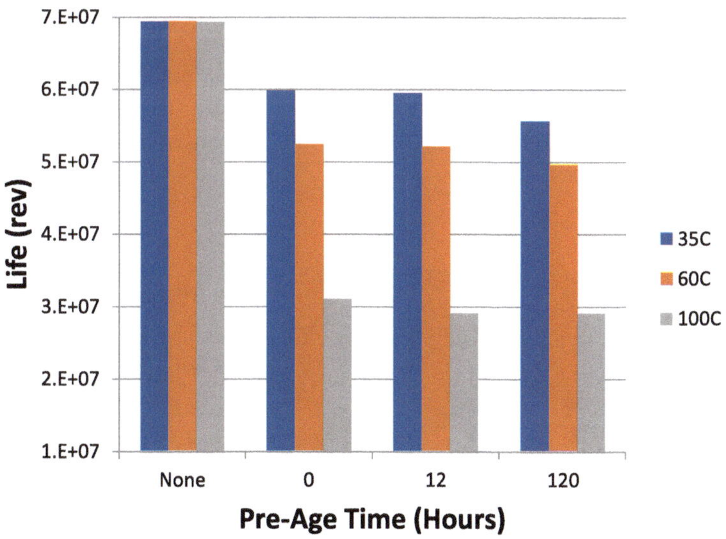

Fig. 62 Runout fatigue life for shortest lived element following aging and impact [166]

occurring 1,000 h of test runout time was 6×10^7. These figures indicate that the oven pre-age exposure time, always less than 120 h, was relatively insignificant compared to the additional exposure time accrued during tire operation, which explains the surprisingly small dependence of life on pre-age time in Fig. 62. The incremental workflows enabled by the scheme of Eq. 63 and Fig. 30, and implemented in the Endurica DT fatigue solver [122], can simulate nearly any mission profile and take account of the combined effects of thermal history, aging, and fatigue.

9.7 Heavy Truck Rubber Bearing

Designers of suspension components face imperatives to maintain or improve durability and vibration isolation functionality whilst simultaneously reducing part cost and weight. In former generations, the development process was painstakingly slow and expensive due to the requirement to build and test physical prototypes. The cost and time involved meant that few design options could be explored. The high risk implicit in a new design meant that many development program iterations were accepted. Simulation now provides an advantageous alternative to physical-only prototyping because the design space can be much more thoroughly explored quickly, leading to the identification of stronger design options and ultimately a winning design in record time. Wang et al. [167] recently made a strong demonstration of the advantages by evaluating and ranking 10 alternative design options for thermomechanical fatigue performance of the rubber bearing shown in Fig. 63.

All designs were made from the same material, with properties reported in Table 14. The simulations showed that out of the 10 design options, 2 (labeled schemes C and G) produced the best results with predicted life improvements of 2.12 and 2.23, and a predicted temperature decrease of 11.0°C for both. A typical set of simulation results is shown in Fig. 64. The original and optimized design options were physically built and tested, with the test results strongly confirming the predicted results for temperature and fatigue life. The observed life improvements were 2.16 and 2.04, see Fig. 65.

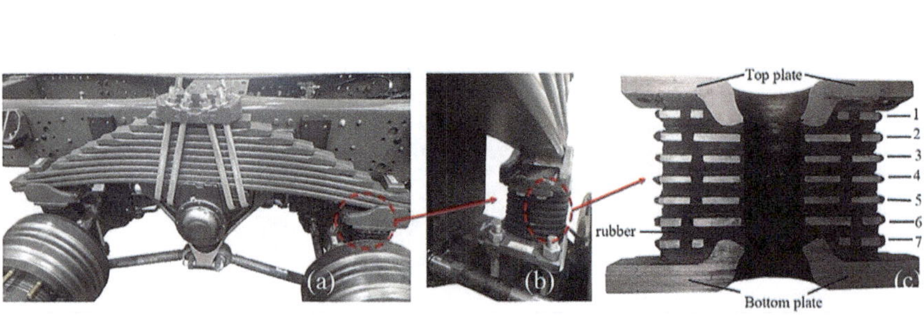

Fig. 63 Heavy truck rubber bearing geometry [167]. Overall suspension structure and location of the rubber bearing (**a**). Installed rubber bearing (**b**). Cross section of the rubber bearing (**c**)

Table 14 Material parameters for heavy truck rubber bearing analysis [167]

Property	Value	Units
Ogden hyperelasticity	1.0941, 0.0086, 1.416×10^{-7}	MPa
μ_1, μ_2, μ_3	−0.6362, 12.0281, −13.3708	Dimensionless
α_1, α_2, α_3		
Thermal transport	0.355	$W\ m^{-1}\ K^{-1}$
Conductivity k	1,473	$J\ kg^{-1}\ K^{-1}$
Specific heat C_p	1.09	$g\ cm^{-3}$
Density ρ		
Thomas crack growth rate law	89.3	$kJ\ m^{-2}$
T_c	1.62	Dimensionless
F	4.505×10^{-4}	mm/cyc
r_c	0.02	mm
c_0		
Strain crystallization	3.71	Dimensionless
F_{exp}		
Self-heating law	6.737×10^{-3}	$mJ\ mm^{-3}\ cyc^{-1}$
h_0	0.10	Dimensionless
ε_0	1.634	Dimensionless
γ	−0.0133	1/°C
r	26.4	°C
θ_0		

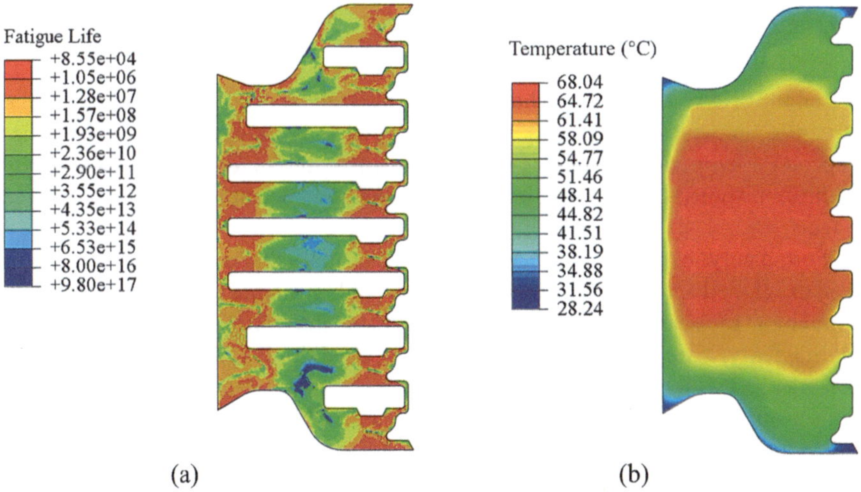

Fig. 64 Fatigue (**a**) and thermal (**b**) analysis results for a heavy truck rubber bearing [167]

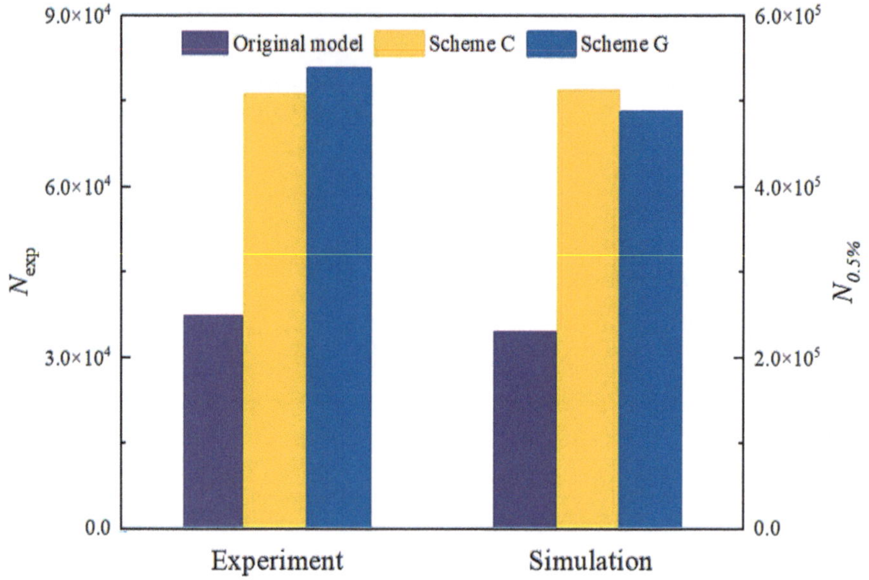

Fig. 65 Comparison of experimental vs simulation life predictions for original design (purple) and design alternatives Scheme C (yellow) and Scheme G (blue) [167]

10 Conclusion

Given the origins of rubber's chemical and mechanical properties in its molecular structure and thermal motions, it is unsurprising to find that most properties depend on temperature. What is more surprising is the diversity and nonlinearity of these responses, as no single rule encompasses the entire range of observed effects.

Fortunately, experimental characterization of the temperature dependencies is well developed and can be made in most cases with available procedures. Analysis/simulation tools are available that can represent the various dependencies with high fidelity. Still, some care must be exercised in judging which effects to consider and which to neglect, in accordance with the requirements and economics of each application. It is the hope of the authors that this review will give readers as full a view as possible at present, so that thermal effects on durability may be properly understood and accounted for in materials selection, design, analysis, and testing.

Acknowledgements The authors thank Endurica LLC for supporting the writing of this review.

References

1. Muhr AH (2005) Modeling the stress-strain behavior of rubber. Rubber Chem Technol 78(3): 391–425
2. Marckmann G, Verron E (2006) Comparison of hyperelastic models for rubber-like materials. Rubber Chem Technol 79(5):835–858
3. He H, Zhang Q, Zhang Y, Chen J, Zhang L, Li F (2022) A comparative study of 85 hyperelastic constitutive models for both unfilled rubber and highly filled rubber nanocomposite material. Nano Mater Sci 4(2):64–82
4. Sasso M, Palmieri G, Chiappini G, Amodio D (2008) Characterization of hyperelastic rubber-like materials by biaxial and uniaxial stretching tests based on optical methods. Polym Test 27(8):995–1004
5. Dedova S, Schneider K, Heinrich G (2018) Biaxial testing system for high static and dynamic loading of elastomers and its application. KGK – Kautschuk Gummi Kunststoffe 6:85–90
6. Treloar L (2005) The physics of rubber elasticity. Oxford University Press, Oxford
7. Mark JE, Erman B (2007) Rubberlike elasticity: a molecular primer. Cambridge University Press
8. Holzapfel GA, Simo JC (1996) Entropy elasticity of isotropic rubber-like solids at finite strains. Comput Methods Appl Mech Eng 132(1–2):17–44
9. Le Cam JB, Samaca Martinez JR, Balandraud X, Toussaint E, Caillard J (2015) Thermomechanical analysis of the singular behavior of rubber: entropic elasticity, reinforcement by fillers, strain-induced crystallization and the Mullins effect. Exp Mech 55:771–782
10. Li X, Dong Y, Li Z, Xia Y (2011) Experimental study on the temperature dependence of hyperelastic behavior of tire rubbers under moderate finite deformation. Rubber Chem Technol 84(2):215–228
11. Fu X, Wang Z, Ma L, Zou Z, Zhang Q, Guan X (2020) Temperature-dependence of rubber hyperelasticity based on the eight-chain model. Polymers 12(4):932
12. Fu X, Wang Z, Ma L (2021) Ability of constitutive models to characterize the temperature dependence of rubber hyperelasticity and to predict the stress-strain behavior of filled rubber under different deformation states. Polymers 13(3):369
13. Mullins L (1969) Softening of rubber by deformation. Rubber Chem Technol 42(1):339–362
14. Diani J, Fayolle B, Gilormini P (2009) A review on the Mullins effect. Eur Polym J 45(3): 601–612
15. Ogden RW, Roxburgh DG (1999) A pseudo–elastic model for the Mullins effect in filled rubber. Proc R Soc Lond Ser A Math Phys Eng Sci 455(1988):2861–2877
16. Elhard JD, Duguid A, Heinrichs M (2017) Research on safety technology verification for materials and pressure high temperature (HPHT) continental shelf (OCS), high corrosions in the US outer material evaluation. technical assessment program report (TAP 767AA) prepared for Bureau of Safety and Environmental Enforcement. Available online https://www.bsee.gov/tap-technical-assessment-program/research-onsafety-technology-verification-for-materials-and. Accessed on 15 Sept 2018
17. Huang LH, Yang XX, Gao JH (2019) Pseudo-elastic analysis with permanent set in carbon-filled rubber. Adv Polym Technol 2019, Article ID 2369329, 8 p. https://doi.org/10.1155/2019/2369329
18. Clément F, Bokobza L, Monnerie L (2001) On the Mullins effect in silica-filled polydimethylsiloxane networks. Rubber Chem Technol 74(5):847–870
19. Mullins L (1948) Effect of stretching on the properties of rubber. Rubber Chem Technol 21(2): 281–300
20. Houwink R (1956) Slipping of molecules during the deformation of reinforced rubber. Rubber Chem Technol 29(3):888–893
21. Thiele JL, Cohen RE (1980) Thermal expansion phenomena in filled and unfilled natural rubber vulcanizates. Rubber Chem Technol 53(2):313–320

22. Kraus G, Gruver JT (1971) Thermal expansion, free volume, and molecular mobility in a carbon black-filled elastomer. Rubber Chem Technol 44(5):1297–1306
23. Wiegand WB, Snyder JW (1935) The rubber pendulum, the Joule effect, and the dynamic stress-strain curve. Rubber Chem Technol 8(2):151–173
24. Farris RJ (1979) Rubber heat engines, analyses and theory. Rubber Chem Technol 52(1): 159–172
25. Greibich F, Schwödiauer R, Mao G, Wirthl D, Drack M, Baumgartner R, Kogler A, Stadlbauer J, Bauer S, Arnold N, Kaltenbrunner M (2021) Elastocaloric heat pump with specific cooling power of 20.9 W g–1 exploiting snap-through instability and strain-induced crystallization. Nat Energy 6(3):260–267
26. Chen B (2002) Influences of thermo-mechanical properties of polymeric cords on tires with and without post-cure inflation. Tire Sci Technol 30(3):156–179
27. Chen B (2004) Material characterization of Tire cords and the effects of cord thermal-mechanical properties on tires. Tire Sci Technol 32(1):2–22
28. Mars WV (2011) Analysis of stiffness variations in context of strain-, stress-, and energy-controlled processes. Rubber Chem Technol 84(2):178–186
29. Dalrymple T, Hurtado JA, Lapczyk I, Ahmadi HR (2015) Parallel rheological framework to model the amplitude dependence of the dynamic stiffness in carbon-black filled rubber, ECCMR IX, Prague
30. Assaad MC, Ebbott TG, Jiang B, Rebel G (2023) Rubber comprehensive constitutive equation and the prediction of Tire temperature and rolling resistance. Accepted for publication Tire science and technology, TSTCA. Presented at the 39th annual meeting and conference of the Tire society, September 2020
31. Aldhufairi H, Olatunbosun R, Essa K (2019) Determination of a Tyre's rolling resistance using parallel rheological framework, SAE Technical Paper, vol 2019-January, 2019-01-5069. https://doi.org/10.4271/2019-01-5069
32. Luchini JR, Peters JM, Arthur RH (1994) Tire rolling loss computation with the finite element method. Tire Sci Technol TSTCA 22(4):206
33. Ebbott TG, Jeusette J-P, Hohman RL, Kerchman V (1999) Tire temperature and rolling resistance analysis with the finite element method. Tire Sci Technol TSTCA 27(1):2
34. Terziyski J, Kennedy R (2009) Accuracy, sensitivity, and correlation of FEA-computed coastdown rolling resistance. Tire Sci Technol 37(1):4–31
35. Williams ML, Landel RF, Ferry JD (1955) The temperature dependence of relaxation mechanisms in amorphous polymers and other glass-forming liquids. J Am Chem Soc 77(14): 3701–3707
36. LeGal A, Yang X, Kluppel M (2005) Evaluation of sliding friction and contact mechanics of elastomers based on dynamic mechanical analysis. J Chem Phys 123:014704-1–014704-11
37. Payne AR (1962) The dynamic properties of carbon black-loaded natural rubber vulcanizates. Part I. J Appl Polym Sci 6(19):57–63
38. Fletcher WP, Gent AN (1954) Nonlinearity in the dynamic properties of vulcanized rubber compounds. Rubber Chem Technol 27(1):209–222
39. Kraus G (1984) Mechanical losses in carbon black filled rubbers. J Appl Poly Sci Appl Polm Symp 39:75–92
40. Ulmer JD (1996) Strain dependence of dynamic mechanical properties of carbon black-filled rubber compounds. Rubber Chem Technol 69(1):15–47
41. Gent AN (2012) Engineering with rubber. Hanser
42. Futamura S (1991) Deformation index concept for hysteretic energy-loss process. Rubber Chem Technol 64(1):57–64
43. Futamura S, Goldstein A (2004) A simple method of handling thermomechanical coupling for temperature computation in a rolling tire. Tire Sci Technol 32(2):56–68
44. Futamura S, Goldstein AA (2016) Prediction and simulation of tire performance characteristics based on deformation index concept. Rubber Chem Technol 89(1):1–21

45. Rivlin RS, Thomas AG (1953) Rupture of rubber. I. Characteristic energy for tearing. J Polym Sci 10(3):291–318
46. Griffith AA (1921) VI The phenomena of rupture and flow in solids. Philos Trans R Soc Lond Ser A Math Phys Character 221(582–593):163–198
47. Lindley PB (1972) Energy for crack growth in model rubber components. J Strain Anal 7(2): 132–140
48. Ebbott TG (1996) An application of finite element-based fracture mechanics analysis to cord-rubber structures. Tire Sci Technol 24(3):220–235
49. Yeoh OH (2002) Relation between crack surface displacements and strain energy release rate in thin rubber sheets. Mech Mater 34(8):459–474
50. Yeoh OH (2006) Strain energy release rates for some classical rubber test pieces by finite element analysis. In: Elastomers and components. Woodhead Publishing, pp 75–89
51. Chang YW, Gent AN, Padovan J (1993) Strain energy release rates for internal cracks in rubber blocks. Int J Fract 60:363–371
52. Thomas AG (1994) The development of fracture mechanics for elastomers. Rubber Chem Technol 67(3):50–67
53. Lake GJ (2003) Fracture mechanics and its application to failure in rubber articles. Rubber Chem Technol 76(3):567–591
54. Lake GJ, Thomas AG (1967) The strength of highly elastic materials. Proc R Soc Lond Ser A Math Phys Sci 300(1460):108–119
55. Andrews EH (1963) Rupture propagation in hysteresial materials: stress at a notch. J Mech Phys Solids 11(4):231–242
56. Ahagon A, Gent AN (1975) Threshold fracture energies for elastomers. J Polym Sci Polym Phys Ed 13(10):1903–1911
57. Gent AN, Tobias RH (1982) Threshold tear strength of elastomers. J Polym Sci Polym Phys Ed 20(11):2051–2058
58. Bhowmick AK, Gent AN, Pulford CTR (1983) Tear strength of elastomers under threshold conditions. Rubber Chem Technol 56(1):226–232
59. Bhowmick AK (1988) Threshold fracture of elastomers. Polym Rev 28(3–4):339–370
60. Lake GJ, Lindley PB (1966) Fatigue of rubber at low strains. J Appl Polym Sci 10(2):343–351
61. Zhang E, Bai R, Morelle XP, Suo Z (2018) Fatigue fracture of nearly elastic hydrogels. Soft Matter 14(18):3563–3571
62. Lake GJ, Yeoh OH (1978) Measurement of rubber cutting resistance in the absence of friction. Int J Fract 14:509 526
63. Robertson CG, Stoček R, Mars WV (2021) The fatigue threshold of rubber and its characterization using the cutting method. Fatigue crack growth in rubber materials: experiments and modelling, pp 57–83
64. Kadir A, Thomas AG (1981) Tear behavior of rubbers over a wide range of rates. Rubber Chem Technol 54(1):15–23
65. Bhowmick AK (1986) Tear strength of elastomers over a range of rates, temperatures and crosslinking: tearing energy spectra. J Mater Sci 21(11):3927–3932
66. Stacer RG, Kelley FN (1985) Criteria for unstable tearing of elastomers. Rubber Chem Technol 58(5):924–938
67. Thomas AG (1958) Rupture of rubber. V. Cut growth in natural rubber vulcanizates. J Polym Sci 31(123):467–480
68. Lake GJ, Lindley PB (1965) The mechanical fatigue limit for rubber. J Appl Polym Sci 9(4): 1233–1251
69. Stoček R (2020) Some revisions of fatigue crack growth characteristics of rubber. In: Heinrich G, Kipscholl R, Stoček R (eds) Fatigue crack growth in rubber materials. Advances in polymer science, vol 286. Springer, Cham. https://doi.org/10.1007/12_2020_72
70. Young DG (1986) Fatigue crack propagation in elastomer compounds: effects of strain rate, temperature, strain level, and oxidation. Rubber Chem Technol 59(5):809–825

71. Young DG, Danik JA (1994) Effects of temperature on fatigue and fracture. Rubber Chem Technol 67(1):137–147
72. Schieppati J, Schrittesser B, Wondracek A, Robin S, Holzner A, Pinter G (2018) Impact of temperature on the fatigue and crack growth behavior of rubbers. Procedia Struct Integr 13: 642–647
73. Heinrich G, Kipscholl R, Stocek R (eds) (2021) Fatigue crack growth in rubber materials – experiments and modeling. Springer
74. Dedova S, Schneider K, Heinrich G (2017) Lion A, Johlitz M (eds) Influence of dissipative specimen heating on the tearing energy of elastomers estimated by global and local characterization methods. Constitutive models for rubber X. CRC Press/Balkema, London, pp 219–224
75. Dedova S, Schneider K, Heinrich G (2019) Energy based characterization of fracture and fatigue behaviour of rubber in complex loading conditions. In: Huneau B, Le Cam J-B, Marco J, Verron E (eds) Constitutive models for rubber XI. CRC Press/Balkema, London, pp 363–367
76. Dedova S, Schneider K, Stommel M, Heinrich G (2020) Dissipative heating, fatigue and fracture behaviour of rubber under multiaxial loading. In: Heinrich G, Kipscholl R, Stoček R (eds) Fatigue crack growth in rubber materials. Advances in polymer science, vol 286. Springer, Cham, pp 421–443. https://doi.org/10.1007/12_2020_75
77. Dedova S, Schneider K (2023) Characterisation of crack propagation in natural rubber under complex loading conditions using thermographic methods. In: Advances in polymer science. Springer, Berlin. https://doi.org/10.1007/12_2023_164
78. Lake GJ, Lindley PB (1964) Cut growth and fatigue of rubbers. II. Experiments on a noncrystallizing rubber. J Appl Polym Sci 8(2):707–721
79. Huneau B, Masquelier I, Marco Y, Le Saux V, Noizet S, Schiel C, Charrier P (2016) Fatigue crack initiation in a carbon black–filled natural rubber. Rubber Chem Technol 89(1):126–141
80. Robertson CG, Tunnicliffe LB, Maciag L, Bauman MA, Miller K, Herd CR, Mars WV (2020) Characterizing distributions of tensile strength and crack precursor size to evaluate filler dispersion effects and reliability of rubber. Polymers 12(1):203
81. Tunnicliffe LB, Robertson CG, Mars WV (2023) A microscopy investigation of rubber compound crack precursors and tensile fracture surfaces. Rubber Chem Technol 96(4): 514–529. https://doi.org/10.5254/rct-23.201163
82. Weibull W (1951) A statistical distribution function of wide applicability. J Appl Mech 18:293
83. Rinne H (2008) The Weibull distribution: a handbook. CRC Press, Taylor & Francis Group, Boca Raton, FL
84. Ignatz-Hoover F, Robertson CG, Basu SK, Ludwig M, Alshuth T (2015) Silica structure in rubber: characterization using X-ray computed tomography, Microscopy and impact on failure properties, paper presented at deutsche Kautschuk-Tagung (DKT), Nürnberg, Germany, June 29–July 2, 2015
85. Huneau B (2011) Strain-induced crystallization of natural rubber: a review of X-ray diffraction investigations. Rubber Chem Technol 84(3):425–452
86. Lindley PB (1973) Relation between hysteresis and the dynamic crack growth resistance of natural rubber. Int J Fract 9:449–462
87. Bekkedahl N, Wood LA (1941) Influence of the temperature of crystallization on the melting of crystalline rubber. Rubber Chem Technol 14(3):544–545
88. Gent AN, Zhang LQ (2002) Strain-induced crystallization and strength of rubber. Rubber Chem Technol 75(5):923–934
89. Candau N, Laghmach R, Chazeau L, Chenal JM, Gauthier C, Biben T, Munch E (2015) Temperature dependence of strain-induced crystallization in natural rubber: on the presence of different crystallite populations. Polymer 60:115–124
90. Schneider K, Zybell L, Domurath J, Heinrich G, Roth S, Rothkirch A, Ohm W (2017) "Investigation of time dependence of dissipation and strain induced crystallization in natural

rubber under cyclic and impact loading", proceeding of the 10th European conference on constitutive models for rubber (ECCMR X), CRC Press, Munich
91. Brüning K, Schneider K, Roth SV, Heinrich G (2012) Kinetics of strain-induced crystallization in natural rubber studied by WAXD: dynamic and impact tensile experiments. Macromolecules 45(19):7914–7919
92. Tosaka M, Kohjiya S, Murakami S, Poompradub S, Ikeda Y, Toki S et al (2004) Effect of network-chain length on strain-induced crystallization of NR and IR vulcanizates. Rubber Chem Technol 77(4):711–723
93. Toki S, Hsiao BS, Kohjiya S, Tosaka M, Tsou AH, Datta S (2006) Synchrotron x-ray studies of vulcanized rubbers and thermoplastic elastomers. Rubber Chem Technol 79(3):460–488
94. Behnke R, Berger T, Kaliske M (2018) Numerical modeling of time-and temperature-dependent strain-induced crystallization in rubber. Int J Solids Struct 141:15–34
95. Mars WV, Fatemi A (2003) A phenomenological model for the effect of R ratio on fatigue of strain crystallizing rubbers. Rubber Chem Technol 76(5):1241–1258
96. Mars WV (2009) Computed dependence of rubber's fatigue behavior on strain crystallization. Rubber Chem Technol 82(1):51–61
97. Harbour RJ, Fatemi A, Mars WV (2007) Fatigue crack growth of filled rubber under constant and variable amplitude loading conditions. Fatigue Fract Eng Mater Struct 30(7):640–652
98. Ruellan B, Le Cam JB, Jeanneau I, Canévet F, Mortier F, Robin E (2019) Fatigue of natural rubber under different temperatures. Int J Fatigue 124:544–557
99. Ruellan B, Le Cam JB, Robin E, Jeanneau I, Canevet F, Mortier F (2019) Influence of the temperature on lifetime reinforcement of a filled NR. In: Fracture, fatigue, failure and damage evolution, vol 6, Proceedings of the 2018 annual conference on experimental and applied mechanics. Springer, pp 45–50
100. Champy C, Le Saux V, Marco Y, Glanowski T, Charrier P, Hervouet W (2021) Fatigue of crystallizable rubber: generation of a Haigh diagram over a wide range of positive load ratios. Int J Fatigue 150:106313
101. Mars WV (2021) Critical plane analysis of rubber. Fatigue crack growth in rubber materials: experiments and modelling, pp 85–107
102. Cunneen JI (1968) Oxidative aging of natural rubber. Rubber Chem Technol 41(1):182–208
103. Baldwin JM, Bauer DR (2008) Rubber oxidation and tire aging-a review. Rubber Chem Technol 81(2):338–358
104. Datta RN, Huntink NM, Datta S, Talma AG (2007) Rubber vulcanizates degradation and stabilization. Rubber Chem Technol 80(3):436–480
105. Ghosh P, Katare S, Patkar P, Caruthers JM, Venkatasubramanian V, Walker KA (2003) Sulfur vulcanization of natural rubber for benzothiazole accelerated formulations: from reaction mechanisms to a rational kinetic model. Rubber Chem Technol 76(3):592–693
106. Arrhenius S (1889) Über die Reaktionsgeschwindigkeit bei der Inversion von Rohrzucker durch Säuren. Z Phys Chem 4(1):226–248
107. Laidler KJ (1984) The development of the Arrhenius equation. J Chem Educ 61(6):494
108. Gillen KT, Bernstein R, Derzon DK (2005) Evidence of non-Arrhenius behaviour from laboratory aging and 24-year field aging of polychloroprene rubber materials. Polym Degrad Stab 87(1):57–67
109. Pazur RJ, Cormier JG, Korhan-Taymaz K (2014) The effect of acrylonitrile content on the thermo-oxidative aging of nitrile rubber. Rubber Chem Technol 87(1):53–69
110. Pazur RJ, Mengistu T (2018) Influence of the crosslink structure on the activation energy calculated under thermo-oxidative conditions. Rubber Chem Technol 91(1):205–224
111. Pett RA, Tabar RJ (1978) The oxidative aging of a compounded natural rubber Vulcanizate. Rubber Chem Technol 51(1):1–6
112. Mott PH, Roland CM (2001) Aging of natural rubber in air and seawater. Rubber Chem Technol 74(1):79–88
113. Paeglis AU (2004) A simple model for predicting heat aging of EPDM rubber. Rubber Chem Technol 77(2):242–256

114. Stokoe AL (1972) Heat ageing of rubbers, ministry of defence, explosive research and development establishment, technical report no. 91. National Technical Information Service, England, 1
115. Ahagon A, Kirino Y (2006) Aging of black filled rubber under deformation. Rubber Chem Technol 79(4):641–652
116. Santoso M, Giese U, Schuster RH (2007) Investigations on initial stage of aging of tire rubbers by chemiluminescence spectroscopy. Rubber Chem Technol 80(5):762–776
117. Soma P, Tada N, Uchida M, Nakahara K, Taga Y (2010) A fracture mechanics approach for evaluating the effects of heat aging on fatigue crack growth of vulcanized natural rubber. J Solid Mech Mater Eng 4(6):727–737
118. Zhong A (2006) Computational fracture mechanics analysis of truck Tire durability, vol 73. Journal of Applied Mechanics, p 799
119. Endurica LLC (2023) *Endurica CL*, 2023-1201, https://endurica.com/fatigue-solver-software/endurica-cl/. Accessed 18 Dec 2023
120. Mars WV (2019) System for tracking incremental damage accumulation. U.S. patent application 16/149,738, filed 4 Apr 2019
121. Mars WV, Suter JD, Bauman M (2018) Computing remaining fatigue life under incrementally updated loading histories (No. 2018-01-0623). SAE Technical Paper
122. Endurica LLC (2023) *Endurica DT*, 2023-1201, https://endurica.com/fatigue-solver-software/endurica-dt/. Accessed 18 Dec 2023
123. Fick A (1855) V. On liquid diffusion. The London, Edinburgh, and Dublin Philosophical Magazine. J Sci 10(63):30–39
124. Daynes HA (1920) The process of diffusion through a rubber membrane. Proc R Soc Lond Ser A Math Phys Character 97(685):286–307
125. Van Amerongen GJ (1964) Diffusion in elastomers. Rubber Chem Technol 37(5):1065–1152
126. Ahagon A, Kida M, Kaidou H (1990) Aging of tire parts during service. I. Types of aging in heavy-duty tires. Rubber Chem Technol 63(5):683–697
127. Gillen KT, Terrill ER, Winter RM (2001) Modulus mapping of rubbers using micro-and nano-indentation techniques. Rubber Chem Technol 74(3):428–450
128. Le Gac PY, Celina MC, Roux G, Verdu J, Davies P, Fayolle B (2016) Predictive ageing of elastomers: oxidation driven modulus changes for polychloroprene. Polym Degrad Stab 130: 348–355
129. Ellwood KR, Baldwin J, Bauer DR (2005) A finite element model for oven aged tires. Tire Sci Technol 33(2):103–119
130. Ellwood KR, Baldwin J, Bauer DR (2006) Numerical simulation of thermal oxidation in automotive tires. Rubber Chem Technol 79(2):249–266
131. Shelton JR, Wherley FJ, Cox WL (1953) Effect of temperature upon rate of oxidation of rubber. Ind Eng Chem 45(9):2080–2086
132. Hands D, Horsfall F (1977) The thermal diffusivity and conductivity of natural rubber compounds. Rubber Chem Technol 50(2):253–265
133. Kerschbaumer RC, Stieger S, Gschwandl M, Hutterer T, Fasching M, Lechner B, Meinhart L (2019) Comparison of steady-state and transient thermal conductivity testing methods using different industrial rubber compounds. Polym Test 80:106121
134. Saxena NS, Pradeep P, Mathew G, Thomas S, Gustafsson M, Gustafsson SE (1999) Thermal conductivity of styrene butadiene rubber compounds with natural rubber prophylactics waste as filler. Eur Polym J 35(9):1687–1693
135. Badawy MM (1996) Effect of sulphur on both thermal and dielectric properties of pre-compressed FEF black-loaded SBR vulcanizates. Polym Test 15(6):507–515
136. Song JP, Tian KY, Ma LX, Li W, Yao SC (2019) The effect of carbon black morphology to the thermal conductivity of natural rubber composites. Int J Heat Mass Transf 137:184–191
137. Zaharescu T, Meltzer V, Vı^lcu, R. (2000) Thermal properties of EPDM/NR blends. Polym Degrad Stab 70(3):341–345

138. Nah C, Park JH, Cho CT, Chang YW, Kaang S (1999) Specific heats of rubber compounds. J Appl Polym Sci 72(12):1513–1522
139. Toth WJ, Chang JP, Zanichelli C (1991) Finite element evaluation of the state of cure in a tire. Tire Sci Technol 19(4):178–212
140. Disch M, Widdecke N, Wiedemann J, Reister H, Weidmann EP (2013) Numerical simulation of the transient heat-up of a passenger vehicle during a trailer towing uphill drive (No. 2013-01-0873). SAE Technical Paper
141. Goodyear Aircraft Tire Care and Maintenance Manual (2020) https://www.goodyearaviation.com/resources/pdf/aviation-tire-care-2020.pdf. Accessed 20 Dec 2023
142. Kerchman V, Shaw C (2003) Experimental study and finite element simulation of heat build-up in rubber compounds with application to fracture. Rubber Chem Technol 76(2):386–405
143. Luo RK, Mortel WJ, Wu XP (2015) Investigation on rubber failure due to heat generation under dynamic loading. Proc Inst Mech Eng L J Mater Des Appl 229(1):77–87
144. Vaněk J, Peter O, Heinrich G, Stoček R (2023) 2D transient thermal analytical solution of the heat build-up in cyclically loaded rubber cylinder, TBD
145. Gent AN (1990) Cavitation in rubber: a cautionary tale. Rubber Chem Technol 63(3):49–53
146. Briscoe BJ, Savvas T, Kelly CT (1994) "Explosive decompression failure" of rubbers: a review of the origins of pneumatic stress induced rupture in elastomers. Rubber Chem Technol 67(3):384–416
147. Lessig ET (1937) The Goodrich flexometer. Ind Eng Chem Anal Ed 9(12):582–588
148. ASTM D 623–99 (1999) Standard test methods for rubber property—heat generation and flexing fatigue in compression
149. Gent AN, Hindi M (1988) Heat build-up and blowout of rubber blocks. Rubber Chem Technol 61(5):892–905
150. Choi SS, Kim JC, Ko JE, Cho YS, Shin WG (2009) Influence of the rubber blend ratio on blowout behaviors of carbon black-reinforced natural rubber/styrene-butadiene rubber. J Appl Polym Sci 112(6):3627–3633
151. Choi SS, Kim JC, Lee HM (2008) Analysis of origin to cause blowout of carbon black-reinforced rubber composites using GC/MS. J Appl Polym Sci 110(5):3068–3072
152. Peter O, Stoček R (2022) Analytical calculation approach of heat-build up in rubber cylinder. In: Constitutive models for rubber XII: proceedings of the 12th European conference on constitutive models for rubber (ECCMR 2022), September 7–9, 2022, CRC Press, Milano, Italy, p 278
153. Peter O, Stoček R, Kratina O (2022) Experimental and numerical description of the heat build-up in rubber under cyclic loading. In: Heinrich G, Kipscholl R, Stoček R (eds) Degradation of elastomers in practice, experiments and modeling. Advances in polymer science, vol 289. Springer, Cham. https://doi.org/10.1007/12_2022_128
154. Kratina O, Stoček R, Zádrapa P, Kipscholl R, Peter O (2023) Kinetics of self-heat build-up in carbon black filled natural rubber caused by cyclic mechanical loading. In: Advances in polymer science. Springer, Berlin. https://doi.org/10.1007/12_2023_167
155. Kyei-Manu W, Tunnicliffe L, Herd C, Akutagawa K, Kratina O, Stoček R, Busfield J (2023) Effect of carbon black on heat build-up and energy dissipation in rubber materials. Adv Polym Sci
156. Mars WV, Castanier M, Ostberg D, Bradford W (2017) Digital twin for tank track elastomers: predicting self-heating and durability. In: proceedings of the 2017 ground vehicle systems engineering and technology symposium (GVSETS), pp 1–10
157. He H, Liu J, Zhang Y, Han X, Mars WV, Zhang L, Li F (2022) Heat build-up and rolling resistance analysis of a solid tire: experimental observation and numerical simulation with thermo-mechanical coupling method. Polymers 14(11):2210
158. United States Department of Transportation National Highway Traffic Safety Association [DOT NHTSA], "Laboratory Test Procedure for FMVSS No. 139 – New Pneumatic Radial Tires for Light Vehicles," TP-139-02, January 6, 2006. Available at https://www.nhtsa.gov/sites/nhtsa.gov/files/tp-139-02.pdf. Accessed 11 Aug 2021

159. United Nations Economic Commission for Europe [UN ECE], "Uniform Provisions Concerning the Approval of Pneumatic Tyres for Motor Vehicles and Their Trailers," United Nations, 29 Mar 2007. Available at https://unece.org/fileadmin/DAM/trans/main/wp29/wp29regs/r030r3e.pdf. Accessed 11 Aug 2021
160. Mars WV, Paudel G, Suter JD, Robertson CG (2021) Incremental, critical plane analysis of standing wave development, self-heating, and fatigue during regulatory high-speed tire testing protocols. Tire Sci Technol 49(3):172–205
161. Warasitthinon N, Robertson CG (2018) Interpretation of the tan d peak height for Particle Filled rubber and polymer nanocomposites with relevance to Tire tread performance balance. Rubber Chem Technol 91:577–594
162. Gobinath G, Ebbott T, Hughes S (2023) Predicting residual casing life of a Tire following an impact event. Tire Sci Technol
163. Norton C (2022) The story of retreading: making sustainability sustainable. Rubber News, 19 Sept 2022, pp 15–16
164. Mars WV, Terrill E, Robertson CG, Stocek R, Kipscholl C (2020) Effects of aerobic and anaerobic ageing on the evolution of upper and lower bounds on tearing energy, 197th fall technical meeting of the ACS rubber division, Cleveland, OH, ISSN:1547-1977
165. Mars WV, Steiner EC (2021) Simulating the effects of post-impact durability of tire ageing, 198th spring technical meeting of the ACS rubber division, Cleveland, OH, ISSN: 1547-1977
166. Mars WV, Steiner EC (2021) Simulating tires and ageing with Endurica DT, fatigue ninja Friday, episode #11, 15 January 2021, https://academy.endurica.com/p/fatigue-ninja-friday-11-ageing-and-tires. Accessed 21 Dec 2023
167. Wang Z, Zhou Y, Liu J, Xu Z, Chen M, Gao R et al (2023) Fatigue life investigation of rubber bearing for heavy trucks: optimal design by using finite element method with experimental verification. Int J Fatigue:108080

Printed by Printforce, the Netherlands